AF878770

Statistical Mechanics in Physics and Biology

MATERIALS RESEARCH SOCIETY
SYMPOSIUM PROCEEDINGS VOLUME 463

Statistical Mechanics in Physics and Biology

Symposium held December 2–5, 1996, Boston, Massachusetts, U.S.A.

EDITORS:

Denis Wirtz
Johns Hopkins University
Baltimore, Maryland, U.S.A.

Thomas C. Halsey
Exxon Research and Engineering Company
Annandale, New Jersey, U.S.A.

MRS
MATERIALS RESEARCH SOCIETY
PITTSBURGH, PENNSYLVANIA

This work was supported in part by the Office of Naval Research under Grant Number ONR: N00014-97-1-0102. The United States Government has a royalty-free license throughout the world in all copyrightable material contained herein.

Single article reprints from this publication are available through University Microfilms Inc., 300 North Zeeb Road, Ann Arbor, Michigan 48106

CODEN: MRSPDH

Copyright 1997 by Materials Research Society.
All rights reserved.

This book has been registered with Copyright Clearance Center, Inc. For further information, please contact the Copyright Clearance Center, Salem, Massachusetts.

Published by:

Materials Research Society
9800 McKnight Road
Pittsburgh, Pennsylvania 15237
Telephone (412) 367-3003
Fax (412) 367-4373
Website: http://www.mrs.org/

Library of Congress Cataloging in Publication Data

Statistical mechanics in physics and biology : symposium held December 2–5, 1996, Boston, Massachusetts, U.S.A. / editors, Denis Wirtz, Thomas C. Halsey
p. cm—(Materials Research Society symposium proceedings ; v. 463)
Includes bibliographical references and index.
ISBN 1-55899-367-3
1. Statistical mechanics—Congresses. 2. Biophysics—Congresses. I. Wirtz, Denis II. Halsey, Thomas C. III. Series: Materials Research Society symposium proceedings ; v. 463.
QC174.7.S742 1997 97-20743
530.13—dc21 CIP

Manufactured in the United States of America

CONTENTS

*Invited Paper

PART IV: STATISTICAL MECHANICS IN BIOLOGY

PART V: MICROMANIPULATION IN PHYSICS AND BIOLOGY

PART VI: FRICTION AND FRACTURE

PART VII: STATISTICAL MECHANICS IN PHYSICS

*Invited Paper

PREFACE

The concepts of statistical mechanics have long been among the most widely applied of physical ideas. The most notable aspect of the symposium "Statistical Mechanics in Physics and Biology" is the sheer range of subjects, including molecular biology, cellular biology, chemical engineering, petroleum engineering, polymer science, and metallurgy, addressed by similar methods.

This symposium could not have taken place without the financial support of the Office of Naval Research, Eastman Kodak Company, and Exxon Research and Engineering Company. In particular, we wish to acknowledge the generous support of Michael F. Shlesinger, Krishnan Chari, and Hans Thomann. We also wish to thank Robert Austin of Princeton University who was pivotal in the organization of the symposium. Finally, we wish to thank the MRS staff for its expert assistance with preparing these proceedings.

Denis Wirtz
Thomas C. Halsey

May 1997

MATERIALS RESEARCH SOCIETY SYMPOSIUM PROCEEDINGS

MATERIALS RESEARCH SOCIETY SYMPOSIUM PROCEEDINGS

Volume 444— Materials for Mechanical and Optical Microsystems, M.L. Reed, M. Elwenspoek, S. Johansson, E. Obermeier, H. Fujita, Y. Uenishi, 1997, ISBN: 1-55899-348-7

Volume 445— Electronic Packaging Materials Science IX, P.S. Ho, S.K. Groothuis, K. Ishida, T. Wu, 1997, ISBN: 1-55899-349-5

Volume 446— Amorphous and Crystalline Insulating Thin Films—1996, W.L. Warren, J. Kanicki, R.A.B. Devine, M. Matsumura, S. Cristoloveanu, Y. Homma, 1997, ISBN: 1-55899-350-9

Volume 447— Environmental, Safety, and Health Issues in IC Production, R. Reif, A. Bowling, A. Tonti, M. Heyns, 1997, ISBN: 1-55899-351-7

Volume 448— Control of Semiconductor Surfaces and Interfaces, S.M. Prokes, O.J. Glembocki, S.K. Brierley, J.M. Woodall, J.M. Gibson, 1997, ISBN: 1-55899-352-5

Volume 449— III–V Nitrides, F.A. Ponce, T.D. Moustakas, I. Akasaki, B.A. Monemar, 1997, ISBN: 1-55899-353-3

Volume 450— Infrared Applications of Semiconductors—Materials, Processing and Devices, M.O. Manasreh, T.H. Myers, F.H. Julien, 1997, ISBN: 1-55899-354-1

Volume 451— Electrochemical Synthesis and Modification of Materials, S.G. Corcoran, P.C. Searson, T.P. Moffat, P.C. Andricacos, J.L. Deplancke, 1997, ISBN: 1-55899-355-X

Volume 452— Advances in Microcrystalline and Nanocrystalline Semiconductors—1996, R.W. Collins, P.M. Fauchet, I. Shimizu, J–C. Vial, T. Shimada, A.P. Alvisatos, 1997, ISBN: 1-55899-356-8

Volume 453— Solid–State Chemistry of Inorganic Materials, A. Jacobson, P. Davies, T. Vanderah, C. Torardi, 1997, ISBN: 1-55899-357-6

Volume 454— Advanced Catalytic Materials—1996, M.J. Ledoux, P.W. Lednor, D.A. Nagaki, L.T. Thompson, 1997, ISBN: 1-55899-358-4

Volume 455— Structure and Dynamics of Glasses and Glass Formers, C.A. Angell, T. Egami, J. Kieffer, U. Nienhaus, K.L. Ngai, 1997, ISBN: 1-55899-359-2

Volume 456— Recent Advances in Biomaterials and Biologically–Inspired Materials: Surfaces, Thin Films and Bulk, D.F. Williams, M. Spector, A. Bellare, 1997, ISBN: 1-55899-360-6

Volume 457— Nanophase and Nanocomposite Materials II, S. Komarneni, J.C. Parker, H.J. Wollenberger, 1997, ISBN: 1-55899-361-4

Volume 458— Interfacial Engineering for Optimized Properties, C.L. Briant, C.B. Carter, E.L. Hall, 1997, ISBN: 1-55899-362-2

Volume 459— Materials for Smart Systems II, E.P. George, R. Gotthardt, K. Otsuka, S. Trolier-McKinstry, M. Wun-Fogle, 1997, ISBN: 1-55899-363-0

Volume 460— High–Temperature Ordered Intermetallic Alloys VII, C.C. Koch, N.S. Stoloff, C.T. Liu, A. Wanner, 1997, ISBN: 1-55899-364-9

Volume 461— Morphological Control in Multiphase Polymer Mixtures, R.M. Briber, D.G. Peiffer, C.C. Han, 1997, ISBN: 1-55899-365-7

Volume 462— Materials Issues in Art and Archaeology V, P.B. Vandiver, J.R. Druzik, J. Merkel, J. Stewart, 1997, ISBN: 1-55899-366-5

Volume 463— Statistical Mechanics in Physics and Biology, D. Wirtz, T.C. Halsey, J. van Zanten, 1997, ISBN: 1-55899-367-3

Volume 464— Dynamics in Small Confining Systems III, J.M. Drake, J. Klafter, R. Kopelman, 1997, ISBN: 1-55899-368-1

Volume 465— Scientific Basis for Nuclear Waste Management XX, W.J. Gray, I.R. Triay, 1997, ISBN: 1-55899-369-X

Volume 466— Atomic Resolution Microscopy of Surfaces and Interfaces, D.J. Smith, R.J. Hamers, 1997, ISBN: 1-55899-370-3

Prior Materials Research Society Symposium Proceedings available by contacting Materials Research Society

Part I

Statistical Mechanics of Proteins

Adhesion of Soft Biological Shells controlled by Bending Elasticity and Macromolecular Networks

R. Simson, A. Albersdörfer, E. Sackmann*
*Physik Department E22, Technische Universität München,
James Franck Strasse, 85748 Garching, Germany.
sackmann@physik.tu-muenchen.de

Abstract

We present an interferometric technique allowing reliable measurements of bending modulus κ, membrane tension Σ and adhesion energy W of cells crawling on substrates. All three parameters are important for cell locomotion and reflect the local balance of attractive and repulsive forces between cell and substratum as well as the internal coupling of cell membrane and the underlying cytoskeleton. Mutants of the Dictyostelium ameba lacking an important cytoskeletal protein, Cortexillin, exhibited a markedly reduced bending modulus and adhesion energy as compared to wild type Dictyostelium. In addition, experiments with model membrane systems suggest that the combination of attractive and repulsive forces results in a local clustering of receptors mediating cell adhesion.

Introduction

Cell adhesion is tuned by a complex interplay of specific (lock and key) and universal forces (Fig. 1). It is also most effectively controlled by the bending elasticity of the soft shells as demonstrated by both theoretical (Seifert et al., 1991) and experimental studies (Albersdörfer et al., submitted). The latter control mechanism is very complex since the plasma membrane is a stratified plate composed of two soft coupled shells, the lipid bilayer and the associated cytoskeleton. The latter can consist of a relatively simple quasi-two dimensional triangular network as in Erythrocytes or of an actin myosin network, about 1 µm thick, called the actin cortex.

Numerous studies of lipid vesicles (Lipowsky and Sackmann, 1995) and erythrocytes showed that bending elasticity and shape changes of stratified membranes are sensitively dependent on the degree of coupling between the two shells (Discher and Mohandas, 1996).Therefore the bending elasticity is determined by the intrinsic stiffness of the cortex and its coupling strength to the bilayer. Measurements of the bending moduli of wild type and mutant Dictyostelium show that cells with a damaged cytoskeleton exhibit a reduced bending stiffness and adhesion energy as compared to normal cells.

Various studies show that cells tend to avoid large contact areas with solid surfaces which results in rapid cell death (Hanein et al., 1994) but attach only strongly through local adhesion plaques (Fig. 1 b,c). This is essential for moving cells since it minimizes loss of membrane material during retraction. In this paper we also present model membrane studies, showing that local clustering of receptors occurs as a natural consequence of the interplay of strong attraction and week repulsion forces.

Materials and Methods

THEORY

We present an interferometric technique allowing reliable measurements of bending modulus κ, membrane tension Σ and adhesion energy W of cells or vesicles

Mat. Res. Soc. Symp. Proc. Vol. 463 © 1997 Materials Research Society

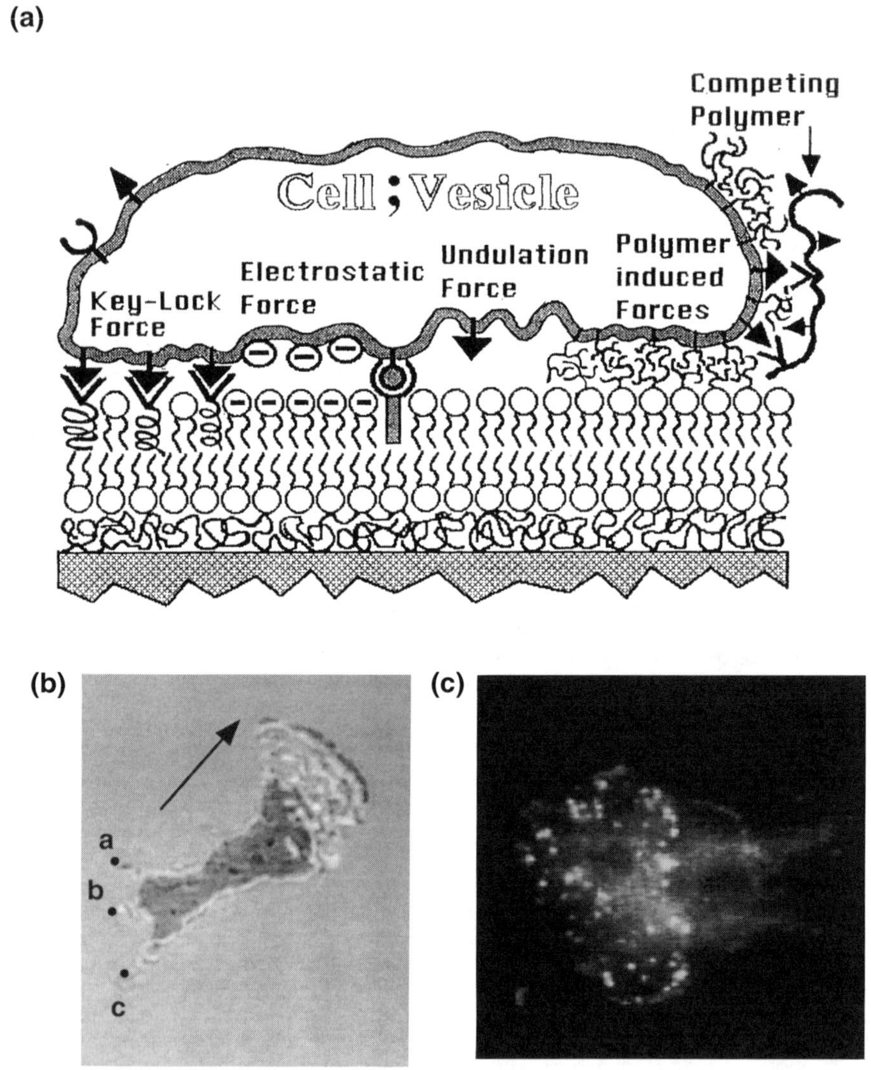

Figure 1
(a) Adhesion is controlled by specific lock and key forces, by a variety of universal forces and by shape and elasticity of the soft shell
(b) Areas of tight local contact (pinning centers) between a Dictyostelium cell and a BSA coated glass substrate can be observed by the tether formation (arrows) at the retracting end.
(c) Indirect demonstration of local contact formation by staining actin left behind on a positively charged surface during retraction. The initial cell adhesion area was labeled with fluorescent phalloidin after the cell moved away.

on substrates. This technique is based on the analysis of the cell contour near the contact zone. Plated cells, when observed with the interference reflection microscope (IRM) yield an interference pattern allowing reconstruction of the cell contour to a height of about 1 μm from the edge of the adhesion area, referred to as the contact line *L* (Rädler at al., 1995; Fig. 2). The mechanical equilibrium at the contact line is determined by the balance of tensions which can be expressed in terms of Youngs law (Bruinsma, 1996),

$$W = \Sigma\,(1-\cos \vartheta_c) \qquad (1)$$

where ϑ_c is the contact angle as defined in Fig. 1c, and by the momentum equilibrium leading to the following relationship between adhesion energy *W* and bending modulus κ (Seifert and Lipowsky,1990).

$$W = \kappa / 2\,R_c \qquad (2)$$

where R_c^{-1} is the contact curvature. A direct measurement of the contact curvature is difficult (Rädler et al., 1995). However, following work by Bruinsma, a simplified procedure can be adopted. Assuming that adhesion does not alter the cell volume (Bruinsma 1996) the elastic energy is dominated by the contribution of the contact zone and can be written as

$$\Delta G = 2\pi u\left[\int_0^{\infty} dx\left\{\frac{1}{2}\kappa\left(\frac{d^2h}{dx^2}\right)^2 + \frac{1}{2}\Sigma\left(\frac{dh}{dx}\right)^2\right\} - \int_{-\infty}^{0} W(x)dx\right] \qquad (3)$$

where the integral is performed along a section through the center of the adhesion disc. Minimizing this energy yields a differential equation of the form

$$\Sigma\frac{d^2h}{dx^2} - \kappa\frac{d^4h}{dx^4} = 0 \qquad (4)$$

This equation introduces a characteristic length scale

$$\lambda = \sqrt{\kappa/\Sigma} \qquad (5)$$

separating different regimes of the cell contour. For distances $\Delta x < \lambda$, measured from the contact line, the elastic energy is dominated by bending elasticity, resulting in a curved bilayer. For distances $\Delta x > \lambda$, the tension dominates and the contour is a straight line forming the contact angle ϑ_c with the substrate. The height profile, satisfying the boundary conditions given in equs. 1 and 2, may be expressed as

$$h(x) = \vartheta x - \vartheta\lambda\left[1-\exp\left(-\frac{x}{\lambda}\right)\right] \qquad x > 0 \qquad (6)$$
$$h(x) = 0 \qquad x \le 0$$

MODEL MEMBRANE SYSTEMS

In order to study the effect of balance of short range attraction forces, mediated by receptors, and long range repulsion forces, based on polymer induced forces or undulation forces, we established the model system shown in Fig.3a. We studied the adhesion of giant lipid vesicles to a supported lipid bilayer using IRM. Both vesicles and supported membranes were composed of 94.9 mole% DOPC, 5 mole% DOPE-PEO_{2000} and 0.1 mole% DOPE-X-biotin and were prepared as described elsewhere [Albersdörfer et al, submitted]. The incorporation of lipopolymers (phospholipids with polyethyleneoxide head groups composed of 45 monomers corresponding to a

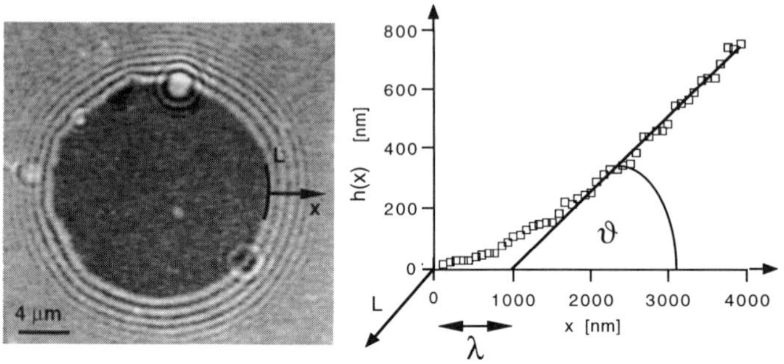

Figure 2
Left: IRM micrograph of the contact zone of an adhering vesicle. Right: Example of the contour h(x) of the adhering vesicle in direction perpendicular to the contact line L as obtained by analyzing the interference pattern of the IRM micrograph. The width of the rim λ and the contact angle ϑ are obtained by analyzing the contact contour along a straight line perpendicular to L , and a locally curved region, at the contact line.

molecular weight of 2000 Da) models the cell glycocalix. The short range attraction was achieved by biotin-streptavidin-biotin coupling.

The vesicle contour at the rim of the adhesion area was calculated from the interference fringe pattern as described above. Performing a linear least square fit to the tension dominated, straight part of the contour one can measure the characteristic length λ and the contact angle ϑ_C. Using equs. 1 and 5 one can locally determine tension Σ and adhesion energy W for a known value of the bending energy κ.

MEASURING TENSION, ADHESION ENERGY AND BENDING MODULUS USING SMALL SHEAR FORCES.

In order to determine both bending energy and tension for crawling cells, we exposed cells to a small laminar shear flow, using a standard parallel plate flow chamber. The flow chamber dimensions (500 µm by 5 mm cross section) were carefully chosen in order to ensure a laminar flow profile, and a syringe pump was used to control flow rate and thus the shear force exerted on the cell. The applied flow rates and resulting shear stresses (0.4 to 1.2 Pa) were small enough to prevent a significant deformation of the cell shape or even detachment. However, the shear stresses were sufficiently large to measurably change contact angle and membrane curvature of the cell at the edge of the adhesion area. When observed with IRM, this change in curvature results in an alteration of the interference fringe pattern which can be quantified (Fig. 4). We compared the cell contour reconstructed from the perturbed fringe pattern with the contour of the undisturbed cell. Due to the bending stiffness of the membrane-cortex complex shear forces acting on the dorsal cell surface result in a change of tension right at the rim of the adhesion disk. At the edge of the cell facing the shear flow, the applied shear stress σ therefore increases the membrane tension at the contact line (Bruinsma, 1996).

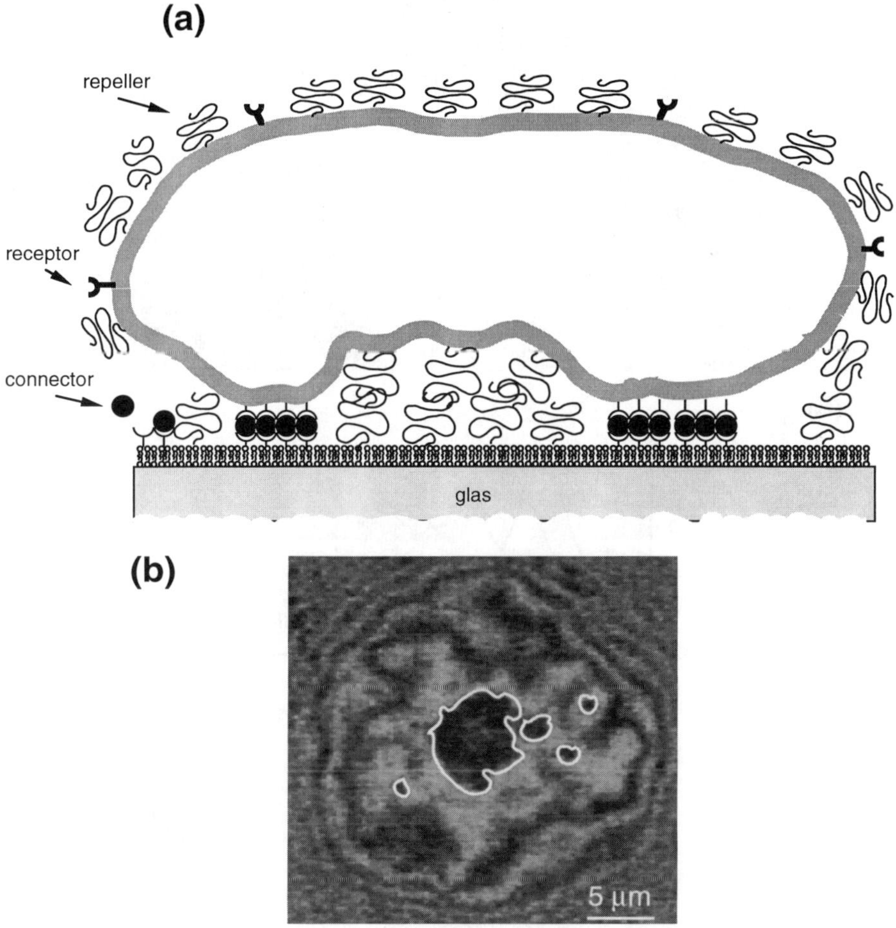

Figure 3
(a) Model system mimicking the interplay of short range attraction and long range repulsion forces. The glycocalix is modeled by incorporation of lipopolymers (phospholipids with polyethyleneoxide head groups composed of 45 monomers corresponding to a molecular weight of 2000 Da), and the short range attraction is achieved by biotin-streptavidin-biotin coupling.
(b) Demonstration of adhesion induced clustering of receptors leading to the formation of local pinning centers. The IRM micrograph shows only the contact zone of the adhering vesicle. The local sites of close contact between vesicle and substrate are outlined with a white border line. Note that dark areas correspond to short interfacial distances.

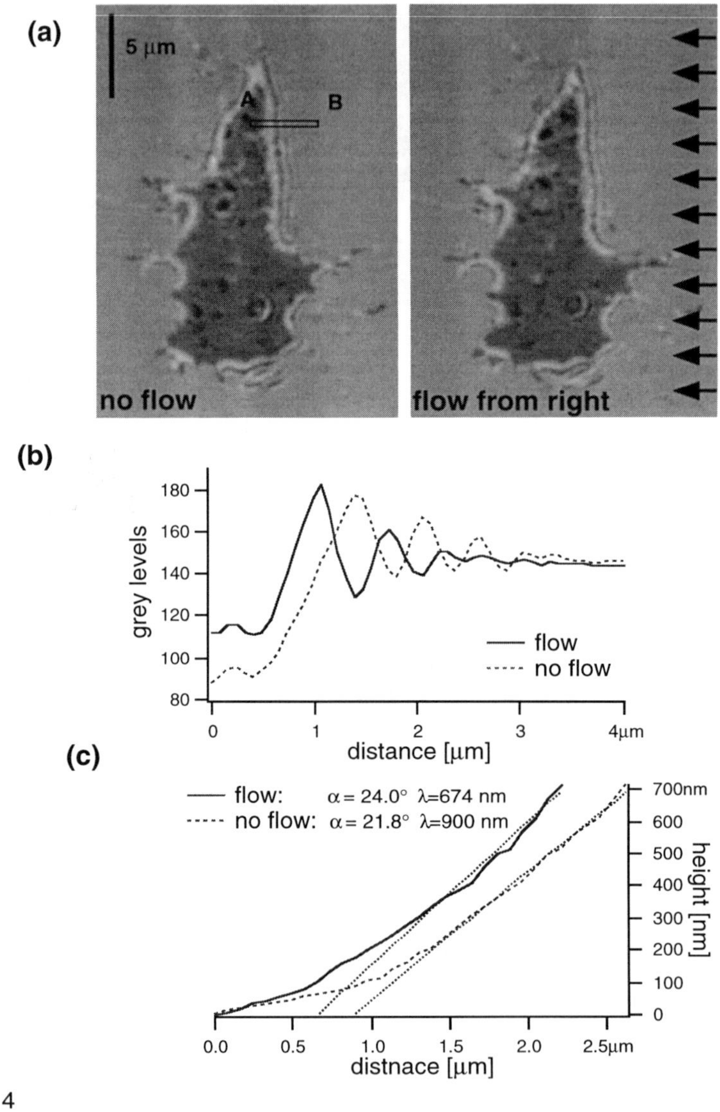

Figure 4
(a) A typical contact area of a wild type Dictyostelium on BSA coated glass observed with IRM. Note the interference pattern around the rim of the adhesion area. On the left side no shear force is exerted. On the right side a shear stress of 1.12 Ps is exposed to the cell.
(b) Distribution of gray values measured for the image shown in 2a from A to B before (dotted line) and at the identical position shortly after (solid line) the laminar shear flow was turned on.
(c) Cell contours reconstructed from the interference patterns shown in 2b. Note that the shear forces result in a reduced value of λ and increased contact angle α.

Assuming that the bending modulus of the cortex remains unaltered by the small shear force , this increase in cell tension results in a measurable decrease of the characteristic length scale λ (Fig. 4). Knowing the flow induced change in tension, both the undisturbed tension and the bending energy can be obtained from equ. 5 For flat cells, the increase in tension $\Delta\Sigma$ due to flow is approximately given by $\Delta\Sigma = \sigma R_c$, where R_c is the diameter of the contact area parallel to the direction of flow. Additional experiments on the laser scanning microscope show that most cells extend less than 5 μm into the volume and the above approximation of a flat cell is valid (data not shown).

For all experiments we used a bovine serum albumin (BSA) coated glass coverslip as substratum. The coverslip was incubated with a 10mg/ml BSA solution of Sörensen phosphate buffer at pH 6.0. As shown in previous studies (Schindl et al., 1995) BSA coated glass provides a very homogeneous and well suited substratum for cell locomotion.

Results and Discussion

MODEL MEMBRANES

As demonstrated in Fig. 3a,b, adhesion leads to the formation of local pinning centers which are separated by regions exhibiting strong flickering. Polymer induced forces as well as undulation forces associated with flickering are strong enough to overcome the Van der Waals attraction (Rädler et al., 1995). In order to form tight connections work has to be performed against these repulsive forces. Single streptavidin-biotin bonds, as used in this study, would be associated with an excessively high energy. Indeed, it has been shown that membrane undulations may mediate attraction forces between strong connectors (Bruinsma et al. 1994). This effect can be attributed to a phase separation of receptors and polymers in the contact area.

LOCAL MEASUREMENTS OF MEMBRANE TENSION AND ADHESION ENERGY

We observed the contact zone of adhering giant vesicles to streptavidin coated supported membranes using IRM (Fig. 5a). The diffraction fringes are remarkably distorted near pinning centers as shown by closer inspection of position 4 in Fig 5a. We calculated membrane tensions and contact angles at different positions of the contact area evaluating the vesicle contours along the sections marked in Fig. 5a. For this calculations we used a bending modulus of 35 k_BT, which was measured for a comparable system (SOPC; Duwe and Sackmann, 1990; Evans and Parsegian, 1983). Three examples of contours are shown in Fig. 5b: one in close proximity to a pinning center, another one in a region which seems to be not influenced by pinning and a third one in an intermediate region. The obvious differences in contact angles (Fig. 5b) indicate a higher tension and an increased adhesion energy in the proximity of a pinning center located close to the contact line. The bent in contour 4d may be attributed to a relaxation of the membrane tension far away from the contact line. Our results (Fig. 5c) show that the local adhesion energy at a pinning center is more than two orders of magnitude larger as compared to energies measured far away from such a center. The same holds for the tension (Fig. 5c).

EFFECT OF MUTATIONS ON CELL BEHAVIOR

An intriguing observation is that chemomechanical processes such as cell division and locomotion are often remarkably insensitive towards depletion of essential proteins regulating the structure of the actin based cytoskeleton. For example, removal of the motor protein myosin which is assumed to be essential for the formation of the cleavage furrow during cell division does not affect division of crawling

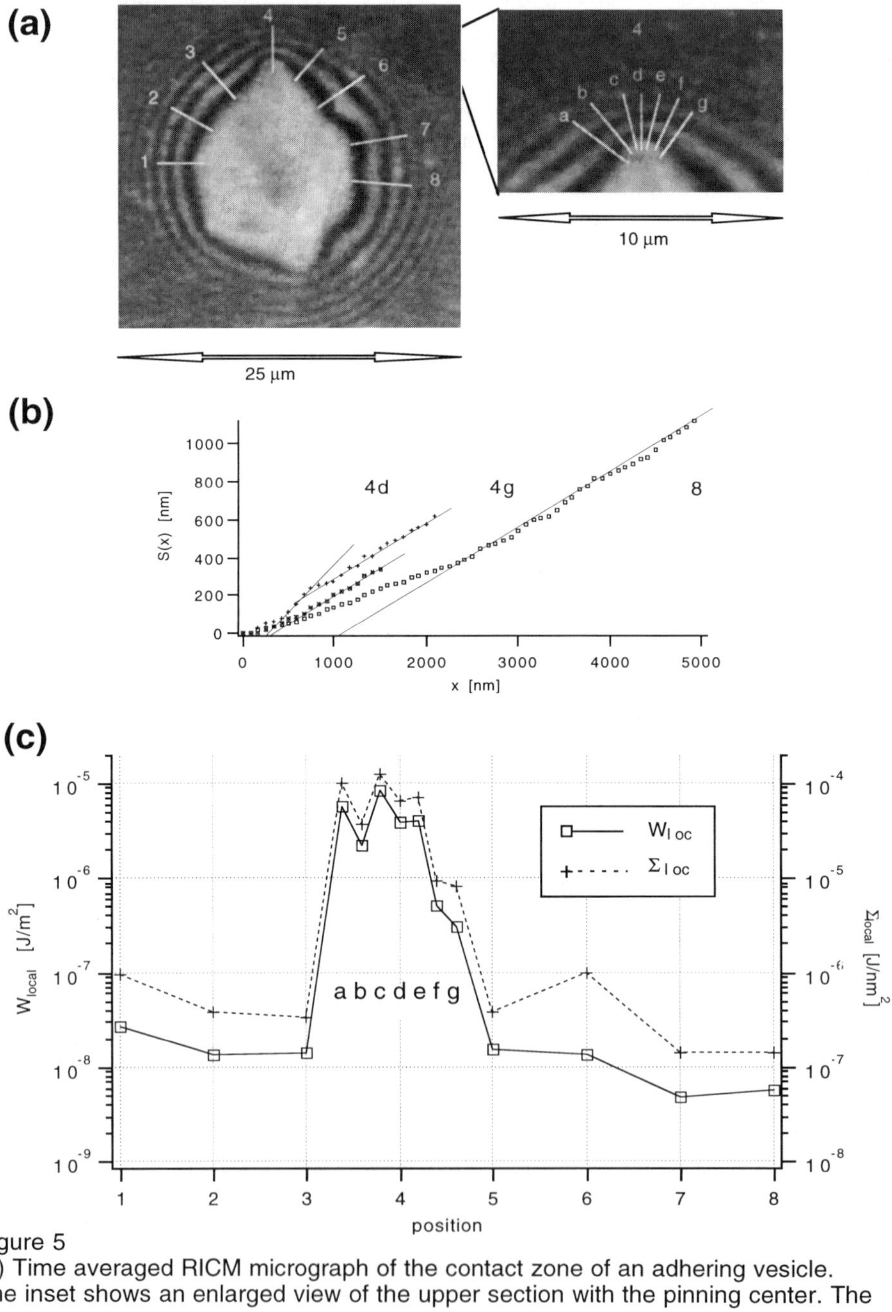

Figure 5
(a) Time averaged RICM micrograph of the contact zone of an adhering vesicle. The inset shows an enlarged view of the upper section with the pinning center. The

white bars mark the directions along which contact contours are analyzed. The diffraction fringes are remarkably distorted near the pinning centers corresponding to increased contact angles ϑ_c and decreased values of λ.
(b) Three examples of contours near contact area: one in the region of a pinning center (position 4d), one in a region which is not influenced by a pinning center (position 8) and one that is situated in between these two regions (position 4g).
(c) Local tensions (dashed line) and adhesion energies (drawn line) plotted versus the position line number marked in Fig. 2a. These values were determined by using the values of ϑ_c and λ which were fitted into the local height profiles h(x). Additionally we used a value of $\kappa \approx 35\ k_BT$ for the bending modulus which was measured for pure SOPC vesicles by other techniques (Duwe and Sackmann, 1990; Evans and Parsegian, 1983).

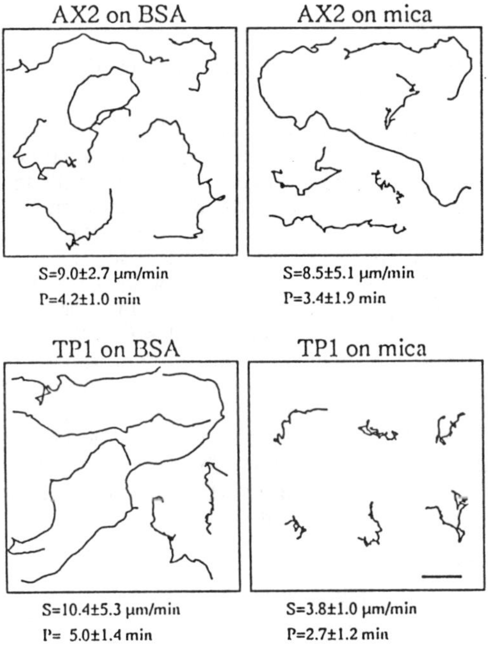

Figure 6
Comparison of random walks of wild type and triple mutant on glass (a) and freshly cleaved mica (b). Note that the mutant locomotion is strongly impeded on the smooth mica surface. Both cells were examined in the elongated state after 6 hours starvation (Weber et al., in press).

cells. Triple mutants of Dictyostelium discoideum, lacking two cross-linking proteins (α-actinin, 120kDa gelation factor) and a severing protein (severin), exhibit unperturbed crawling on substrates exerting strong adhesion forces such as albumin coated glass. However, their advancement is strongly impeded on weakly adhesive surfaces as freshly cleaved mica (Fig. 6).

Experiments show that mutations of the cytoskeleton can be overcome by strong adhesion. The strikingly different behavior of mutants on glass and mica points to an inability of mutants to stabilize adhesion of newly formed pseudopodia. Moreover, the quasi-periodic contact area oscillations, with periods of 3 to 6 min, (Schindl, 1995) suggest that the pseudopod formation, although occurring in random directions, is driven by an intracellular process. This process could in turn be associated with lateral contractions and expansions of the actin-myosin cortex. The cell can only advance, first if pseudopod spreading is followed by retraction of the cell near the opposite pole and second, if the newly formed pseudopod is attached strong enough to the substrate for transmitting momentum. If the adhesion is too weak the pseudopod is retracted again if a second pseudopod is formed in a different direction, as a consequence of the bilayer incompressibility.

EVALUATION OF EFFECTS OF MUTATIONS ON BENDING MODULUS OF THE CELL CORTEX AND ADHESION ENERGY.

Using small shear forces (Materials and Methods) we measured bending modulus, tension and adhesion energy for wild type and a mutant of *Dictyostelium. discoideum* lacking a protein called Cortexillin I. This recently found protein belongs to a new class of actin binding proteins. Cortexillin I has been shown to promote the formation of actin bundles that associate into meshworks and plays a key role in cytokinesis and maintaining the cell shape (J. Faix et al., 1996). It's presumed importance for mechanical cell stability could be confirmed by our measurements. With a bending energy of $94 \pm 18\ k_BT$ the Cortexillin I- mutant appears to be much softer than the wild type cell which exhibited a bending modulus of $386 \pm 158\ k_BT$. Also the membrane tension is higher for the wild type as compared to the mutant. Interestingly, the undisturbed contact angles of wild type and mutant cells do not differ at all. Only their response to an external force reveals the dramatic differences in their elastic properties. A difference could be observed for the adhesion energies, where the wild type adhered stronger (2.2 ± 1.2 e-7 J/m^2) than the mutant (1.5 ± 1.0 e-7 J/m^2). This reduction in adhesion energy my be attributed to an increased repulsion based on stronger undulation forces for the softer mutant.

Acknowledgments

The authors thank Prof. Robin Bruinsma for helpful discussion. We are also grateful to Dr. Jan Faix and Prof. Günther Gehrisch at the MPI Martinsried for providing the Dictyostelium cell lines.

List of abbreviations

DOPC	dioleoylphosphatidylcholine
DOPE-X-biotin	biotin-X-dioleoylphosphatidylethanolamine
DOPE-PEO$_{2000}$	dioleoylphosphatidylethanolamine-polyethyleneoxide
SOPC	stereoylphosphatidylcholine

References

Albersdörfer, A., T. Feder, and E. Sackmann. submitted (1996).

Bruinsma, R., M. Goulian, and P. Pincus. *Biophys. J.* 67: 746-750 (1994).

Bruinsma, R. NATO, ASI, Winterschool: Physics of Biomaterials, Plenum Press, in press (1996).

Duwe, H.P., and E. Sackmann. *Physica A.*163: 410-428 (1990).

Discher, E.D. and N. Mohandas. *Biophys. J.* 71: 1680-1694 (1996).

Evans, E., and V.A. Parsegian. *Ann. N.Y. Acad. Sci.* 416: 13-33 (1983).

Faix, J., M. Steinmetz, H. Boves, R.A. Kammerer, F. Lottspeich, U. Mintert, J. Murphy, A. Stock, U. Aebi and G. Gehrisch. *Cell* 86: 631-642 (1996).

Hanein, D., B. Geiger and L. Addadi. *Science* 263: 1413-1416 (1994).

Lipowsky, R. and E. Sackmann. Structure and Dynamics of Membranes, Elsevier Amsterdam (1995).

Rädler, J., T.J. Feder, H.H. Strey and E. Sackmann. *Phys. Rev. E.* 51(5): 4526-4536 (1995).

Seifert, U., and R. Lipowsky. 1990.*Phys, Rev. A.* 42: 4768-4771.

Seifert, U., L. Miao, H. Döbereiner and M. Wortis in The structure and conformation of amphiphilic membranes. Springer Proceedings in Physics 66 (1991).

Weber, I., E. Wallraff, R. Albrecht and G. Gehrisch. in press.

LOCAL VISCOELASTICITY OF BIOPOLYMER SOLUTIONS

B. SCHNURR*, F. GITTES*, P.D. OLMSTED**, C.F. SCHMIDT*, F.C. MACKINTOSH*
* *Dept. of Physics & Biophys. Res. Div., University of Michigan, Ann Arbor, MI 48109-1120*
** *Dept. of Physics, University of Leeds, Leeds, LS2 9JT, United Kingdom*

Abstract

We describe a new, high-resolution technique for determining the local viscoelastic response of polymer gels on a micrometer scale. This is done by monitoring thermal fluctuations of embedded probe particles. We derive the relationship between the amplitude of fluctuations and the low-frequency storage modulus G', as well as the relationship between the fluctuation power spectrum, measured between 0.1Hz and 25kHz, and the complex shear modulus $G(\omega)$. For both, semiflexible F-actin solutions and flexible polyacrylamide (PAAm) gels we observe high-frequency power-law dependence in the spectra, which reflects the behavior of the shear modulus. However, we observe distinctly different scaling exponents for $G(\omega)$ in F-actin and PAAm gels—presumably due to the semiflexible nature of the actin filaments.

Introduction

Synthetic polymer solutions and gels are common and technologically important materials. They exhibit rich equilibrium and non-equilibrium behavior, particularly in their response to shear stress not unlike their biological analogues. Plant and animal cells contain a multitude of filamentous protein, especially in the *cytoskeleton*, which is a complex network of biopolymers. A principal component of the cytoskeleton is polymeric actin (F-actin), which is largely responsible for the viscoelastic response of plant and animal cells [1]. To understand the mechanical properties and dynamics of living cells, the viscoelastic properties of F-actin solutions *in vitro* have been measured by a number of groups [2-7], using macroscopic rheology. These measurements focused on the elastic shear modulus, a quantity of particular significance for the elastic response of cells. The microscopic properties of actin filaments lead to fundamental differences between the actin cortex and synthetic polymer systems under similar conditions. Actin filaments are rather rigid on the scale of the characteristic mesh size of the networks. This appears to be responsible for the anomalously large shear moduli of actin solutions even at low concentrations of order 1 mg/ml (or about 0.1% volume fraction)[6]. Microscopic measurements of the viscoelasticity of actin gels and other polymer systems have been reported recently. Ziemann et al. [8,9] and Amblard et al. [10] used the manipulation of micron size magnetic beads by field gradients and video displacement detection. Other important experiments have employed multiple light scattering [11].

In this paper, we describe a new method to measure local viscoelasticity based on the high-resolution observation of thermal fluctuations of probe particles (silica beads). The bead displacements can be measured to better than 1nm resolution at frequencies up to 50kHz. The experiments with F-actin and PAAm gels are aimed at understanding the fundamental differences between semiflexible and flexible polymer gels, as well as the basic viscoelastic properties of polymer systems on a local scale. The method extends rheology to a frequency range that has not been studied before.

Theory

Consider the motion of a spherical bead of radius R in a viscoelastic medium. The drag force in a purely viscous fluid of viscosity η is given by the Stokes formula [12]: $f = 6\pi\eta R v$, where v is the bead velocity. This result also applies to oscillatory motion, provided that the viscous penetration depth $\delta = \sqrt{2\eta/(\rho\omega)}$ is large compared with R, where ρ is the fluid density. Inertial effects can be neglected at low frequencies. For a micron size bead in water this holds for

Mat. Res. Soc. Symp. Proc. Vol. 463 © 1997 Materials Research Society

frequencies less than about 1 MHz. For oscillatory motions below this frequency, the motion of the fluid is described by an amplitude which is everywhere independent of the frequency. Thus, the stress is independent of ω, apart from the dependence on the velocity v. The analogous relationship between the applied force and particle displacement x in a viscoelastic medium is[13]

$$f_\omega = 6\pi G(\omega) R x_\omega, \tag{1}$$

where $G(\omega) = G'(\omega) + iG''(\omega)$ is the (complex) shear modulus. The Stokes formula assumes an incompressible medium. For a solution of polymer and solvent, this means no *relative* motion of the two. The full expression [14,15] for the force at zero frequency on a (rigid) bead of radius R displaced by a distance x in a continuum elastic medium is given by

$$f = Kx = \frac{24\pi(1-v)}{5-6v} G'(0) R x. \tag{2}$$

where K is the elastic spring constant. Taking the Poisson ratio $v = 1/2$ (for an incompressible medium) this general expression reduces to Eq. (1).

The extent to which the polymer solution behaves as an incompressible medium depends on the frequency of bead motion. At very low frequencies the stress in the solution is due entirely to the compressible elastic network. Both shear and compression moduli are finite, and the Poisson ratio need not be 1/2. At high frequencies, on the other hand, the medium behaves as an incompressible viscoelastic fluid. This can be seen within a "two-fluid" model [16,17], in which an elastic network is viscously coupled to the solvent. This coupling leads to a stress on the network given by a force per unit network volume $\eta v / \xi^2$, where η is the solvent viscosity, ξ is the mesh size of the network, and v is the relative velocity of the solvent with respect to the network. The elastic force on a volume element is given by $G\nabla^2 u$, where u is the network displacement field. This is of order Gx / R^2 for a bead of radius R, displaced by x. Thus, the viscous coupling is dominant for $\omega \geq G(\xi / R)^2 / \eta$. For G of order 1Pa and $\xi \approx 0.1R$ the lower limit is of order 10Hz; for higher frequencies the two fluids move as one, while the Poisson ratio may fall below 1/2 for lower frequencies.

Given the force-displacement relationship in Eq. (1), the power spectrum can be determined from the fluctuation-dissipation theorem [18]. The result for the power spectral density (PSD) is

$$\langle x_\omega{}^2 \rangle = \frac{2kT}{\omega} \mathrm{Im}\left(\frac{1}{6\pi G(\omega) R} \right) = \frac{kT}{3\pi\omega R} \frac{|G''(\omega)|}{|G(\omega)|^2}. \tag{3}$$

The complex shear modulus $G(\omega)$ usually exhibits three distinct frequency regimes in polymer solutions [19]. For a non-crosslinked solution the behavior is essentially that of a viscous liquid at frequencies lower than $1/\tau_R$, where τ_R is the reptation time. At intermediate frequencies a rubber-like plateau appears, for which the dominant response is elastic, and nearly independent of frequency. Above this, the storage (G') and loss (G'') moduli increase with frequency as power laws: $G', G'' \propto \omega^Z$ [19]. In this regime, Eq. (3) implies a power spectrum scaling as ω^{-1-Z}.

Materials and Instrumentation

Actin was purified from chicken skeletal muscle to about 95% purity following standard recipes [20]. Actin concentration was determined both by staining (BioRad) and by UV absorption at 290nm. Monomeric actin (G-actin) was stored at -85°C. Samples were prepared by mixing a small number of silica beads with G-actin, adding concentrated polymerization buffer (F-buffer: 2mM HEPES, pH 7.5, 2mM $MgCl_2$, 50mM KCl, 1mM EGTA, 1mM ATP) and immediately transferring the solution into a glass sample chamber (15mm x 3mm x 70µm). The chambers were sealed and the actin polymerized at room temperature for at least one hour under slow rotation. The samples were used within one day.

PAAm gels were prepared according to a standard gel electrophoresis recipe [21], with a relative concentration of 3% bis-acrylamide. Solutions were degassed under vacuum, and polymerization started by adding TEMED and ammonium persulfate. The polymerizing solutions with silica beads were immediately transferred into sample chambers (15mm x 6mm x 140μm), which were sealed and slowly rotated at room temperature for at least one hour before starting experiments. All experiments were done with PAAm gels of a final concentration of 2% (w/v).

The microscope we used is a custom built instrument, constructed on an optical rail system and mounted on a vibration isolated optical bench. To detect the thermal motion of beads imbedded in the gels we used an interferometric method employing near infrared laser illumination as described previously [22,23]. The laser power was typically about 0.6mW in the specimen, low enough to make optical forces negligible. The laser light is detected by two photodiodes. The normalized difference of the signals from the two diodes provides, due to a differential phase shift, a measure for the position of the bead. The linear range of the detector is about ±200nm for 0.5μm beads and increases with bead size. The amplified position signal is anti-alias filtered at the Nyquist frequency and then digitized. For all experiments reported here, the data acquisition rate was 50kHz and the filter frequency 25kHz, the total duration of a run was typically 20 seconds. Fast Fourier Transforms (FFT) were performed using a Bartlett window. Power spectra were smoothed by averaging spectra obtained from windows smaller than the full data set.

The detector was calibrated for each observed bead, immediately following the thermal motion recording, by moving the sample with a piezo-actuated stage with a triangular signal, producing between 150nm and 1μm peak-to-peak displacement at a frequency of 0.1 Hz. The driving voltage was corrected for the non-linearity and hysteresis of the piezo actuators. Piezo voltage and detector response were digitized and recorded for about 40 seconds. These time series were analyzed to obtain an averaged displacement vs. voltage curve for each individual bead. The slopes of linear fits (typical error 20%) were used to convert signal voltages to nm.

Results

Thermal fluctuations of entrapped silica beads (diameters: 0.5, 0.96, 1.8, and 5.0μm) were observed in actin gels of 1 and 2mg/ml concentrations. The spring constant K for elastic displacements of the bead in the gel can be obtained from the total variance in bead position $\langle x^2 \rangle$ (see [24] for a general discussion of micromechanical measurements). By Eq. (2), K determines the macroscopic zero-frequency shear modulus G_0' of the network. The variance could, in principle, be directly obtained from the time series data. Low frequency noise, however, can dominate this variance. Therefore we estimate the variance by summing the displacement power spectrum from a lower cut-off frequency of 1Hz. This cut-off was chosen by inspecting the dependence of the apparent variance on the lower cut-off frequency (data not shown). A sharp increase was typically observed with cut-offs below 1Hz. In Fig. 1 we plot the inverse of these high-pass variances of bead position in an actin gel, as a function of bead size. According to Eq. (2) the reciprocal variance is expected to be proportional to the bead radius. Some of the data points obviously deviate from a slope of 1 (line in Fig. 1). The 0.96μm and 1.8μm beads used in these experiments were aminopropyl and bromopropyl derivatized respectively. We believe that surface effects on the polymerization of the surrounding gel account for the deviations.

Data from both 0.5 and 5.0 μm beads in 1 and 2mg/ml gels are consistent with a slope of 1 with amplitudes corresponding to shear moduli of $G' \cong 0.2$Pa and $G' \cong 0.4$Pa. Because of the lower cut-off frequency discussed above, these are upper bounds for the shear moduli. At least part of the scatter in the observed total variances for each size beads in the same sample reflects local inhomogeneities in the gels. Smaller beads sample inhomogeneities on smaller scales and should therefore show more scatter. This trend is clearly visible in the data. The absolute values of G' are consistent with other reported experiments [3,7,8], although discrepancies among labs persist [4].

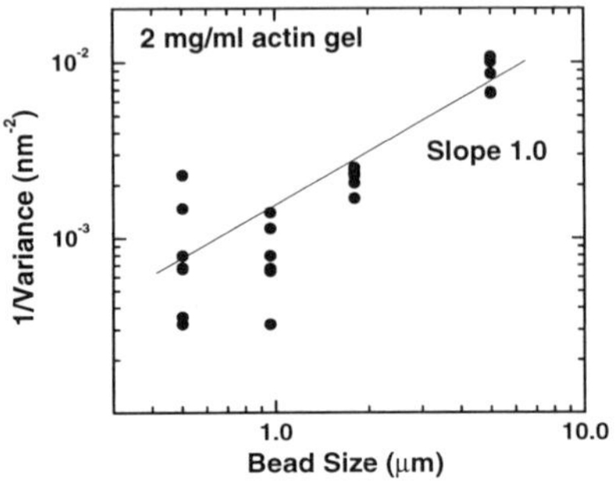

FIGURE 1: The reciprocal variance $1/\langle x^2 \rangle$ in bead position for silica beads in an actin gel of concentration 2mg/ml at room temperature, obtained by summing the PSD for frequencies above a 1Hz cutoff.

The power spectral densities (PSD) for silica beads in an actin gel of 2mg/ml concentration are shown in Fig. 2. The downturn at about 10kHz is due to electronic anti-alias filtering. The PSD for the largest bead diameter, 5.0μm, exhibits power law behavior over about three orders of magnitude in frequency, with an exponent of about -1.75. The ratio of bead diameter to mesh size, ξ, is about 20 ($\xi \cong 0.2\mu m$, [2,25]), so the continuum model of Eq. (3) should apply. Assuming that $G'(\omega)$ and $G''(\omega)$ scale, as for conventional gels [19], with the same power at high frequencies, Eq. (3) implies a scaling of $G'(\omega), G''(\omega) \propto \omega^{3/4}$. The different scaling exponent for flexible polymers can be understood using the Rouse or the Zimm models [19]. No existing model is consistent with our data on semiflexible polymers [26]. For smaller beads, the amplitude of fluctuations increases, as described by Eq. (3), and the spectra increasingly deviate from power law behavior both at low and high frequencies. At low frequencies the assumption of incompressibility may break down, as explained above. This effect is expected when the bead size becomes comparable to the mesh size. The smallest beads are only about 2x larger than the average mesh, although they remain well trapped. To check for slow diffusion due to reptation of actin filaments, we observed 1.8μm beads by video for two hours and estimated an upper limit for the long range diffusion coefficient of $4 \times 10^{-14} cm^2/s$. Reptation is therefore not relevant on the time scale of our experiments. Note that the deviation of the power spectra from power law behavior is bead-size dependent, unlike what we observe for PAAm gels (see below), and can therefore not be the onset of the elastic plateau regime. We cannot explain deviations at high frequencies. Inertial effects are only relevant at frequencies higher than 1 MHz (see above).

The shape of the PSD can be compared to direct measurements of $G'(\omega)$ and $G''(\omega)$ in actin networks. Refs. [3] report data suggesting $\omega^{1/2}$ scaling, using macroscopic rheology. Recent microrheology data from the same lab [8], however, may be

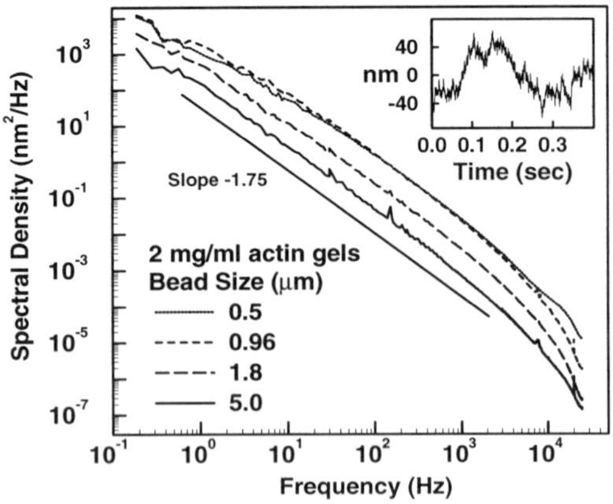

FIGURE 2: The power spectrum for beads in actin gels of concentration 2mg/ml. The line indicates a log-log slope of -1.75. Inset: Portion of the time series data for bead size 1.8 μm.

consistent with $\omega^{3/4}$ scaling. In both cases, the observed range of frequencies is very close to the transition from the plateau regime. A different interpretation of the time dependence of bead motion in actin gels has been given in Ref. [10]. These authors observe particle dynamics consistent with our $\omega^{-7/4}$ scaling of the PSD, but suggest an explanation based on single filament dynamics. We believe that at our higher concentrations, for which the mesh size is substantially smaller than the bead diameter, a continuum elastic approach such as we have described is correct.

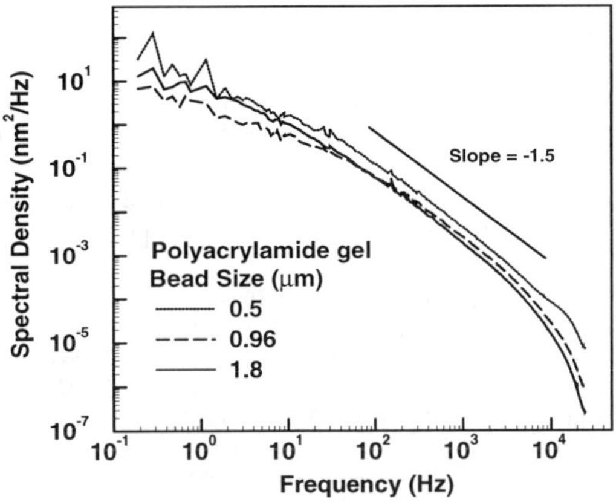

FIGURE 3: The PSD for beads in 2% polyacrylamide gels. The line indicates a log-log slope of -1.5.

The PSD of thermally fluctuating silica beads in 2% PAAm gels for three different bead sizes are shown in Fig. 3. The downturn at about 10 kHz is again due to anti-alias filtering. The log-log curves are less linear than the actin spectra for all beads, but are in their mid range tangential to the line drawn in Fig. 3 with a slope of -1.5. This result is consistent with internal chain dynamics described by the Rouse model [19], which predicts a scaling behavior for the complex modulus at high frequencies: $G'(\omega), G''(\omega) \propto \omega^{1/2}$. This translates via Eq. (3) into a slope of -1.5 in the log-log power spectrum.

The mesh size of a 2% PAAm gel is on the order of 50Å [27], but relatively large variations are possible because this concentration is at the gelation threshold. As in actin, we observed no long-range diffusion, showing that the polymer is indeed cross-linked into a solid. The smallest bead (0.5μm) is still about a factor of 100 larger than the mesh, and the continuum model should apply. At low frequencies, the slope of the spectra again decreases, but in contrast to the actin spectra in a bead-size independent way. This is consistent with the onset of a plateau, where $G(\omega)$ becomes weakly dependent on frequency [19]. Eq. (3) then predicts a PSD proportional to ω^{-1}, consistent with the observed slope. This provides further evidence for the validity of our continuum viscoelastic interpretation.

By integrating the spectra, from a lower cut-off frequency of 1Hz, we again obtain a spring stiffness K which is roughly proportional to the bead radius, and calculate a zero-frequency shear modulus of 4.6 Pa. The measured value for G' appears low compared with conventional rheology results [28]. Those are typically reported for higher gel concentrations, but extrapolating results assuming scaling as $G' \propto c^{9/4}$ predicts $G' \cong 200$Pa. PAAm gels are known to be complex and the deviation is most probably due to the proximity to the gelation threshold. On the other hand, the gels appear homogeneous on the scale of the bead size, judging from the high reproducibility of the spectra at various locations within one sample.

Conclusions

We have demonstrated that thermal fluctuations of micron size inclusions can be used to measure the viscoelastic properties of polymer solutions locally. The total variance of fluctuations determines the zero frequency shear modulus. The power spectrum, on the other hand, reflects the frequency dependence of the complex shear modulus $G(\omega)$. In particular, we find that the high frequency scaling of $G(\omega)$ for actin differs from that of PAAm, as well as from the predictions of the Rouse model. This is in contrast to previous claims of Rouse-like scaling for actin [3]. Our continuum-elastic model should be valid for inclusions of diameter large compared with the mesh size of the polymer network. This is true for our largest beads in actin solutions, and for all of our experiments on PAAm. Judging from the appearance of the spectra, the continuum model appears to be a reasonable approximation even for the smaller beads. For PAAm we clearly see the onset of the rubber plateau for lower frequencies.

---

1 T.P. Stossel, Sci Am **271**, 54-5, 58-63 (1994).

2 J. Kas, H. Strey, J.X. Tang *et al.*, Biophysical Journal **70**, 609 (1996).

3 O. Muller, *et al.*, Macromolecules **24**, 3111 (1991); R. Ruddies, *et al.*, Eur Biophys J **22**, 309 (1993).

4 P.A. Janmey, S. Hvidt, J. Kas *et al.*, J Biol Chem **269**, 32503 (1994).

5 P.A. Janmey, *et al.*, Biochemistry **27**, 8218 (1988); P.A. Janmey, J Biochem and Biophys Meth **22**, 41 (1991).

6 F.C. MacKintosh, J. Kas, and P.A. Janmey, Physical Review Letters **75**, 4425 (1995).

7 T.D. Pollard, I. Goldberg, and W.H. Schwarz, J Biol Chem **267**, 20339 (1992); D.H. Wachsstock, W.H. Schwartz, and T.D. Pollard, Biophys J **65**, 205 (1993); **66**, 801 (1994); M. Sato, *et al.*, J Biol Chem **260**, 8585 (1985); K.S. Zaner and J.H. Hartwig, J Biol Chem **263**, 4532 (1988); J. Newman, *et al.*, Biophys J **64**, 1559 (1993).

8 F. Ziemann, J. Radler, and E. Sackmann, Biophys J **66**, 2210 (1994).

9 F.G. Schmidt, F. Ziemann, and E. Sackmann, Eur Biophys J **24**, 348 (1996).

10 F. Amblard, *et al.*, Physical Review Letters **77**, 4470 (1996).

11 T.G. Mason and D.A. Weitz, Physical Review Letters **74**, 1250 (1995).

12 L.D. Landau and E.M. Lifshitz, *Fluid mechanics* (Pergamon Press, Reading, MA, 1959).

13 The shear stress σ and displacement field u in a viscoelastic medium are related in the same way as σ and the velocity field in a viscous fluid, provided that both are incompressible.

14 L.D. Landau and E.M. Lifshitz, *Theory of elasticity*, 2d ed. (Pergamon Press, New York, 1970).

15 B. Schnurr, *et al.*, to be published.

16 F. Brochard and P.G. de Gennes, Macromolecules **10**, 1157 (1977).

17 S.T. Milner, Physical Review E **48**, 3674 (1993).

18 L.D. Landau, E.M. Lifshitz, and L.P. Pitaevskii, *Statistical physics* (Pergamon Press, New York, 1980).

19 M. Doi and S.F. Edwards, *The Theory of Polymer Dynamics* (Clarendon Press, Oxford, 1988).

20 J.D. Pardee and J.A. Spudich, in *Structural and Contractile Proteins (PartB: The Contractile Apparatus and the Cytoskeleton)*, ed. by D W Frederiksen and L W Cunningham (Academic Press, San Diego, 1982).

21 Bio-Rad Laboratories, US/EG Bulletin **1156.**

22 W. Denk and W.W. Webb, Applied Optics **29**, 2382 (1990).

23 K. Svoboda, C.F. Schmidt, B.J. Schnapp *et al.*, Nature **365**, 721 (1993).

24 F. Gittes and C.F. Schmidt, in *Laser tweezers in cell biology*, ed. M. P. Sheetz (Academic Press, San Diego, 1997).

25 C.F. Schmidt, *et al.*, Macromolecules **22**, 3638 (1989).

26 In analogy with the Rouse model, one might have expected a scaling $G'(\omega), G''(\omega) \propto \omega^{1/4}$ for semiflexible polymers. This would result in a $\omega^{-5/4}$ dependence of the power spectrum.

27 J.S. Fawcett and C.J.O.R. Morris, Separation Science **1**, 9 (1966).

28 Y. Cohen, *et al.*, Journal of Polymer Science, Part B (Polymer Physics) **30**, 1055 (1992).

Hydration of Biological Macromolecules: From Small Solutes to Proteins and Nucleic Acids

Shekhar Garde,[1,2] Gerhard Hummer,[1] Michael E. Paulaitis,[2,3] and Angel E. García[1,*]

[1] *Theoretical Biology and Biophysics T-10, MS K710, Los Alamos National Laboratory, Los Alamos, NM 87545*

[2] *Center for Molecular and Engineering Thermodynamics, Department of Chemical Engineering, University of Delaware, Newark, DE 19716*

[3] *Department of Chemical Engineering, Johns Hopkins University, Baltimore, MD 21218.*

We present a method that uses two- and three-particle correlation functions between solute atoms and water molecules to approximate the density profile of water surrounding biomolecules. The method is based on a potential of mean force expansion and uses X-ray crystallography, NMR, or modeling structural input information on the biomolecule. For small hydrophobic solutes, we have calculated entropies of hydration using the predicted water densities that are in good agreement with experimental results. We have also predicted the hydration of the *catabolite activator protein*-DNA complex. The method is extremely efficient and makes possible the study of hydration of large biomolecules within CPU minutes.

1 Introduction

The unique balance of forces underlying biological processes such as protein folding, molecular recognition, and formation of biological membranes owes its origin in part to the surrounding aqueous medium.[1] Hydrophobic interactions, in particular, are believed to contribute to the overall stability of folded proteins, protein-ligand complexes, and biological membranes. The hydrophobic driving force for self-assembly is mainly entropic in origin, and is directly connected to the inhomogeneous structure of water in the vicinity of hydrophobic solutes.[2] Thus, quantifying structural hydration of biomolecules and relating water structure to the key thermodynamic properties of hydration constitute important steps towards understanding of the role of water in biomolecular structure-function relationship.

Experimental information about the hydration of biomolecules at atomic resolution is obtained mainly from X-ray and neutron diffraction,[3] and NMR spectroscopic [4] techniques. Even though these techniques are widely used, unambiguous assignment of water molecules on the surface and in the interior of biomolecules is complicated due to methodological difficulties. Further, the disorder of water molecules hydrating biomolecules in different unit cells of crystals presents difficulties in the interpretation of experimental results.[5,6] This disorder is quantified best by an inhomogeneous density of water molecules in space. The experimental analysis is however limited to identifying the regions of high localization of water molecules.

Theoretical methods for the detailed study of the structural hydration of biomolecules rely mostly on the molecular dynamics (MD) and Monte Carlo (MC) computer simulation techniques.[7] Computer simulations of hydration of small molecules have advanced our fundamental understanding of solvation phenomena. However, the computational effort needed to obtain satisfactory statistical precision increases significantly when applied to the study of biomolecular hydration. Further, analysis of systems like protein-DNA complexes would need simulations of the individual protein, the DNA, and the complex to understand the role of hydration in complexation.

Mat. Res. Soc. Symp. Proc. Vol. 463 © 1997 Materials Research Society

We present here an alternative approach to predicting position-dependent water densities near biomolecules. Our method is based on the potential-of-mean-force (PMF) expansion [8] applied previously to hydration studies of B-DNA,[9] an α-helix,[10] t-RNA, the photosynthetic reaction center, an antigen-antibody complex,[11,12] and a host of other systems. We will show that the water densities surrounding biomolecules can be approximated by a limited number of well-defined, lower-order correlation functions accurately calculated from simulations of small molecules in water. The efficiency of the method allows analysis of hydration of biomolecules in CPU minutes rather than weeks. For purely hydrophobic molecular solutes we relate the calculated water densities to the entropies of hydration *via* an entropy expansion to show the feasibility of accurately estimating macroscopic thermodynamic properties from water density profiles. Finally, we analyse of the hydration of the interface of the *catabolite activator protein* (CAP)-DNA complex.

2 Theory and Computer Simulations

The hydration of biomolecules can be quantified by the equilibrium position-dependent density of water molecules surrounding them. The biomolecule is characterized by a set of atomic coordinates $\{\mathbf{s}_{i\alpha}\}$, where α indicates the type of atom, $\alpha = 1,\ldots,M$, and $i = 1,\ldots,N_\alpha$. The conditional density of water molecules at a position $\mathbf{r}$ (given that biomolecule is fixed in a given configuration $\{\mathbf{s}_{i\alpha}\}$) is then given by,[11]

$$\rho(\mathbf{r}|\,\{\mathbf{s}_{i\alpha}\}) = \rho_0 \frac{g^{(1;\{N_\alpha\})}(\mathbf{r},\{\mathbf{s}_{i\alpha}\})}{g^{(\{N_\alpha\})}(\{s_{i\alpha}\})}. \tag{1}$$

The n-particle distribution function, $g^{(n)}$, is related to the n-particle potential of mean force, $W^{(n)}$, by $g^{(n)} = \exp(-W^{(n)}/kT)$, and ρ_0 is the bulk density of water. Successful prediction of density thus depends on how accurately one can approximate the n-particle PMF. For strictly nonpolar molecular solutes, water structure is found to be only locally sensitive to the structural details of the solute. Thus, the water density around nonpolar solutes can be approximated simply by its correlations with one or two nearest solute sites.[13,14] Namely,

$$\rho^{(n,1)}(\mathbf{r}|\mathbf{r}_1,\ldots,\mathbf{r}_n) \approx \rho_0 g(\mathbf{r}|\mathbf{r}_j)\ , \tag{2}$$
$$\rho^{(n,1)}(\mathbf{r}|\mathbf{r}_1,\ldots,\mathbf{r}_n) \approx \rho_0 g(\mathbf{r}|\mathbf{r}_j,\mathbf{r}_k)\ , \tag{3}$$

with j and k such that $|\mathbf{r}-\mathbf{r}_j| = \min_{l=1,..,n}|\mathbf{r}-\mathbf{r}_l|$ and $|\mathbf{r}-\mathbf{r}_k| = \min_{m=1,..,n;m\neq j}|\mathbf{r}-\mathbf{r}_m|$. Eqs. (2) and (3) are defined, respectively, as the one- and two-site proximity approximations.

In contrast to nonpolar molecules, biomolecules are a complex mix of a variety of sites. In this case, we formally expand $W^{(n)}$ in terms of multiparticle potentials of mean force. Difficulties in calculating correlation functions of order four and higher necessitates truncation of this expansion at the three-particle level. Further, geometric and energetic considerations of interactions of nitrogen, oxygen, and sulphur atoms with water molecules allow us to effectively group these atoms into one class of solute sites. That is, their effect on ordering nearby water molecules is assumed to be similar to that of water oxygens. Also, the effect on water structure of nonpolar carbon atoms, represented as methyl [Me] sites, is treated using the one-site proximity approximation. That these approximations are reasonable for applications to biomolecules has been shown elsewhere.[9-12] These approximations render the complex problem of biomolecular hydration tractable. The following approximate expression for the conditional water density is obtained.

$$\rho(\mathbf{r}|\mathbf{s}_1,\ldots,\mathbf{s}_n) \approx \rho_0 \prod_{j=1}^{n} g^{(2)}(\mathbf{r},\mathbf{s}_i) \prod_{j=1}^{n-1} \prod_{k=j+1}^{n} \frac{g^{(3)}(\mathbf{r},\mathbf{s}_j,\mathbf{s}_k)}{g^{(2)}(\mathbf{r},\mathbf{s}_j)g^{(2)}(\mathbf{s}_j,\mathbf{s}_k)g^{(2)}(\mathbf{s}_k,\mathbf{r})} \times g_{Me-O}(\mathbf{r},\mathbf{s}_{\text{C-prox}}) \,. \tag{4}$$

Additional effects, such as polar hydrogens forming directional hydrogen bonds, can also be incorporated into this expression.[11]

Application of eqn. (4) to biomolecules requires the calculation of the O-O-O triplet correlation function, $g^{(3)}$, as well as Me-O pair correlation function g_{Me-O}. The $g^{(3)}$ is calculated from simulations of SPC water.[9,18] Simulations were also performed for a single Me and the Me-Me dimer in water where Me-Me dimer was held fixed at separations between 0.12 to 0.66 nm. [14] These simulations were used to calculate Me-O pair, and Me-Me-O triplet correlations required in the proximity approximations, as well as eqn. (4).

3 Results and Discussion

3.1 Hydration of nonpolar molecules

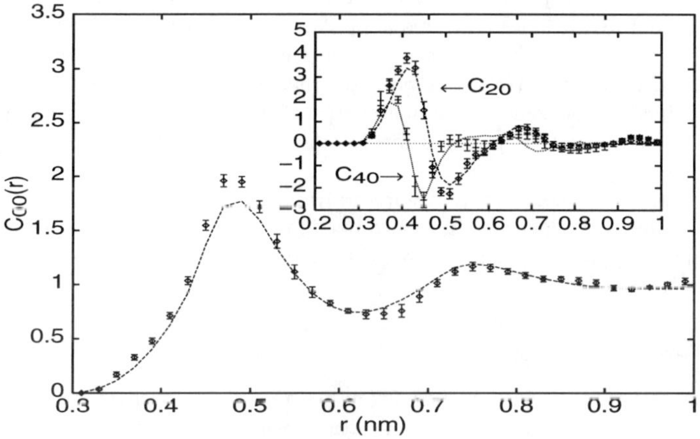

Figure 1: Spherically averaged water-oxygen density in units of bulk density as a function of distance r from the center of benzene calculated from simulation (symbols) and the two-site proximity approximation (dashed line). Inset shows the C_{20} and C_{40} coefficients in the spherical harmonic expansion of the density.[14]

Figure 1 shows a comparison of water-oxygen densities calculated from MC simulation of benzene (represented by six Me sites in a planar configuration) with densities calculated from the two-site proximity approximation. [14] The positions and heights of both peaks corresponding to the first and the second hydration shells, respectively, are well reproduced. The symmetry of the density profile about the plane of benzene, captured by the higher order coefficients in the spherical harmonic expansion, is also very well reproduced by the two-site proximity approximation. Application to other model solutes also gives similar results.[14]

This suggests a key feature of the hydration of nonpolar solutes is that water structure in their vicinity is only locally sensitive to the structural details of the nonpolar solute.

Table 1. Standard entropies of hydration calculated using an entropy expansion.

	$\Delta S^{*}_{\mathrm{hydration}}/(\mathrm{cal/mol/K})$	
solute	calculated	expt
ethane	-18.0	-20.1
propane	-21.7	-22.8
n-butane[a]	-23.3	-26.0
benzene	-28.8	-21.2

[a]56/44 % *trans/gauche* mixture.
The calculated entropies include two terms corresponding to positional and orientational correlations of water molecules with nonpolar solute approximated using two- and one-site proximity approximations, respectively. The latter amounts to an orientational entropy of 0.35 cal/mol/K per water molecule in the first hydration shell.[15]

We have used the water density profiles obtained from proximity approximations to calculate standard entropies of hydration for various nonpolar solutes.[13,16,17] The results are given in Table 1. The entropies of hydration are large and negative and are in good agreement with experimental results for alkanes. A value significantly more negative than the experimental value is obtained for benzene. Note, however, that the description of benzene as six overlapping Me sites is a severe approximation for this molecule which neglects its complex electronic structure.[19]

The PMF between two methane molecules in water and water contribution to the conformational equilibria of n-butane and n-pentane have also been predicted from water densities obtained using the proximity approximations to show that hydration favors the more compact conformations, as expected. [15]

3.2 Ice-water interface

Eq. (4) was applied to the ice/water interface. The ice phase is described by fixing SPC water oxygens at ideal hexagonal-ice I_h lattice positions while allowing for molecular reorientations representing thermal disorder in the ice phase. The basal plane of the ice layer is covered by liquid water. Figure 2 shows a comparison of the oxygen density profiles predicted by the PMF expansion truncated at the level of pair and triplet correlations with calculations from explicit MC computer simulations.[9] It is seen that truncation of the PMF expansion at the level of pair correlations results in water densities much higher than those observed in MC simulations in the region close to the basal plane ($Z< 0.2$nm). Including triplet correlations improves the agreement. The position as well as the height of the first peak are well reproduced. The agreement at longer distances is qualitative. Clearly, information on triplet correlations is needed for identifying high density regions accurately. We thus include the terms up to triplet correlations in the PMF expansion eqn. (4) for the applications to the biomolecular hydration.

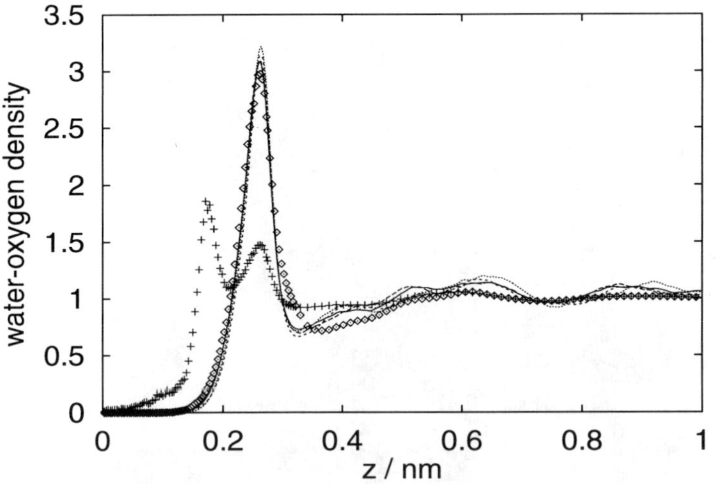

Figure 2: Water-oxygen density at the ice/water interface in units of the bulk- water density, ρ_0. Z is the distance from the closest plane of oxygen atoms in the ice phase. (+) PMF expansion including only pair correlations; ($\diamond$) PMF expansion including both pair and triplet correlations. Solid lines: explicit MC simulations using different system sizes and methods for Coulombic interactions.[9]

3.3 CAP-DNA complex

The catabolite activator protein (CAP), also known as cyclic adenosine monophosphate receptor protein (CRP), plays an important role in mediating transcription activation of several genes in *E. coli*. The CAP-cAMP complex binds, as a dimer, to the specific sequences in several *E. coli* promoter regions.[20] The activation process is complex, involving not only the binding of cAMP, but also the interactions of CAP with the DNA [21, 22] and with the RNA polymerase.[23] Atomic resolution structure of CAP bound to the consensus DNA sequence has been determined by X-ray crystallographic studies.[21, 22] Upon binding, the DNA molecule is bent at approximately 90° near the position where helix-turn-helix motif of each CAP monomer binds. Previous crystallographic studies of protein-DNA complexes have shown that the sequence recognition can also be mediated by water molecules bridging between hydrogen-bond acceptor groups in the protein and the DNA.[24] A recent moderately high resolution (2.7 Å) crystal structure of the CAP-DNA complex also shows water mediated interactions. To elucidate the extent of water mediated recognition, we have studied the hydration of the regions in the vicinity of the recognition sites by means of a local water density calculation using the PMF expansion. Here we present the results for one of the recognition motifs.

Points corresponding to a high local density of water molecules near the binding region in the complex are indicated in the panels A and B. The binding region is seen to be considerably hydrated. We have also calculated water densities near the isolated protein and DNA molecule. Both the amino acids (panel D) as well as the DNA bases (panel C) comprising the recognition sites are considerably hydrated. Thus, water molecules corresponding to high

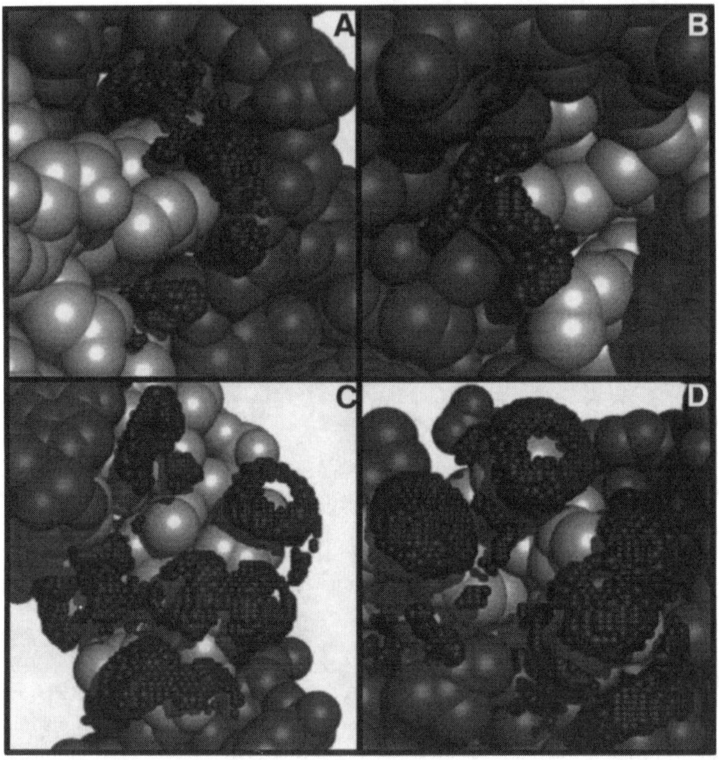

Figure 3: Hydration of the CAP-DNA complex. Regions of water density $\rho(r)/\rho_0 > 3$ and $\rho(r)/\rho_0 > 5$ are identified in each panel by small and medium sized dark gray spheres. Panel (A) and (B) show two different views of the CAP (white) and DNA (gray) complex at the recognition motif (5'-TGTGA-3'/3'-ACACT-5'). Panel (C) shows the hydration of the DNA recognition site in the absence of CAP. (white: DNA bases comprising the recognition site, gray: other bases). Panel (D) shows the hydration of the CAP recognition site in absence of the DNA. (white: amino acids comprising recognition site, gray: other amino acids.) The crystal structure from pdb file 1ber [22] was used for the calculations.

local densities near the isolated recognition sites have to be removed upon binding. The release of bound water represents an entropic gain upon binding while the water molecules that hydrate the interface of the complex (as shown in panel A and B) give a favorable enthalpic contribution, *e.g.*, by mediating hydrogen bonds.

Our results demonstrate that we can efficiently investigate medium-to-low resolution structures (> 0.2 nm crystallographic resolution) where water molecules are difficult to assign. Further, we get the density profiles of water hydrating biomolecules which represents a more complete picture compared to assignment of only individual water molecules. We have studied the hydration of the recognition sites of the CAP and the DNA for which the binding free energies have been measured previously.[25] Based on our calculations, it is our hypothesis that water-mediated interaction in this complex can partly account for the observed specificity.

4 Conclusions

We have presented a method for studying the hydration of biological macromolecules. Detailed simulations of the hydration of purely hydrophobic solutes and the ice-water interface indicate that the water structure is only locally sensitive to the structure of hydrophobic solutes and that information at the level of three-particle correlations is essential for reproducing water density profiles observed in simulations of polar interface. These results have been incorporated into a truncated PMF expansion for describing biomolecular hydration.

An advantage of this approach is that the database of well-defined, lower order correlation functions required for the determination of water densities around a macromolecule must be obtained only once. This makes the application of the PMF expansion extremely efficient. The direct connection to the statistical mechanics also allows us to include any additional information on correlations (*e.g.* directional hydrogen bonding interactions) in an unambiguous, systematic manner.

The PMF expansion formalism can be applied to hydration studies in both crystals or solutions to give average density profiles rather than assigning individual water molecules. The method can also be applied to medium-to-low resolution X-ray structures for the assignment of water molecules in the process of refinement.

The water density profiles surrounding hydrophobic molecules have also been used to calculate entropies of hydration. Similar application to an antigen-antibody complex was performed recently using water densities calculated from the PMF expansion. Qualitative correlation with experimental entropies of binding was observed.[11,26] Accurate calculations of free energies of binding are complicated by the orientational component in the entropy and the electrostatic part in the enthalpy. Thus, more detailed calculations on biological systems are needed before quantitative agreement with experiments and possibly predictions of thermodynamic quantities can be achieved. However, our results for the CAP-DNA complex presented here underscore the importance of water in biomolecular interactions.

Acknowledgements

G.H. would like to thank Dr. D. M. Soumpasis for many discussions on the role of water and ions in the biomolecular systems.

References

[1] J. A. Rupley and G. Careri, Adv. Protein. Chem., **41**, 37, (1991).

[2] M. E. Paulaitis, S. Garde, and H. S. Ashbaugh, Curr. Opin. Coll. Int. Sci., **1**, 376, (1996).

[3] H. Savage and A. Wlodawer, Meth. Enzymol., **127**, 162, (1986).

[4] G. Otting, E. Liepinsh, and K. Wüthrich, Science, **254**, 974, (1991).

[5] E. Westhof, Annu. Rev. Biophys. Biophys. Chem., **17**, 125, (1988).

[6] X-J. Zhang and B. W. Matthews, Protein Sci. **3**, 1031, 1994.

[7] C. L Brooks III and M. Karplus, Meth. Enzymol., bf 127, 369, (1986).

[8] D. M. Soumpasis, in Computation of Biomolecular Structures, edited by D. M. Soumpasis and T. M. Jovin (Springer, Berlin, 1993) p. 223.

[9] G. Hummer and D. M. Soumpasis, Phys. Rev. E., **49**, 591, (1994); ibid. **50**, 5085, (1994).

[10] G. Hummer and A. E. García, Prot. Struct. Funct. Gen., (1996), in press.

[11] G. Hummer, A. E. García, and D. M. Soumpasis, Faraday Discuss., **103**, (1996), in press.

[12] G. Hummer, A. E. García, and D. M. Soumpasis, Biophys. J., **68**, 1639, (1995).

[13] H. S. Ashbaugh and M. E. Paulaitis, J. Phys. Chem., **100**, (1996).

[14] S. Garde, G. Hummer, A. E. García, L. R. Pratt, and M. E. Paulaitis, Phys. Rev. E., **53**, R4310, (1996).

[15] S. Garde, G. Hummer, and M. E. Paulaitis, Faraday Discuss., **103**, (1996), in press.

[16] T. Lazaridis and M. E. Paulaitis, J. Phys. Chem., **96**, 3847, (1992); ibid. **98**, 635, (1994).

[17] M. E. Paulaitis, H. S. Ashbaugh, and S. Garde, Biophys. Chem., **51**, 349, (1994).

[18] H. J. C. Berendsen, J. P. M. Postma, W. F. van Gunsteren, and J. Hermans, in Intermolecular Forces: Proceedings of the 14th Jerusalem Symposium on Quantum Chemistry and Biochemistry, edited by B. Pullman, (Reidel Dordrecht, The Netherlands), p. 331, (1981).

[19] W. L. Jorgensen and D. L. Severance, J. Am. Chem. Soc., **112**, 4768, (1990).

[20] J. L. Boshford and J. G. Harman, Microbiol. Review, **56**, 100, (1992).

[21] S. Schultz, G. Shields, and T. Steitz, Science, **253**, 1001, (1991).

[22] G. Parkinson, C. Wilson, A. Gunasekera, Y. W. Ebright, R. H. Ebright, and H. M. Berman, J. Mol. Biol. **260**, 395, (1996).

[23] R. H. Ebright, Mol. Microbiol., **8**, 797, (1993).

[24] B. F. Luisi, W. X. Xu, Z. Otinowski, L .P. Freedman, K. R. Yamamoto, and P. B. Sigler, Nature, **352**, 495, (1991).

[25] A. Gunasekera, Y. W. Ebright, and R. H. Ebright, J. Biol. Chem., **21**, 14713, (1992).

[26] T. N. Bhat, G. A. Bentley, G. Boulot, M. I. Green, D. Tello, W. Dall'Acqua, H. Souchon, F. P. Schwartz, R. A. Mariuzza, and R. A. Poljak, Proc. Nat'l. Acad. Sci. USA., **91**, 1089, (1994).

Part II

Statistical Mechanics of DNA

STRETCHED, TWISTED, SUPERCOILED AND BRAIDED DNA

John F. Marko
The University of Illinois at Chicago, Department of Physics,
845 West Taylor Street, Chicago IL 60657

ABSTRACT

The DNA double helix is a semi-flexible polymer with twist rigidity. Its bending elasticity gives rise to entropic polymer elasticity, which can be precisely studied in single-molecule experiments. DNA's twist rigidity causes it to wrap around itself, or 'supercoil', when it is sufficiently twisted; thermal fluctuations destabilize supercoiling for DNAs twisted fewer than once per twist persistence length. Twisted DNAs under tension, braided DNAs, and the internal dynamics of supercoiled DNAs are discussed. The interplay between braiding and supercoiling free energy is argued to be important for the decatenation of duplicated DNAs in prokaryote cells.

I. DNA BENDING, TWISTING AND STRETCHING ELASTICITY

The DNA double helix has a bending persistence length of about $A = 50$ nm in 'physiological' aqueous solution (300 K, ≈ 0.15 M univalent ions)[1]. It also has twisting elasticity, described by a twist persistence length $C \approx 75$ nm[2]. Elasticity theory may be used to describe deformation of the double helix on such length scales, since A and C are well separated from the DNA diameter (2 nm) the helix repeat (3.5 nm), and the spacing of successive base-pairs (0.34 nm; note 1000 bp = 1 kb = 0.34 μm). DNAs that encode genes in prokaryote cells (cells without nuclei and other organelles, e.g. bacteria) are 10 to 1000 microns in length, indicating that thermal conformational fluctuations will play a role in large-scale DNA organization in such cells.

Several experimental groups have stretched[3,4,5] and twisted[6] single DNA molecules of lengths 10 to 30 microns, further motivating the study of DNA from the point of view of polymer statistical mechanics. As a polymer, DNA offers unique possibilities for experiments: it is available in essentially monodisperse form, and enzymatic techniques can be used to make precise changes to its structure.

The bending elasticity, when thermal fluctuations are taken into account, leads to entropic elasticity. Idealizing a linear DNA as a worm-like chain of length L, a force must be applied to separate its ends by a distance z:

$$f = \frac{k_B T}{A}\left[z/L + \frac{1}{4(1-z/L)^2} - 1/4\right] \tag{1}$$

When subjected to the force $k_B T/A \approx 0.1$ piconewtons (pN), a DNA is about half-extended. Over the range $f < 10$ pN experimental data fit Eq. 1 very well[7], indicating that for such forces the DNA contour length is essentially fixed.

For forces larger than 10 pN, the double helix starts to stretch, and at ≈ 60 pN DNA undergoes an abrupt 'overstretching' transition to a new structure 1.6 times the normal double-helix length[5]. This lengthening is consistent with the straightening of the normally helical sugar-phosphate backbones, while the force scale is consistent with the work that must be done to overcome the $\approx 10 k_B T/\text{nm} = 40$ pN of free energy which binds DNA in its double helix structure. It is plausible that the double helix untwists during overstretching, and it may be argued from symmetry that DNA stretching and twisting degrees of freedom must be linearly coupled[8]. The strength of this coupling might be determined in an experiment which measures how much the double helix is lengthened when it is twisted, or equivalently, how much it twists in response to stretching.

Mat. Res. Soc. Symp. Proc. Vol. 463 © 1997 Materials Research Society

A simple model for the elastic energy per length of DNA which accounts for all the above effects is

$$\frac{\mathcal{E}}{k_B T} = \frac{A}{2}\kappa^2 + \frac{C}{2}\Omega^2 + \frac{\gamma}{2}u^2 + g\Omega u + \frac{w}{2}\left(\frac{du}{ds}\right)^2 - \frac{q}{3}u^3 + \frac{r}{4}u^4 + \cdots \qquad (2)$$

Here the elastic degrees of freedom are curvature κ, the twist Ω (radians per length) in excess of the 1.8 radians/nm of the relaxed double helix, and the longitudinal strain u. The stretching elastic constant is $\gamma \approx 300/\text{nm}$ [5]. Recent experiments [6,9] suggest that the twist-stretch coupling is on the order of $g \approx 25$[8]. These linear elastic constants are consistent with the assumption that DNA is an elastic medium with a Young modulus of roughly $100 k_B T/\text{nm}^3 = 4 \times 10^8$ Pa.

The nonlinear-u terms of (2) define the extension of the 'overstretched' state, and the force on the 'plateau'. Fig. 1 shows the force-distance response of this model with $q = 500$ nm^{-1}, $|u| < 0.8$ (enforcing this constraint eliminates the need for higher-order nonlinear terms), and $w = 11$ nm^{-1}. This last constant gives energy to 'domain walls' between overstretched and normal DNA, and controls the 'width' of the plateau.

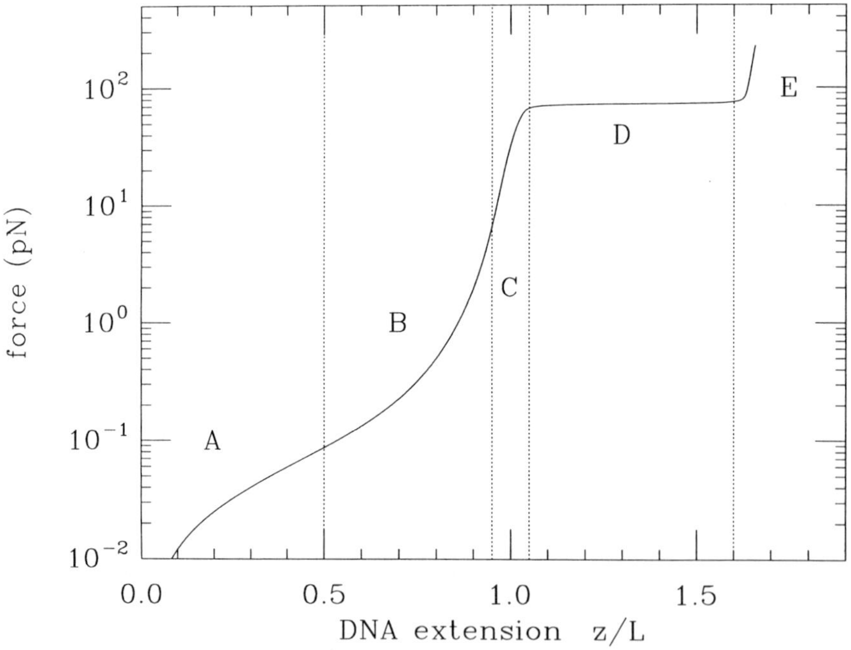

Figure 1: Elasticity of double-helix DNA. For forces $< k_B T/A \approx 0.1$ pN, one observes linear polymer elasticity (A); for forces between 0.1 pN and 10 pN, the elasticity is that of an inextensible worm-like chain (B); for forces > 10 pN the double helix itself starts to stretch (C); there is a sharp transition (D) to an 'overstretched' state at about 40 pN (E). In experiments, either the DNA or the ligands used to anchor it break at ≈ 100 pN.

II. SUPERCOILING OF DNA

The DNA elasticity described above suggests that a twisted DNA will wrap around itself in the manner of a twisted telephone cord. If the ends of a double helix are joined together, a circular DNA with a fixed integer *linkage number* Lk of its two strands is obtained. Lk is the number of times one strand is wrapped around the other, and is a topological property of a closed circular double-stranded DNA. A circular DNA of length L is relaxed when $\mathrm{Lk}=\mathrm{Lk}_0 = L/(3.5\ \mathrm{nm})$, and may be considered to be a (closed) random coil. Change in linkage number away from the relaxed state is conveniently measured as a fractional linkage number change, $\sigma \equiv \mathrm{Lk}/\mathrm{Lk}_0 - 1$. Most of the DNA in prokaryote cells is supercoiled (a typical circular 'plasmid' from E. coli is shown in Fig. 2a), with a typical linkage deficit of $\sigma \approx -0.05$. Supercoiled DNA has thus been the focus of intense experimental study [10,11].

The simple physical problem of the conformation of a closed DNA with fixed linkage σ is therefore of interest. Remarkably little work has been done on the statistical mechanics of supercoiled DNAs except using numerical MC simulation (Fig. 1b) [12]; next it will be shown how simple statistical-mechanical calculations give insight into many of the features of experiment and simulation [13].

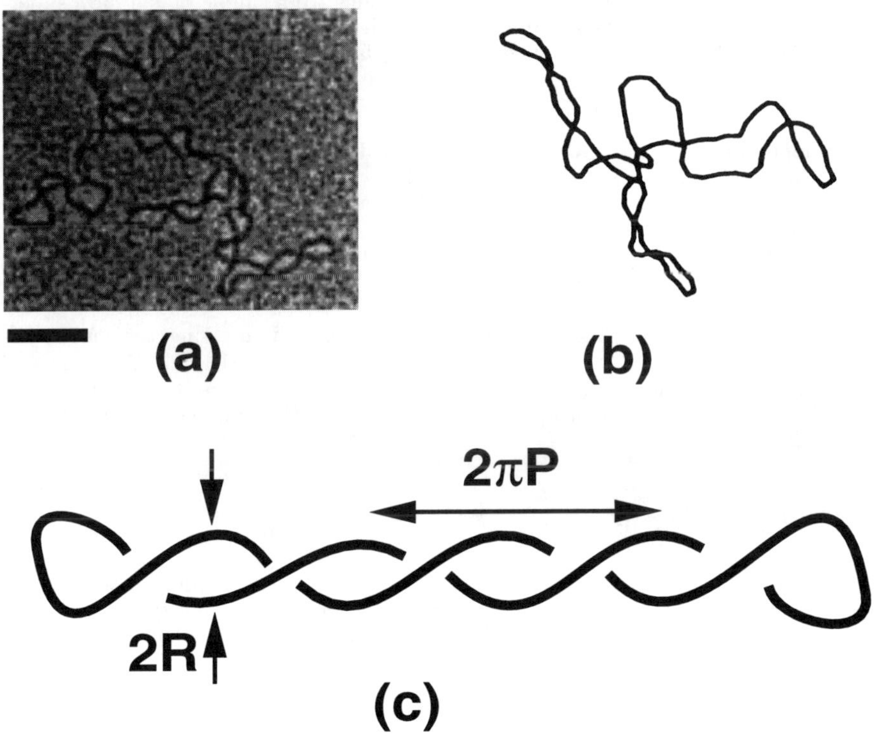

Figure 2: (a) electron-microscope picture of 7 kb (2.4 μm) supercoiled plasmid with linkage number 3.3% below that of the relaxed double helix ($\sigma = -0.033$) [11], bar is 100 nm; (b) MC simulation snapshot of 3.5 kb (1.2 μm) supercoiled plasmid with $\sigma = |0.03|$[12]; (c) regular plectonemic superhelix of radius R and pitch P.

First, one must impose the constraint of fixed σ. The linkage number of the two DNA strands is a nonlocal function of the conformations C_1 and C_2 of the two strands,

$$\mathrm{Lk} = \oint_{C_1} \oint_{C_2} \frac{d\mathbf{r}_1 \times d\mathbf{r}_2 \cdot (\mathbf{r}_1 - \mathbf{r}_2)}{|\mathbf{r}_1 - \mathbf{r}_2|^3} \tag{3}$$

Thanks to the fact that the two DNA strands are close to each other, the linkage number can be broken into (a) a contribution from the DNA twist Ω, which measures the local crossings of one strand over the other, and (b) a contribution from the nonlocal crossings of the double helix over itself. These two contributions are termed Tw (twist) and Wr (writhe), and take the form[10,14]:

$$\mathrm{Lk} = \mathrm{Tw} + \mathrm{Wr} \tag{4}$$

$$\mathrm{Tw} = \mathrm{Lk}_0 + \frac{1}{2\pi} \int_0^L ds\, \Omega(s) \qquad \mathrm{Wr} = \frac{1}{4\pi} \oint_C \oint_C \frac{d\mathbf{r} \times d\mathbf{r}' \cdot (\mathbf{r} - \mathbf{r}')}{|\mathbf{r} - \mathbf{r}'|^3}$$

where C is a curve running down the center of the double helix.

Motivated by experiment and simulation (Fig. 2a and 2b) a simple model for a supercoiled DNA is a regular interwound superhelix (or 'plectoneme') of radius R and pitch P (Fig. 2c). The writhe per length of such a structure is [10]

$$2\pi \mathrm{Wr}/L = \mp P/(P^2 + R^2) \tag{5}$$

where the upper/lower signs are for right-/left- handed superhelicies. In concert with (4), (5) allows the constraint of fixed σ to be imposed.

The next challenge is to take into account the entropy of thermal fluctuations. If DNA forms an interwound structure, there will be a loss of conformational entropy. A simple estimate for this entropy cost is that of confinement of a worm-like chain in a tube of radius R. For $R < A$, the free energy cost of confinement is on the order of $k_BT/(AR^2)^{1/3}$.

Taking the non-stretching linear elastic terms of (2), and inserting the curvature-squared of a regular superhelix $\kappa^2 = R^2/(R^2 + P^2)^2$, eliminating the twist Ω in favor of σ via (5), adding the confinement free energy, and finally minimizing the resulting free energy with respect to the undetermined parameters R and P, the free energy per length of a supercoiled DNA is

$$\frac{F_\mathrm{p}}{k_BTL} = \min_{R,P} \left[\frac{A}{2} \frac{R^2}{(R^2 + P^2)^2} + \frac{C}{2} \left(\sigma\omega_0 \pm 2\pi \frac{P}{R^2 + P^2} \right)^2 + \frac{1}{(AR^2)^{1/3}} + w(R) \right] \tag{6}$$

Here $\omega_0 = 2\pi \mathrm{Lk}_0/L = 1.8\ \mathrm{nm}^{-1}$, the radians per length of twist (or link) of the relaxed double helix. The term $w(R)$ is included to take into account short-ranged repulsions between the double helicies, and might include hard-core and screened Coulomb interactions of range ≈ 1 nm [13].

To understand the 'phase diagram' of (6), first consider its behavior without the third entropic repulsion term (i.e. in the absence of thermal fluctuations). For any nonzero linkage σ, R and P may be chosen so that the second twisting energy term is zero. Then if R tends to zero, the curvature energy is reduced essentially to zero as well; for finite hard-core interaction $w(R)$, R will be essentially determined by the range of that interaction. This is a reasonable description of a twisted telephone cord. It is clear that the sign chosen for the writhe is determined by the sign of σ; positive σ leads to left-handed plectonemes, while negative σ generates right-handed ones.

For real DNA, the effect of the entropic repulsion term must be considered: for large enough σ, its effect is minimal since it is overwhelmed by the very large elastic energy of the plectoneme. On the other hand, for small enough σ, it will overwhelm the elastic energy, and destabilize the interwound supercoil. The point at which a transition from unsupercoiled DNA to plectonemic supercoiling should occur can be estimated by supposing that $A \approx C$ and $P \approx R$, and examining the scaling behavior of (6),

$$\frac{F_{\mathrm{p}}}{k_B T L} \approx \min_R \left\{ \frac{C}{R^2} + C(\sigma\omega_0)^2 - \frac{C|\sigma|\omega_0}{R} + \frac{1}{(CR^2)^{1/3}} \right\} \tag{7}$$

Due to the slow $1/R^{2/3}$ decay of the entropic repulsion, the $R = \infty$ unsupercoiled state is always locally stable. However, for sufficiently large $|\sigma|$, (7) develops a minimum for finite R. At some point, that minimum becomes lower in free energy than the unsupercoiled state, corresponding to stable plectonemic supercoiling. From (7) it is evident that the critical value of linkage at which this transition occurs is $|\sigma| \approx 1/(C\omega_0) \approx 0.01$ for DNA; this is close to the experimentally observed threshold for plectonemic supercoiling[10].

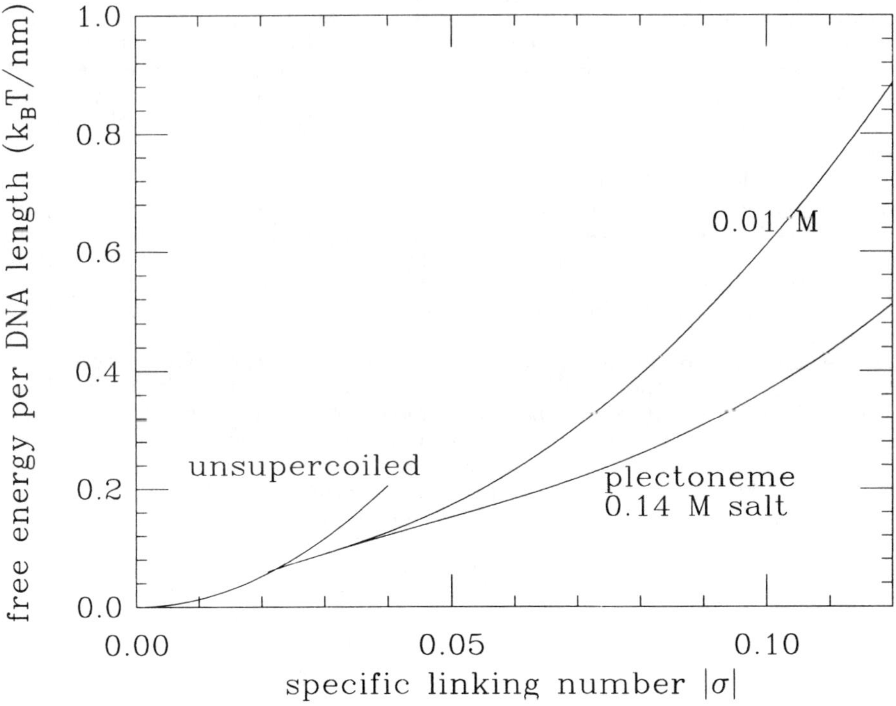

Figure 3: Free energy of plectonemic and unsupercoiled states obtained from Eq. 6. For $|\sigma| < 0.02$, unsupercoiled DNA is stable, while for $|\sigma| > 0.02$, plectonemically supercoiled DNA is stable. The free energy is approximately linear for $0.02 < |\sigma| < 0.08$.

At the point $|\sigma| \approx 1/(C\omega_0)$ the number of excess links per length, $\sigma\omega_0$, is on the order of one per twist or bend persistence length, comparable to the random fluctuation of twist and writhe per persistence length along a linear chain. More detailed analysis of (6) gives $R \approx A^{1/2}/(\sigma\omega_0)^{3/2}$ and $P \approx 1/(\sigma\omega_0)$; the results for R and P obtained by numerically minimizing (6) are in good agreement with experimental data [10].

The free energy of the supercoiled state relative to the unsupercoiled state is shown in Fig. 3. As can be seen, the plectoneme becomes lower in free energy than the unsupercoiled polymer for $|\sigma| \approx 0.02$. For this model, the free energy of the plectoneme is approximately $F \approx k_B T(\mathrm{Lk} - \mathrm{Lk}_0)$ in the regime where thermal fluctuations dominate the free energy (for 0.15 M NaCl, roughly $0.02 < |\sigma| < 0.07$). In this regime, each added link essentially adds a turn of the superhelix, thus forcing one more constraint on the supercoiled configuration. Thus each turn of supercoil may be considered to be a correlation 'blob'. For $|\sigma| > 0.07$, the hard core interaction keeps R from decreasing further, and the free energy starts to increase quadratically with σ; in this regime thermal fluctuations are not important.

In Fig. 2, it is clear that supercoiled DNAs are not linear plectonemes; they have branch points in their structure. Creation of a Y-shaped branch point costs on the order of $k_B T$ of free energy since a turn of supercoil must be disrupted. However, there is an entropy gain on the order of $k_B T \log L/(R^2 + P^2)^{1/2}$ since that branch point may be placed at any superhelical turn. Assuming no correlations between the branches, one expects one branch point per every few turns of supercoil; a more detailed calculation based on the free energy difference of supercoiled and unsupercoiled states of (6) suggests one branch point per ≈ 1 kb of DNA [13], in rough agreement with experimental (1 branch/kb) and simulation data (0.5 branch/kb) [15]. Thermal fluctuations will cause long supercoiled DNAs to take the shape of branched polymers with annealed branch structure.

III. STRETCHING TWISTED DNA

The previous section discussed supercoiling of a closed double-helical DNA: how can the same ideas be applied to the problem of a twisted DNA under tension? Experimentation with a twisted telephone cord suggests that tension can bring an extended state into 'coexistence' with plectonemic supercoiling (Fig. 4). Once again, there are two main problems: imposing the linkage number constraint, and taking account of thermal fluctuations.

Linkage number is of on the face of it, undefined for a linear DNA. However, any experiment expecting to probe DNA twisting elasticity must anchor both strands at both ends of the molecule (e.g. one end to a microscope slide, the other end to a colloidal particle to which tension and torque are applied). Once attachments to surfaces are made, the linkage number is well defined *if* one assumes that those surfaces are infinite in extent, so as to keep the polymer contour from passing 'behind' either end. Under such circumstances, $\mathrm{Lk} - \mathrm{Lk}_0$ is the number of twists applied to e.g. the colloidal particle, starting from a torsionally relaxed state (i.e. where $\mathrm{Lk} = \mathrm{Lk}_0$).

The thermodynamics of a twisted, stretched DNA may be treated by supposing a fraction $1-x$ of its length is extended DNA, leaving a fraction x as plectonemic supercoil. Thanks to the fact that the start and end points of the plectonemic 'domains' are near to one another, it is reasonable that the total linkage is just the sum of the extended DNA linkage, and that of the plectonemic region. In terms of fractional linkage changes, this means

$$\sigma = x\sigma_\mathrm{x} + (1-x)\sigma_\mathrm{p} \tag{8}$$

where σ is the linkage of the whole DNA, while σ_x and σ_p are the linkage in the extended and plectonemic domains.

Taking the extension of the entire molecule to be z, the extension per length of the extended state is $z/(xL)$. The free energy of a stretched twisted DNA is then computed from the free energies of extended and supercoiled 'pure phases' (using the plectoneme result of (6)):

$$F = \min_{x,\sigma_X} \{ xF_X(\sigma_X, z/[xL]) + (1-x)F_P(\sigma_P) \} \tag{9}$$

where (8) is used to eliminate σ_P.

The only remaining problem is to construct $F_X(\sigma_X, y)$, the free energy of the extended state as a function of its linkage and fractional extension y. Since some writhing of the extended state is expected, one can begin with a uniformily writhed state, i.e. a helical supercoil, with added thermal fluctuations [13,16]. For forces larger than $f > (C\sigma_x\omega_0)^2 k_BT/4A$, the writhing is expelled in favor of twisting. Therefore, in that

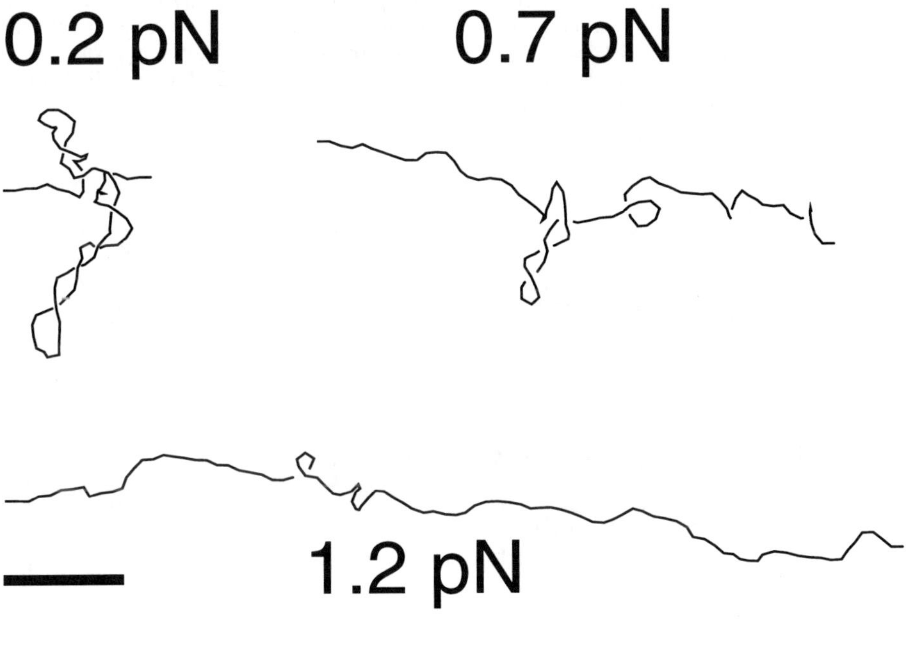

Figure 4: MC simulation configurations for DNA under tension after twisting [17]. The fixed linkage ($|\sigma| = 0.043$) leads to a plectonemically supercoiled conformation at low force ≈ 0.2 pN (upper left), a fully extended conformation at high force ≈ 1.2 pN (lower), and a mixed state of the two conformations at intermediate forces ≈ 0.7 pN (upper right). Bar is 100 nm; total DNA length is 1000 nm.

regime the free energy is just that corresponding to the force law (1), plus the twist energy for Tw = Lk, or $\Omega = \sigma_X \omega_0$:

$$\frac{F_X(\sigma_X, y)}{k_B T L} = \frac{1}{A}\left[\frac{y^2}{2} - \frac{y}{4} + \frac{1}{4(1-y)} - \frac{1}{4}\right] + \frac{C(\sigma_X \omega_0)^2}{2} \tag{10}$$

For lower forces, one can either make the approximation that there is no writhe and thus continue to use (10), or one can construct a free energy model for the writhed state [13,16]; this choice does not crucially alter the results.

The main results of this model are as follows. When the linkage is well below $|\sigma| \approx 0.02$, the force vs distance is unperturbed from that of unsupercoiled DNA (Eq. 1). In this regime, the added linkage per persistence length is less than the random thermal fluctuations of twist and writhe in each persistence length. For $|\sigma| \approx 0.02$, the force needed to obtain a given extension starts to increase with $|\sigma|$ due to the appearance of plectonemic domains along the chain. As $|\sigma|$ is increased, progressively more of the molecule is bound up as plectonemic supercoil at zero force. Finally a regime is reached where at zero force, all of the molecule is plectonemically supercoiled (Fig. 5).

For larger $|\sigma|$ than this, a finite force must be applied to obtain any extension at all, corresponding to the finite amount of work per length that must be done to unwind the plectonemic supercoil[13,16]. A rough estimate for this critical force is given by the twist energy per length, or $f^* \approx C k_B T(\sigma\omega_0)^2$. Near the threshold of supercoil stability $|\sigma| \approx 0.02$, this force is roughly $f^* \approx 0.1$ pN.

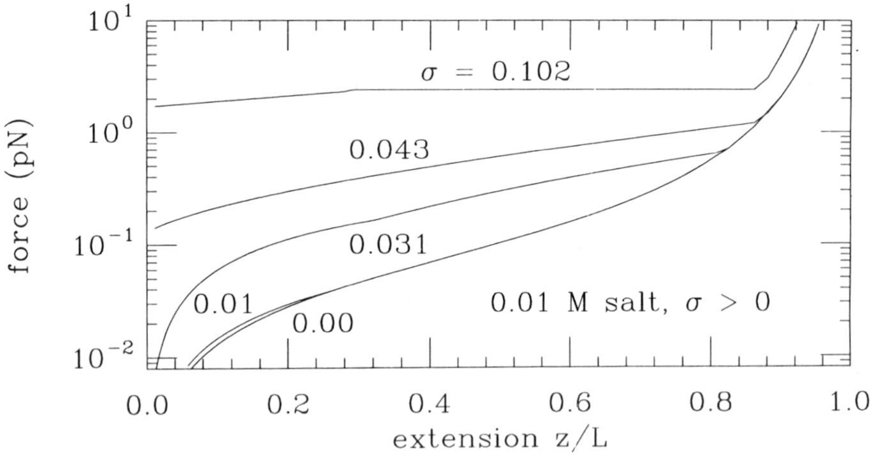

Figure 5: Coexisting extended-plectoneme model for force versus extension of twisted DNA (0.01 M univalent salt). For low $|\sigma| \leq 0.02$, the elasticity is essentially that of untwisted DNA; increasing $|\sigma|$ increases the force required to obtain a given extension, and for $|\sigma| \approx 0.04$ a finite force is required to obtain any extension at all, while at larger forces ≈ 1 pN, enough work has been done on the molecule to alter the double-helix structure (force plateau).

An important result of this model is that as the molecule is extended, the linkage number is highly concentrated in the plectonemic domain, leading to a buildup of torsional stress in the DNA. This will tend to destabilize the double helix, and one can expect transitions to alternate double helix structures for $\sigma < -0.04$ and $\sigma > 0.08$ at relatively low forces ≈ 1 pN [13] (such transitions occur more readily for unwinding than for overwinding [16]). Fig. 5 shows this for the case where for $|\sigma| = 0.12$, the DNA strands inside the plectonemic domains start to separate (i.e. normal plectonemic supercoiling coexists with regions where $|\sigma_{\mathrm{p}}| = 1$). At the point where the DNA starts to change structure, this model predicts a force 'plateau' (Fig. 5)[13].

Strick et al [6] have twisted and stretched a 50 kb DNA, with both strands anchored at each of its ends using a magnetic bead to apply torque and force to one end. The results of their study are in good agreement with the extended-plectoneme coexistence model [13,16]: they observe a force at fixed extension that appreciably increases for $|\sigma| > 0.02$, they observe a force threshold for extension for sufficiently large σ, and they observe 'plateaus' which plausibly indicate transitions to alternate 'melted' DNA states. Finally, we note that MC simulations of twisted DNAs by Vologodskii et al [17] are in good agreement with both experiment and the extended-plectoneme domain coexistence theory.

IV. INTERNAL DYNAMICS OF SUPERCOILED DNA

Plectonemic supercoiling introduces the possibility of a new type of dynamics, the sliding around of the molecule inside the supercoil. This motion has been proposed to be important to the juxtaposition of distant sequences along DNA in prokaryote cells (such juxtapositions are known to be required for recombination of distant regions along the E. coli chromosome)[19].

For an unbranched plectonemically supercoiled DNA of length L, such a process is bound to be slow, since slithering involves the 'reptation' of the molecule inside the supercoil. The typical time required for an encounter, starting from a random initial condition will be roughly

$$\tau_{\mathrm{sl}} \approx \frac{\eta L^3}{k_B T} \qquad (12)$$

where η is the solution viscosity. For a 1 μm (3 kb) DNA, this time is 250 msec.

Branches modify this scaling law for large L; the motion of branches along their trunks accelerates the rate at which distant sequences can communicate. Numerical results for a model of this process indicate that $\tau_{\mathrm{sl}} \propto L^2$ for very large L.

The chemical kinetics of site-specific recombination of 4 kb and 8 kb supercoiled DNAs has been studied in the test tube[19], and characteristic times for the initial communication process are consistent with (12), although the predicted strong length effect has not been observed [19]. Recent experiments on the same enzyme system operating in a prokaryote cell [20] over distances ranging from 10 kb to 100 kb show a strong length effect, in surprisingly good agreement with (12).

V. BRAIDED DNAs

If two double-helix DNAs are wrapped around one another, one can ask what the free energy of the resulting 'braid' is. Such braids naturally occur in prokaryote cells in replicated DNAs, due to incomplete removal of all the twists of the 'parent' double stranded DNA. The linkage number of the two DNAs is called the 'catenation' number, Ca.

For large $|\mathrm{Ca}| > L/A$, the DNAs are confined to a well-defined helical braid. Using the free energy functional (6) without the twist energy, and fixing instead

$|\mathrm{Ca}| = L/[2\pi(R^2 + P^2)^{1/2}]$ (the number of braid turns), one obtains

$$\frac{F}{k_BT} \approx 4\pi(2/3)^{3/4}\ k_BT\ |\mathrm{Ca}| \approx 20\ k_BT\ |\mathrm{Ca}| \qquad (13)$$

in the regime where the DNAs are kept far enough away from one another by entropic repulsion that the hard-core interaction is not important. Eq. 13 is in accord with each additional catenation forcing one more constraint on the polymers; intriguingly there is a large numerical constant. As the catenation is increased, the fluctuations of the DNAs are progressively decreased; the tension generated in the DNAs is $f_{\mathrm{braid}} = (k_BT/A)(2\pi \mathrm{Ca}A/L)^2$.

For such large $|\mathrm{Ca}|$ that the hard cores are interacting, the free energy becomes dominated by static bending energy of the braid, and begins to increase $\propto \mathrm{Ca}^2$. In terms of the specific catenation $\sigma_c = \mathrm{Ca}/\mathrm{Lk}_0$, this upturn in the catenation free energy occurs for $|\sigma_c| > 0.07$ under 0.15 M ionic conditions. Overall, the catenation free energy as a function of σ_c has nearly the same shape as the plectonemic supercoiling free energy shown in Fig. 3.

For small $|\mathrm{Ca}|$, the DNAs can be considered to be random-coil polymers; Rudnick and Hu[18] have computed the free energy to be

$$\frac{F}{k_BT} \approx \mathrm{Ca}^2/\log(L/A) \qquad (14)$$

The logarithmic elastic constant is due to the wandering apart of the two chains between successive turns of the braid.

An interesting possibility follows from considering how to connect the loose and tight braiding regimes. Consider $\sigma_c \sim \mathrm{Ca}/L$ fixed, and take $L \to \infty$: note that the loose-braid $dF/d\mathrm{Ca} \sim L/\log L$, while the tight-braid $dF/d\mathrm{Ca}$ is L-independent. This indicates that for any nonzero σ_c, for sufficiently large L, the loose braids will be unstable to the formation of tight braids. For long DNAs, if $|\mathrm{Ca}| < L/A$, one can expect 'braid separation', with a domain of tight braiding at the terminus of essentially unbraided chains.

If a DNA braid is viewed as a 'double helix' (of double-helix DNAs), then the braiding free energy can be considered to be the 'twist energy' of that object. This twist energy is lower than that of DNA itself due to the lack of any permanent connection between the two DNAs in the braid. The bending energy of such a braid will be larger than that of DNA since two DNAs must be bent. Both these effects disfavor supercoiling of the braid, but for large enough $|\mathrm{Ca}|$, the twist energy does become large enough for the braid to itself supercoil! [16] For 0.15 M NaCl, this point is reached for $|\mathrm{Ca}|/\mathrm{Lk}_0 \approx 0.07$; for 0.1 M NaCl, the longer range of the electrostatic repulsion of the nearby DNAs reduces this point to about 0.03. The force-distance behavior of braided DNAs can be computed using a extended-plectoneme coexistence model similar to that discussed in Sec. V.

VI. DECATENATION OF REPLICATED DNAs IN PROKARYOTES

During replication of DNA, not all of the twists of the 'parent' DNA are removed, leaving on the order of one catenation per kb between 'sibling' DNAs (each sibling carries one strand from the parent)[21]. Removal of these catenations is an important biophysical problem. Enzymes which can pass one DNA through a transient gap in another are implicated in this process, but the question remains of how those locally acting enzymes are guided to remove *all* the Ca. Once Ca is reduced below one per 10 persistence lengths, it is implausible that the free energy of catenation drives decatenation.

Experiment provides a crucial hint: *efficient decatenation of small circular plasmids requires the presence of an enzyme (gyrase) which supercoils DNA* [22]. Consider the free energy of supercoiled sister DNAs, each with excess linkage $\text{Lk} - \text{Lk}_0$, with Ca catenations between them. Following the approach outlined above, this free energy is of the form

$$F(\text{Lk} - \text{Lk}_0, \text{Ca}) \approx F_\text{p}(|\text{Lk} - \text{Lk}_0| + |\text{Ca}|) \tag{15}$$

where F_p is the supercoil free energy described in Sec. II. Thus, by cranking up $|\text{Lk} - \text{Lk}_0|$, the free energy per Ca liberated can be kept large, even when $\text{Ca} \to 0$: in this model, supercoiling-generated stress guides the removal of catenations betweeen the catenated siblings. A simple dynamic model based on (15) indicates that this process can efficiently remove sibling catenations.

This invites the speculation that supercoiling of DNA was an advantageous variation for early prokaryote cells, allowing longer DNAs to be efficiently segregated following their replication.

ACKNOWLEDGEMENTS

A large part of this work was done in collaboration with E. D. Siggia and A. V. Vologodskii. The author thanks D. Bensimon, C. Bustamante, P. Cluzel, D. Chatenay, N. Cozzarelli, V. Croquette, M. Frank-Kamenetski, S. E. Halford, P. Higgins and P. Nelson for many useful discussions.

REFERENCES

1. P. J. Hagerman, Annu. Rev. Biophys. Biophys. Chem. **17**, 265 (1988).

2. D. M. Crothers, J. Drak, J. D. Kahn and S. D. Levene, Meth. Enzymology **212**, 3 (1992).

3. S. B. Smith, L. Finzi and C. Bustamante, Science **258**, 1122 (1992);

4. T. T. Perkins, S. R. Quake, D. E. Smith and S. Chu, Science **264**, 822, (1994); T. T. Perkins, D. E. Smith, R. G. Larson and S. Chu, Science **268**, 83 (1995).

5. P. Cluzel, A. Lebrun, C. Heller, R. Lavery, J. L. Viovy, D. Chatenay and F. Caron, Science **271**, 792 (1996); S. B. Smith, Y. Cui and C. Bustamante, Science **271**, 795 (1996).

6. T. R. Strick, J.-F. Allemand, D. Bensimon, A. Bensimon and V. Croquette, Science **271**, 1835 (1996).

7. C. Bustamante, J. F. Marko, E. D. Siggia and S. Smith, Science **265**, 1599 (1994); A. V. Vologodskii, Macromolecules **27**, 5623 (1994); J. F. Marko and E. D. Siggia, Macromolecules **28**, 8759 (1995).

8. R. E. Kamien, T. C. Lubensky, P. Nelson and C. S. O'Hern, preprint (1996); J. F. Marko, preprint (1996).

9. P. Cluzel, Ph.D. thesis, University of Paris (1996).

10. N. R. Cozzarelli, T. C. Boles and J. White, in *DNA Topology and its Biological Effects*, edited by N. R. Cozzarelli and J. C. Wang (Cold Spring Harbor Laboratory, Cold Spring Harbor NY, 1990) Ch. 4;

11. S. D. Levene, C. Donahue, T. C. Boles and N. R. Cozzarelli, Biophys. J. **69**, 1036 (1995);

12. K. V. Klenin, A. V. Vologodskii, V. V. Anmshelevich, A. M. Dykhne and M. D. Frank-Kamenetskii, J. Mol. Biol. **217**, 413 (1991); A. V. Vologodskii, S. D. Levene, K. V. Klenin, M. Frank-Kamenetskii and N. R. Cozzarelli, J. Mol. Biol. **227**, 1224 (1992).

13. J. F. Marko and E. D. Siggia, Science **265**, 506 (1994); Phys. Rev. E **52**, 2912 (1995).

14. G. Calagareau, Rev. Math. Pur. et Appl. **4**, 58 (1959); Czech. Math. J. **4** 588 (1961); J. H. White, Am. J. Math. **91**, 693 (1969); F. B. Fuller, Proc. Natl. Acad. Sci. USA **68**, 815 (1971); A. V. Vologodskii, *Topology and Physics of Circular DNA* (CRC, Boca Raton FL, 1990) Ch. 4.

15. A. V. Vologodskii and N. R. Cozzarelli, Annu. Rev. Biophys. Biomol. Struct. **23**, 609 (1994).

16. J. F. Marko, Phys. Rev. E (in press, 1996).

17. A. V. Vologodskii and J. F. Marko (preprint, 1996).

18. J. Rudnick and Y. Hu, Phys. Rev. Lett. **60**, 712 (1988).

19. C. N. Parker and S. E. Halford, Cell **66**, 781 (1991); M. Oram, J. F. Marko and S. E. Halford (preprint, 1996).

20. N. P. Higgins, X. Yang, Q. Fu and J. Roth, J. Bact. **178**, 2825 (1996).

21. D. E. Adams, E. M. Shekhtman, E. L. Zechiedrich, M. B. Schmidt and N. R. Cozzarelli, Cell **71**, 277 (1992).

22. E. L. Zechiedrich and N. R. Cozzarelli, Genes and Devel. **9**, 2859 (1995).

TWIST-STRETCH ELASTICITY OF DNA

RANDALL D. KAMIEN, T.C. LUBENSKY, PHILIP NELSON, COREY S. O'HERN*
Department of Physics and Astronomy, University of Pennsylvania,
Philadelphia, PA 19104
* ohern@lubensky.physics.upenn.edu

ABSTRACT

The symmetries of the DNA double helix require a new term in its linear response to stress: the coupling between twist and stretch. Recent experiments with torsionally-constrained single molecules give the first direct measurement of this important material parameter. We extract its value from a recent experiment of Strick *et al.* and find rough agreement with an independent experimental estimate recently given by Marko. We also present a very simple microscopic theory predicting a value comparable to the one observed.

INTRODUCTION

In this paper we will study the response of DNA to mechanical stress using the methods of classical elasticity theory[1]. While many elements of DNA function require detailed understanding of specific chemical bonds (for example the binding of small ligands), still others are quite nonspecific. Moreover, since the helix repeat distance of $l_0 \approx 3.4$nm involves dozens of atoms, it is reasonable to hope that this length-scale regime would be long enough so that the cooperative response of many atoms would justify the use of a continuum, classical theory, yet short enough that the spatial structure of DNA matters.

Since various important biological processes involve length scales comparable to l_0 (notably the winding of DNA onto histones), the details of this elasticity theory should prove important. Yet until recently little was known about the relevant elastic constants. Extensive experimental work yielded fair agreement on the values of the bend and twist persistence lengths, though the former was plagued with uncertainties due to the polyelectrolyte character of DNA [2]. A simple model of DNA as a circular elastic rod gives a reasonable account of many features of its long-scale behavior, for example supercoiling [3].

Recently, techniques of micromanipulation via optical tweezers and magnetic beads have yielded improved values for the bend stiffness from the phenomenon of thermally-induced entropic elasticity [4][5][6], as well as a direct measurement of a third elastic constant, the stretch modulus[7][8]. Significantly, the relation between bending stiffness, stretch modulus, and the diameter of DNA turned out to be roughly as predicted from the classical theory of beam elasticity [7][8][9], supporting the expectations mentioned above.

Still missing, however, has been any direct measurement of the elastic constants reflecting the *chiral* (*i.e.* helical) character of DNA. One such constant, a twist-bend coupling, was investigated by Marko and Siggia [10], but no direct experimental measurement has yet been devised. We will introduce a new chiral coupling, the twist-stretch energy. Electrostatic effects do not complicate the analysis of this coupling. We will explain why our term is needed, extract its value from the experiment of Strick *et al.* [11], and compare it to the prediction of a microscopic model to see that its magnitude is in line with the expectations of classical elasticity theory. J. Marko has independently introduced the same coupling and estimated its value from different experiments[12]; our values are in rough agreement.

Mat. Res. Soc. Symp. Proc. Vol. 463 © 1997 Materials Research Society

EXPERIMENT

DNA differs from simpler polymers in that it can resist twisting, but it is not easy to measure this effect directly due to the difficulty of applying external torques to a single molecule. The first single-molecule stretching experiments constrained only the locations of the two ends of the DNA strand. The unique feature of the experiment of Strick *et al.* was its added ability to constrain the *orientation* of each end of the molecule.

We will study Fig. 3a of ref. [11]. In this experiment, a constant force of 8pN was applied to the molecule, and the end-to-end length $z_{\rm tot}$ was monitored as the terminal end was rotated through ΔLk turns from its relaxed state (which has Lk_0 turns). In this way the helix could be over- or undertwisted by as much as $\pm 10\%$. Over this range of imposed linkage $z_{\rm tot}$ was found to be a linear function of σ:

$$\epsilon = \epsilon_{\sigma=0} - 0.15\sigma, \tag{1}$$

where $\sigma \equiv \Delta Lk/Lk_0$ and $\epsilon \equiv (z_{\rm tot}/z_{\rm tot,0}) - 1$. Thus σ is the fractional excess link, and ϵ is the extension relative to the relaxed state. Eq. (1) is the experimentally observed twist-stretch coupling.

THEORY

Phenomenological Model

A straight rod under tension and torque will stretch and twist. We can describe it using the following reduced elastic free energy per equilibrium length $z_{\rm tot,0}$ of the rod:

$$f_1(\sigma, \epsilon) \equiv \frac{F_1(\sigma, \epsilon)}{k_B T z_{\rm tot,0}} = \frac{\omega_0^2}{2}\left[\bar{C}\sigma^2 + \bar{B}\epsilon^2 + 2\bar{D}\epsilon\sigma\right]. \tag{2}$$

The twist persistence length is $\bar{C} \approx 75\,\text{nm}$ [2], while the helix parameter $\omega_0 = 2\pi/l_0 = 1.85/\text{nm}$. We will take $\bar{B} \approx 1100pN/\omega_0^2 k_B T \approx 78\text{nm}$ [8]. In the experiment under study, there is an applied reduced force $\tau = 8\text{pN}/k_B T \approx 1.95/\text{nm}$. For a circular beam made of isotropic material, the cross-term $\bar{D}$ is absent [9] because twisting is odd under spatial inversion while stretching is even. For a helical beam, however, we must expect to find this term.

Setting $\tau = \partial f_1/\partial\epsilon|_\sigma$, we find

$$\epsilon = \epsilon_{\sigma=0} - (\bar{D}/\bar{B})\sigma\ . \tag{3}$$

Comparing to Eq. (1), we obtain the desired result: $\bar{D} = 12\,\text{nm}$. To compare this to Marko's analysis, we note that his dimensionless g equals our $\bar{D}\omega_0$, so that we get $g = 22$. The rough agreement with Marko's result $g = 35$ [12] indicates that the data show a real material parameter of DNA and not some artifact. We do not expect exact agreement, since Marko studied the nonlinear overstretching transition of [7][14]; our value came from the linear regime of small strains.

Microscopic Model

To gain further confidence in our result, we will now see how the expected twist-stretch coupling emerges from a simple elastic model for DNA. Fig. 1 reviews the relevant geometric

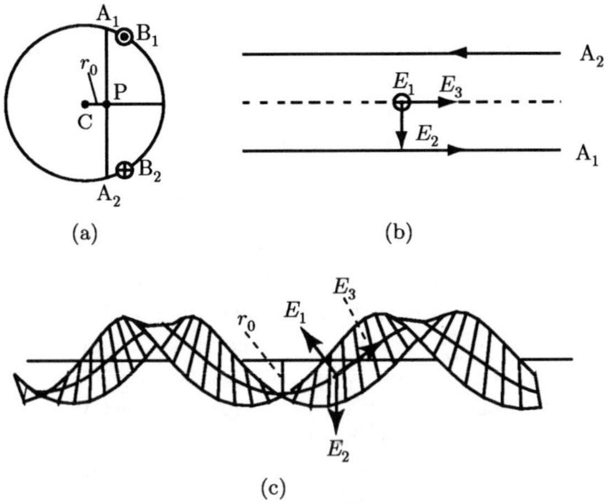

Figure 1: (a) Schematic representation of a cross section of DNA showing its intersections (B_1 and B_2) with the phosphate backbones, its intersection (line A_1PA_2) with the parameterizing helical ribbon, and its helical center C. (b) A section of ribbon showing oppositely oriented edges. At each point along the ribbon's center (dotted curve), there is a triad of orthonormal vectors, E_1, E_2, and E_3. E_3 is parallel to the center line, E_2 points to one of the edges, and E_1 is perpendicular to the ribbon. (c) Representation of the helical ribbon.

properties of DNA. Base pairs are connected by inequivalent sugar-phosphate backbones whose twisting pattern defines the major and minor grooves. These backbones intersect a given cross section of DNA at two points. Lines parallel to the line connecting these two points sweep out ribbon-like surfaces as the DNA twists. We will parameterize the DNA structure by one of these ribbons, whose center is a distance r_0 from the central helical axis. Any choice of r_0 is acceptable. We will see, however, that a small value of r_0 can explain the measured value of $\overline{D}$.

Our ribbon is described by the triad of unit vectors E_1, E_3, and $E_2 = E_3 \times E_1$, where E_3 is the unit tangent vector to the center of the ribbon and E_2 points from the center of the ribbon to one of its edges. The triad varies as we move along the arc length s of the ribbon. The motion is described by

$$\frac{dE_i}{ds} = -\epsilon_{ijk}\Omega_j E_k. \tag{4}$$

The parameter s labels each point along the central axis of the ribbon in its unstressed state and runs from 0 to L. The actual arc length along the distorted central axis of the ribbon will not be ds but rather $[1 + \epsilon_2(s)]ds$ where ϵ_2 is the intrinsic strain. Therefore, the total length for constant ϵ_2 is $L' = (1 + \epsilon_2)L$. The intrinsic strain allows the spacing between successive phosphate groups to change.

The edges of our ribbon, like the two sugar-phosphate backbones in DNA, are distinguishable and point in opposite directions. This symmetry can be incorporated into our microscopic model by considering a rotation of 180° about the vector E_1 followed by $s \to -s$. Under this transformation E_2 and E_3 change sign, but s derivatives of these vectors do not. Also, E_1 does not change sign, but dE_1/ds does. Therefore, the free energy should remain unchanged upon changing the sign of Ω_1 but not of Ω_2 and Ω_3[10]. The most general reduced free energy per length of ribbon relative to that of the flat unstretched ribbon up to second order in Ω_i and ϵ_2 is

$$\begin{aligned} f_{DNA} &= \frac{1}{2}[A'{\Omega_1}^2 + A(\Omega_2 - \Omega_{20})^2 + C(\Omega_3 - \Omega_{30})^2 + B{\omega_0}^2\epsilon_2^2 + \\ & 2D\omega_0\Omega_3\epsilon_2 + 2G\Omega_2\Omega_3 + 2K\omega_0\Omega_2\epsilon_2 - A{\Omega_{20}}^2 - C{\Omega_{30}}^2]. \qquad (5) \end{aligned}$$

ϵ_2 does not change sign under $s \to -s$, and so it can appear in combination with Ω_2 and Ω_3 in Eq. (5). This model is the simplest semi-microscopic model that incorporates all of the symmetries of DNA. It is an expansion to harmonic order in first-order derivatives of the vectors E_i (*e.g.* of $\Omega_3 = -E_1 \cdot \frac{dE_2}{ds}$). Thus, it is a model with quantitative predictive power so long as the Ω_i are slow on a scale set by the distance $a = 0.6$ nm between successive phosphate groups, (*i.e.* so long as $\Omega_i a \ll 1$). In the ground state $\Omega_3 = \Omega_{30} \approx \omega_0 = 1.85/\text{nm}$, so that $\Omega_{30}\ a \approx 1.1$ is not small. This implies that higher derivative terms (*e.g.* $(\frac{d\Omega_3}{ds})^2$, etc.) are needed for a quantitative theory. Nevertheless, our simple semi-microscopic model captures the essential symmetry of the DNA structure and allows us to address questions like the nature of the twist-stretch coupling.

One can easily show that the center of the ribbon describes a helix in the ground state of Eq. (5). We will assume that D, G, and K can be made small by an appropriate choice of r_0. Then, to keep the model as simple as possible, we will simply set these parameters equal to zero for this choice of r_0. We parameterize the helical ribbon using three angles ψ, γ, and ϕ:

$$\begin{aligned} E_3 &= \sin\gamma\hat{z} + \cos\gamma\hat{\phi} \\ E_2 &= \cos\psi(-\sin\gamma\hat{\phi} + \cos\gamma\hat{z}) - \sin\psi\hat{\rho} \\ E_1 &= -\sin\psi(-\sin\gamma\hat{\phi} + \cos\gamma\hat{z}) - \cos\psi\hat{\rho}, \qquad (6) \end{aligned}$$

where $\hat{\rho}$ and $\hat{\phi}$ are cylindrical unit vectors spinning at frequency $\dot{\phi}$. If $\dot{\phi}$ is a constant and $\psi = 0$, then the ribbon wraps around a cylinder of length z_{tot} and radius r.

$$z_{\text{tot}} = L(1 + \epsilon_2)\sin\gamma \quad \text{and} \quad r = \frac{\cos\gamma}{\omega}. \qquad (7)$$

In its ground state the helix has the following properties: $\Omega_1 = \Omega_{10} = 0$, $\Omega_2 = \Omega_{20} = \omega_0 \cos\gamma_0$, $\Omega_3 = \Omega_{30} = \omega_0 \sin\gamma_0$, $\dot{\phi} = \omega_0 = \sqrt{{\Omega_{20}}^2 + {\Omega_{30}}^2}$, and $\psi = \psi_0 = 0$. Also, the ground state length and radius of the molecule are, respectively, $z_{\text{tot},0} = L\sin\gamma_0$ and $r_0 = \cos\gamma_0/\omega_0$.

We now consider deviations in the ground state length z_{tot} and twist rate $\dot{\phi}$ of the molecule. Since the total twist is $\dot{\phi}L$, the excess twist is σL where $\sigma = (\dot{\phi} - \omega_0)/\omega_0 = \delta\dot{\phi}/\omega_0$. Using Eq. (7) we find that changes in length are produced both by intrinsic strain ϵ_2 and by changes in γ:

$$\epsilon = \frac{z_{\text{tot}}}{z_{\text{tot},0}} - 1 = \epsilon_1 + \epsilon_2 \quad \text{where} \quad \epsilon_1 = \cot\gamma_0\delta\gamma. \qquad (8)$$

The energy of harmonic deviations from equilibrium are obtained by expanding f_{DNA} to second order in $\delta\gamma(s)$, $\delta\psi(s)$, and $\delta\phi(s)$. The ground state is a periodic helix implying these variables can be expressed in terms of Fourier modes in different Brillouin zones defined by ω_0. Rotations of the helix about space-fixed axes x, y, and z are described by the variables $\delta\theta_x$, $\delta\theta_y$, and $\delta\theta_z$. The variables $\delta\gamma$, $\delta\psi$, and $\delta\phi$ can be expressed in terms of these variables:

$$
\begin{aligned}
\delta\gamma &= \cos\omega_0 s\,\delta\theta_x + \sin\omega_0 s\,\delta\theta_y \\
\delta\psi &= \frac{1}{\cos\gamma}(\sin\omega_0 s\,\delta\theta_x - \cos\omega_0 s\,\delta\theta_y) \\
\delta\phi &= \tan\gamma(\sin\omega_0 s\,\delta\theta_x - \cos\omega_0 s\,\delta\theta_y) + \delta\theta_z.
\end{aligned} \tag{9}
$$

Thus, variations of $\delta\theta_x$ and $\delta\theta_y$ in the first BZ give rise to variations in $\delta\gamma$ and $\delta\psi$ in the second BZ. A complete long-wavelength theory can, therefore, be expressed in terms of the first BZ components of $\delta\theta_x$, $\delta\theta_y$, $\delta\gamma$, $\delta\psi$, and $\delta\phi$ (whose 1st BZ component is equal to that of $\delta\theta_z$).

Using the relations for the Ω_i obtained from Eq. (4) and Eq. (6), we find

$$
\begin{aligned}
\Omega_2 - \Omega_{20} &= \cos\gamma_0\delta\dot{\phi} - \omega_0\sin\gamma_0\delta\gamma \\
\Omega_3 - \Omega_{30} &= \sin\gamma_0\delta\dot{\phi} + \omega_0\cos\gamma_0\delta\gamma.
\end{aligned} \tag{10}
$$

Then, using these expressions in f_{DNA}, integrating out ψ, and remembering that length along the pitch axis is a factor of $\sin\gamma_0$ less than the ribbon length, we find that the effective reduced free energy per unit length of pitch axis is

$$
f = f_B + f_{TS}, \tag{11}
$$

where

$$
f_B = \frac{1}{4\sin\gamma_0}(A' + A\sin^2\gamma_0 + C\cos^2\gamma_0)(\dot{\theta}_x^2 + \dot{\theta}_y^2) = \frac{1}{2}\overline{A}(\dot{\theta}_x^2 + \dot{\theta}_y^2) \tag{12}
$$

is the bending energy[16] and f_{TS} is the twist-stretch energy defined by

$$
f_{TS} = \frac{{\omega_0}^2}{2\sin\gamma_0}[C_{\sigma\sigma}\sigma^2 + B_{\epsilon_1\epsilon_1}{\epsilon_1}^2 + B{\epsilon_2}^2 + 2D_{\epsilon_1\sigma}\epsilon_1\sigma], \tag{13}
$$

where $C_{\sigma\sigma} = (A\cos^2\gamma_0 + C\sin^2\gamma_0)$, $B_{\epsilon_1\epsilon_1} = (A\frac{\sin^4\gamma_0}{\cos^2\gamma_0} + C\sin^2\gamma_0)$, and $D_{\epsilon_1\sigma} = (C - A)\sin^2\gamma_0$. The twist-stretch energy can be expressed in terms of the total strain by setting $\epsilon_1 = \epsilon - \epsilon_2$ and integrating over ϵ_2. The result is that f_{TS} has the same form as Eq. (2) with macroscopic elastic constants $\bar{B}$, $\bar{C}$, and $\bar{D}$ expressed in terms of our microscopic parameters A, B, and C. The stretch moduli associated with ϵ_1 and ϵ_2 add in parallel to yield a total stretch modulus $\bar{B} = \frac{1}{\sin\gamma_0}(\frac{1}{B_{\epsilon_1\epsilon_1}} + \frac{1}{B})^{-1}$. The twist-stretch coupling is

$$
\bar{D} = \frac{B}{(B_{\epsilon_1\epsilon_1} + B)}(C - A)\sin\gamma_0. \tag{14}
$$

Thus, a description of DNA in terms of a helical ribbon with an axis offset from the central helical axis generates a twist-stretch coupling even if the bare twist-stretch coupling (D in Eq. (5)) is zero. We can estimate the offset r_0 necessary to produce the measured $\overline{D}$ assuming it arises entirely from Eq. (14). If we assume $A' \approx A$ and $\omega_0 r_0 \ll 1$, then $B \approx \overline{B} =$ 75nm, $C \approx \overline{C} =$ 78nm, and $A \approx \overline{A} =$ 40nm, and we find $\overline{D} \approx (\omega_0 r_0)^2(\overline{B}/\overline{A})(\overline{C} - \overline{A})$ and $(\omega_0 r_0)^2 \approx 0.176$ or $r_0 \approx 0.23$nm[15]. Corrections to this estimate are of order $(\omega_0 r_0)^4 \sim 0.03$.

CONCLUSION

We have pointed out a strong twist-stretch coupling in torsionally constrained DNA stretching experiments, evaluated it, argued that it reflects intrinsic elasticity of the DNA duplex, and shown that the value we obtained is consistent with elementary considerations from classical elasticity theory. A greater challenge remains to predict this coupling from the wealth of available crystallographic information on the conformation of short oligomers.

ACKNOWLEDGMENTS

We would like to thank D. Bensimon, S. Block, and J. Marko for their help and for communicating their results to us prior to publication, and W. Olson for discussions. RK, TL, and CO were supported in part by NSF grant DMR96–32598. PN was supported in part by NSF grant DMR95–07366.

REFERENCES

1. R. Kamien, T. Lubensky, P. Nelson, and C. O'Hern, "Direct Determination of DNA Twist-Stretch Coupling", (1996), preprint.
2. M. Record, S. Mazur, P. Melancon, J. Roe, S. Shaner, and L. Unger, Annu. Rev. Biochem. **50**, 997 (1981).
3. C. Benham, Biopolymers **22**, 2477 (1983).
4. S. Smith, L. Finzi, and C. Bustamante, Science **258**, 1122 (1992).
5. C. Bustamante, J. Marko, E. Siggia, and S. Smith, Science **265**, 1599 (1994).
6. J. F. Marko and E. D. Siggia, Macromolecules **28**, 8759 (1995).
7. S. Smith, Y. Cui, and C. Bustamante, Science **271**, 795 (1996).
8. M. D. Wang, H. Yin, R. Landick, J. Gelles, and S. M. Block, "Stretching DNA with optical tweezers", Biophys. J., (1997), in press.
9. L. Landau and E. Lifshitz, Theory of Elasticity, 3rd ed. (Pergamon, London, 1986), pp. 59-86.
10. J. F. Marko and E. D. Siggia, Macromolecules **27**, 981 (1994).
11. T. Strick, J. Allemand, D. Bensimon, A. Bensimon, and V. Croquette, Science **271**, 1835 (1996).
12. J. Marko, "Stretching must twist DNA", 1996, preprint.
13. C. Calladine and H. Drew, Understanding DNA: the molecule and how it works (Academic, London, 1992).
14. P. Cluzel, A. Lebrun, C. Heller, R. Lavery, J.-L. Viovy, D. Chatenay, and F. Caron, Science **271**, 792 (1996).
15. W. Saenger, Principles of Nucleic Acid Structure (Springer-Verlag, New York, 1984), pp. 225-226.
16. J. F. Marko and E. D. Siggia, Phys. Rev. E **52**, 2912 (1995).

UNCOILING OF DNA BY DOUBLE CHAINED CATIONIC SURFACTANTS

H. H. PARADIES* AND S. F. CLANCY**
*Biotechnology & Physical Chemistry, Märkische Fachhochschule, Iserlohn, D- 58644, Germany,HParadies@aol.com
**Witco Corporation, Safety, Health, & Environmental Affairs, One American Lane, Greenwich, CT 06831– 2559, USA, clancsh@witco-corp.com

ABSTRACT

Distearyldimethylammonium (DSDMA) X (X^- = OH, H_2PO_4) interact with double stranded T4 DNA (166 kilobase pairs) below and above the CMC ($1.5x10^{-6}$). Below the CMC of either DSDMA X where the cationic double-chained surfactants are in the monomeric state, T4 DNA and $DSDMA^+$ form a compact complex where all surfactant molecules are bound. Close to the CMC, particularly for DSDMA OH but also in the presence of $H_2PO_4^-$, T4 DNA exhibits a condensed (globule) conformation, though some DNA molecules are in the more extended DNA conformation. Above the CMC of DSDMA X a plateau is reached (up to 3.0×10^{-6} M surfactant) revealing a penetration of DSDMA molecules into the DNA globules resulting in a loosening of the tightened state of DNA. The various stages of the transition from the condensed coil to the extended state of T4 DNA with changing DSDMA X concentrations was monitored by static and inelastic light scattering experiments which were supplemented by small-angle X-ray scattering measurements.

INTRODUCTION

The polyamine-induced co-operative transition of the DNA helix from an extended conformation to a highly condensed structure *in vitro* is of extreme significance in molecular biology, transcription and bending of DNA, charge distribution, and polymerase-DNA interactions, since nucleic acids often occur tightly packed *in vivo* and require polyamines to achieve and stabilise the compact form [1]. Calculations of the energy changes due to conformational changes of the packing of the DNA, performed by Riemer & Bloomfield [2] using the encapsulation of DNA by bacteriophage T4 as a model system, it became obvious that electrostatic repulsions dominate the other forces which oppose the collapse of DNA into an extended conformation. The process of collapsing can be facilitated by neutralisation of the high charge density within the condensed DNA domain. The purely electrostatic counterion condensation theory of polyelectrolytes [3,4] and neutralisation depend on the condensation of counterions close to the backbone of the DNA and/or the cationic double chained micelles (which are in equilibrium with their monomers) in the outer layer in a process which is driven by the charge density of the biopolymer, the character of the counterions (including their valence, structure and chirality), and as studied here, the physical solution properties of the double-chained distearyldimethylammonium (DSDMA) X salts. The extent of neutralisation under specified ionic and DSDMA X conditions can be estimated from theory which provides the means to determine critical conditions beyond which the collapse of T4 DNA becomes spontaneous. However, these conditions conceivably depend not only on the charge of the counterion, its structure [5], or the N-ionic surfactant concentration (concentrations of charged species vs. total surfactant concentration where the extended T4 DNA is the polyelectrolyte), but also on the geometry of the collapse and the DSDMA induced mechanism leading to a collapse, which is balanced between DNA-DNA contacts and DNA ≈ DSDMA contacts (including solvent). The possible role of the counterions (including those having a chiral center), in stabilising various DSDMA X forms in

Mat. Res. Soc. Symp. Proc. Vol. 463 © 1997 Materials Research Society

aqueous solutions (e.g. unilamellar vesicles, threadlike micelles & spherical micelles [6]) is an additional factor in stabilising the compact (globule) form of the DNA. Furthermore, cationic lipids are known as efficient non-viral reagents to transfect outer DNA into animal cells *in vitro* [7] through liposome like DNA complexes. Anti-viral and anti-microbial activities *in-vitro & in-vivo* of single and double chained N-cationic surfactants have been reported and are being used in medical applications [8]. In addition, the environmental concerns of double chained N cationic surfactants of the DSDMA X type (including their biodegradability which is related to the interactions between DNA (or RNA & proteins) and the surfactant at very low surfactant concentrations) prompted us to study the interactions of DSDMA X with T4 DNA since there are few in this area [9].

This paper reports on the conformation changes of double-stranded T4 DNA in the presence of DSDMA X (X= OH, $H_2PO_4^-$). Applying inelastic (QELS) and static (LS) light scattering measurements we were able to examine the transition of T4 DNA from discrete DNA transition states with respect to higher order structures between an uncoiled state (elongated coil state) and the compact ("packaged") globule state by addition of DSDMA X. Additional information has been obtained from small-angle X-ray-scattering experiments supporting the conclusions drawn from QELS & LS measurements of this complex system.

EXPERIMENTAL

Materials

T4 phage was prepared by the method of Thomas & Abelson [10]. DNA preparation was performed by repeated phenol extractions (20 °C) in 0.1 M Na citrate, pH 7.1. In order to eliminate the polyamines inherent to T even phages, the ionic strength was increased to 2.5 M NaCl in the course of the phenol treatment. The presence of 0.5% SDS during the T4 DNA preparations has been avoided since it has been found that the T4 DNA, in the presence of SDS, interferes with the binding experiments of DNA and DSDMA X due to binding of residual SDS to DNA to DSDMA X. The purity of the T4 DNA was proved by measuring the A_{260} / A_{280} and A_{260} / A_{230} ratios, which were determined to be 1.86 and 2.4, respectively and are close to those reported in the literature [11], suggesting pure preparations of T4 DNA. The absence of fragmented T4 DNA was confirmed by agarose gel electrophoresis (Pharmacia, Phase System). Acetylated T4 DNA was prepared by acetylation of T4 DNA under mild conditions and characterised according to Gerhard & Warren [12]. The concentrations of T4 DNA were determined spectrophotometrically by applying an extinction coefficient ε_{250} = 6,650 cm^{-1} M^{-1} nucleotides. DSDMA X was prepared as described in [6], and further purified by HPLC (reversed phase C_{16}, Bondapak).

Sample Preparations

T4 DNA solutions were diluted into experimental buffer (1mM TRIS-H_2PO_4 or TRIS-OH, containing 1 mM NaCl, pH 7.1) to twice the desired final concentration (0.55 - 1.5 μM DNA-H_2PO_4), and centrifuged at 3,000 x g (20 °C) for 5 min. About 0.25 mL from the top was added to the scattering cylindrical scattering cells containing the counterion solutions (10 μM DSDMA OH or DSDMA-H_2PO_4) at pH 7.1. Mixing was performed through gentle tipping in order to avoid high local concentrations of DNA or DSDMAX, minimising the possibility of multi-molecular condensation or premature collapse to the globule state [13]. The samples were permitted to equilibrate for 2 h, centrifuged for 15 min at 2,000g (20 °C), and analysed. The total scattering intensities of buffer alone, of T4 DNA in experimental buffer, and of DSDMA X solutions at various concentrations were monitored with each set of data to provide proper controls. The correction for light absorption by DSDMA X was negligible even at the highest concentrations of DSDMA X. The solutions contained 1 μg/mL DNA in the standard buffer.

METHODS.

QELS and LS experiments were performed and the data processed as described in [6]. The laser was a He-Ne laser (NEC 50 mW) and an ALV-5000 multi-τ-correlator and a computer controlled stepping motor driven variable angle detector system was used. The cell holder was filled with index-matching fluid (toluene) and the temperature equilibrated at 20 °C (Lauda 15). Small Angle X-ray scattering (SAXS) experiments including data collection, processing, smoothing procedures were performed as described in [14] also at 20 °C.

RESULTS

LS and SAXS. In order to characterize the conformational transition of T4 DNA in the presence of DSDMA X, a series of experiments were performed by changing the DSDMA X concentration from 0.50-3.5 x 10^{-6} M (CMC=1.3x10^{-6} M) at constant DNA concentration. Assuming that the flexibility of T4 DNA is reflected in its radius of gyration, it is important that the data recorded are in the "linear region" before the correct equations can be applied for analysis. However, using LS data where $\theta \leq 15°$ and those from SAXS experiments, representative $(HC_{DNA}/R_\theta)_{C\to 0}$ for T4 DNA in 1mM TRIS-H_2PO_4, 1mM NaCl, pH 7.1(20 °C) can be obtained. Applying the determined values for $(\partial n/\partial c_{DNA})_{T,c} = 0.165$ mL g^{-1}, $n_0 = 1.325$ and λ= 632.8 nm, the radius of gyration was calculated to be $< R_G > = 1{,}327 \pm 58$ A (0.37 μm). The value obtained from SAXS experiments is $< R_G > = 1{,}320 \pm 40$ Å, the weight-average molecular weight of our T4 DNA preparations was found to be 14x10^6 by LS and 13.5x10^6 by SAXS. Thus, for a chain of 160 kilobase pairs, the contour length, L_c, is estimated to be 520,000 Å or 52 μm. Therefore, we calculate that the contour distance below which the T4 DNA is rodlike in solution ($<r^2>_0 \geq 0.85\, L_c^2$ or $<r^2>_0^{1/2} \geq 0.95\, L_c$) corresponds to a value of $(1/3)(L_c/a) = 186$ with a= 0.28 μm. From this we can estimate that $L_c = 0.3a - 16.5$ μm which is equivalent to $<r^2>_0^{1/2} = 52.3 \pm 0.5$ μm or 2,730 base pairs. However, as L_c increases, $<r^2>_0 < L^2_c$ as the effect of curvature becomes significant. So at very high values of L_c the ratio $<r^2>_0/L_c^2 = 2a/L_c$, which means that $<r^2>_0$ from the value expected for a rod is determined by the ratio of the persistence length to the contour length, hence bending of the DNA becomes manifest and has to be taken into account. Moreover, since $L_c << a$ it follows that $< r^2 > \cong L_c^2$, which is the unperturbed end-to-end distance of the structure in solution (accessible by SAXS) and is equal to the contour length. These changes in the ratios are actually seen by the addition of DSDMA X in the various concentration ranges and monitored by SAXS and QELS experimentally. The persistence length, λ, of the T4 DNA molecule is determined to 0.08 μm under the assumption of a Gaussian chain. The number of Kuhn segments of this giant DNA molecule under our experimental conditions is approximately $L_c/2x0.08 \cong 344$.

The effects of DSDMA OH or DSDMA $H_2PO_4^-$ on the total intensity of scattered light from T4 DNA samples at scattering angle 90° are shown in Fig.1.

All of the DNA samples tested, including acetylated T4 DNA, reveal a sharp increase in the intensity accompanying the co-operative condensing of the DNA into a compact conformation. The transition range is determined between DSDMA X concentrations of 1.0 x 10^{-6} M for $DSDMA^+$ OH^- and 1.15 x 10^{-6} M for $DSDMA^+$ $H_2PO_4^-$. Although the concentration difference between the DSDMA hydroxide and the phosphate is small and the transition range is broader for the phosphate, background scattering and total scattering intensity after condensation are closely comparable as the T4 DNA collapses by a different mechanism in the presence of DSDMA phosphate.

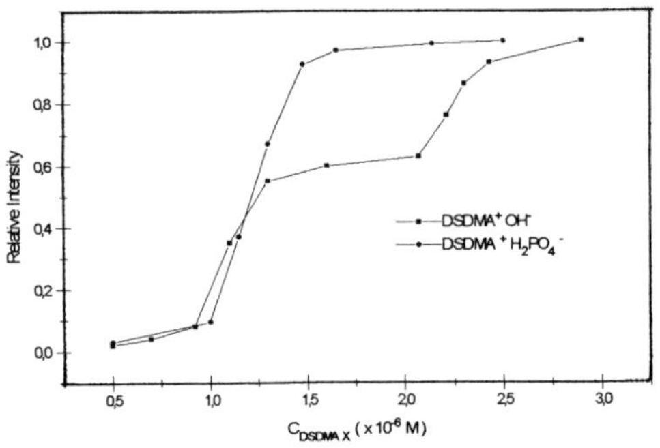

Fig. 1 Normalized relative intensity of scattered light, $(I - I_{sample}) / (I_{max} - I_{blank})$ vs. DSDMA X (X = OH^-, $H_2PO_4^-$) concentrations at scattering angle of 90° (20 °C) for solutions of 1 μM DNA in standard buffer.

Furthermore, there is no consideration of effects related to charge densities and possible non-homogenous distribution of charge density leading to long-range effects along the DNA helix. Fig.2 shows the dependence of the determined hydrodynamic radii, R_H, obtained from SAXS experiments according to $< R_G^2 > = < R^2_H > / 6$ of the T4 DNA upon addition of DSDMA X. We noticed three different areas where the T4 DNA changes conformation with surfactant concentration.

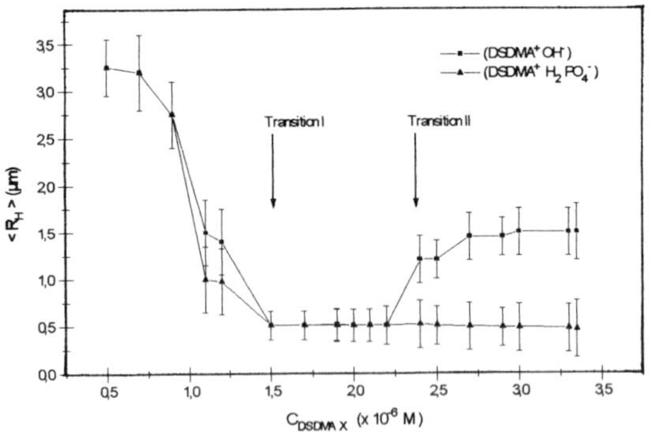

Fig. 2 $< R_H >$ of T4 DNA molecules vs. $C_{DSDMA\ X}$ (X = OH^-, $H_2PO_4^-$) added to a 1μM DNA solution in standard buffer at 20 °C.

At DSDMAX concentrations below 0.5×10^{-6} M to 0.9×10^{-6} M all DNA found to be in the coil state and no changes of the apparent size with increase of DSDMA X concentrations appear to occur. However, between 1.0×10^{-6} M to 1.8×10^{-6} DSDMA X only condensed (globular) T4

DNA are observed. This particular region of DNA≈DSDMA X complexes tend to be excluded from the bulk aqueous phase when raising the concentration of DNA. At DSDMA X concentrations above 2.5×10^{-6} M, which is well above the CMC, an increase of L can be noticed with a transition region between C_{DSDMAX} = (1.9-2.5) x 10^{-6} M. This phenomenon can be reconciled by the penetration of DSDMA cations inside the tightly packed DNA globules due to attraction of the surfactant with both the negative charges of the DNA chain and the interference with the base pairing of the C_{18} chains as seen in CD-spectra. Furthermore, the aggregation in this region can be enhanced in the presence of additional DNA molecules which also results in an increase of the size of the globules and denser packing. This is also consistent with CD-spectra using different T DNAs having different GC and AT compositions. Single-chained surfactants having the equivalent chain length and the same head group show a very narrow transition region but magnitudes higher than $C_{surf.} = 2.5 \times 10^{-6}$ M, and no additional uncoiling region which levels off at $C_{DSDMAX} = 2.8 \times 10^{-6}$ M.

Calculation of Shape-Dependent Hydrodynamic Dimensions of T4 DNA in the presence of DSDMA X. - QELS studies directly yield the diffusion coefficient, D_T, of particles in solution. Table I lists the diffusion coefficients of all the condensed DNA species as well as the ones in the uncoiled (extended) state.

Table I

Hydrodynamic Properties of T4 DNA at 90° Scattering Angle in the Presence of DSDMA X (X^- = OH, $H_2PO_4^-$) at 20 °C.

Conditions (x 10^{-6} M)		D_T (x 10^{-8} $cm^2 s^{-1}$) [a.)]		Outside Diameter of Toroids, o (Å) [b.)]	
		OH^-	$H_2PO_4^-$	OH^-	$H_2PO_4^-$
T 4 DNA C_{DSDMA} =	0.01	1.05 ± 0.20	1.02 ± 0.20	4150 ± 50	4200 ± 50
T 4 DNA C_{DSDMA} =	0.57	1.15 ± 0.25	1.25 ± 0.20	3790 ± 50	3700 ± 50
T 4 DNA C_{DSDMA} =	1.50	4.00 ± 0.15	4.15 ± 0.20	1240 ± 50	955 ± 25
T 4 DNA C_{DSDMA} =	1.90	4.15 ± 0.15	4.20 ± 0.25	1050 ± 25	785 ± 24
T 4 DNA C_{DSDMA} =	2.50	3.95 ± 0.20	4.25 ± 0.25	1350 ± 30	751 ± 24
T 4 DNA C_{DSDMA} =	3.00	2.25 ± 0.25	4.40 ± 0.30	1940 ± 25	720 ± 20
T 4 DNA, acetyl C_{DSDMA} =	2.00	4.35 ± 0.25	4.20 ± 0.30	1075 ± 25	784 ± 24
T 4 DNA C_{DSDMA} =	3.50	2.01 ± 0.25	4.95 ± 0.25	2100 ± 35	650 ± 25
T 4 DNA $C_{SPERMIDINE}$ =	100	4.15 ± 0.15	-	1240 ± 42	-
T 4 DNA [c.)] C_{CTAB} =	640	4.00 ± 0.25	4.19 ± 0.25	1250 ± 50	1150 ± 50

[a.)] QELS measurements; [b.)] according to eq. 1; [c.)] CTAB = cetyltrimethylammonium Bromide.

Since T4 DNA appears under the electron microscope as toroidal in shape [18], D_T, the translational diffusion coefficient of the studied samples can be related to the outer diameter, o, of toroids according to eqn.1 [15]:

$$(1) \qquad D_T \cong \frac{2k_B \cdot T}{3\pi\eta(o+i)} \cdot \left(\frac{2h^2}{3\pi}\right) \cdot (0.127h + 0.908)^{-1}$$

where i is the inner toroidal diameter, h = (o + i)/ (o - i), and η is the viscosity of the solution. Applying i = 0(h = 1) rests on the assumption that D_T is remarkable insensitive to i what has been found in the present work. The outer diameter of the tori that diffuse with D_T are listed in Table I also. The diameters calculated from eqn.1, vary from 4150 Å to 1050 Å for DSDMA OH or DSDMA $H_2PO_4^-$, respectively.

The angular dependence of the diffusion coefficient under these saline conditions was minimal and random in all samples studied and reflects monodisperse suspensions and no interference with dust contamination. At DSDMA X concentrations above 3.9×10^{-6} M perturbation of the spectra with time has been noticed since forming of DSDMA X unilamellar vesicles started to occur. This interfered with the T4 DNA because of intercalation of DNA by the DSDMA X vesicles in the form of a liposome-DNA complex which can be isolated and physically characterised revealing very different hydrodynamic parameters. Three major trends can be observed: (1) DNA condenses in the presence of monomeric DSDMA OH or DSDMA $H_2PO_4^-$ almost at the same concentrations; (2) the diffusion coefficient of the T4 DNA increases with DSDMA concentration regardless of OH^- or $H_2PO_4^-$ counterions as long as the ionic strength is low or not significantly changed during the experiments. However, this is only true until a concentration of 2.0×10^{-6} M of DSDMA X is maintained; (3) the tightening is again loosened at $C_{surf} = 3.0 \times 10^{-6}$ M or higher. The additional trend in uncondensing of T4 DNA is not observed in the presence of $H_2PO_4^-$ as the counterion of $DSDMA^+$. A similar trend is observed for the acetylated form of T DNA. The changes in magnitude either by R_H or ξ_H obtained from the translational diffusion coefficient, as well as of the outer toroidal diameter, o, before and after the transition, is found to be of the order of three and higher. Since the spatial density of DNA is in the coil state and quite low, ξ_H is somewhat smaller than for the extended state of the coil because ξ_H holds only for nearly spherical molecules with relative high density. Comparisons of relative sizes of DNA with surfactant concentration are only meaningful if the samples have the same basic shape and are almost monomolecular. However, the low ionic strength and low DNA concentrations used provide conditions that minimize interchain interactions. To ensure that the observed figures are the results of changes in tightening of DNA, and not intermolecular arrangements, the D_T values were measured at various DSDMA X concentrations in the transition region. For all cases D_T slightly increased by $\cong 8\%$ as the transition progressed and leveled off at DSDMA X concentrations at which the complex reaction was complete. This is in accord with the view that the condensed T4 DNA≈DSDMA X complex were loosely folded at the onset of the transition (Fig.1), but tightened up as DSDMA X concentration was increased. The reversibility of tightening along the transition curve was provided by incremental additions of NaCl or $MgCl_2$ (used as an additional control for macromolecular collapse in the presence of DSDMA X). Furthermore, a comparison of the torus volume (Tab. I) to the total volume of hydrated DNA helix of diameter d = 25 Å and length L = 55 μm yields 2.8×10^{-16} and 3.1×10^{-16} cm, respectively, strongly suggesting that the observed torus only accommodate a single DNA.

CONCLUSION

The double-chained surfactant DSDMA X below its CMC exhibits a discrete transition between a coil and globular state at μM concentrations. This transition is reversible upon addition of NaCl or $MgCl_2$ resulting in a sort of uncoiling of the condensed state. The preliminary results reveal that a double-chained cationic surfactant having "point charge" characteristics not only could cause tightening of DNA but also seems to exhibit great efficiency: It took less than one-tenth as much a single chain length or as spermidine to induce DNA condensation. The particles in the collapsed state show a drastically smaller hydrodynamic radii than in the presence of spermidine or CTAB, respectively.

REFERENCES

[1] S.S.Cohen, Adv. Polyamine Res.,1, 1-10, (1978); L.C.Gosule & Schellman, Nature, 259, 33, (1976); L.C.Gosule & J.A.Schellman, J.Mol.Biol., 121, 311, (1979); R.W.Wilson & V.A.Bloomfield, Biochemistry, 18, Biochemistry, 18, 2192, (1979)

[2] S.C.Riemer & V.A.Bloomfield, Biopoymers, 17, 785, (1978)

[3] G.S.Manning, Q.Rev.Biophys., 11, 179, (1978)

[4] G.S.Manning, J.Phys.Chem., 85, 870, (1981)

[5] H.H.Paradies, M.Thies, S.F.Clancy, this symposium "Statistical Mechanics in Physics & Biology, MRS, eds. D.Wirtz, T.C.Halsey, J.van Zanten, (1996)

[6] S.F.Clancy, P.H.Steiger, D.A.Tanner, M.Thies, H.H.Paradies, J.Phys.Chem., 98, 11143, (1994); M.Thies, S.F.Clancy, H.H.Paradies, J.Phys.Chem., 100, 9881, (1996); H.H.Paradies, J.Phys.Chem., 90, 5956, (1986)

[7] P.L.Felgner et al., Proc. Natl. Acad. Sci. (USA), 84, 7413, (1987)

[8] H.H.Paradies, US-Pat. # 4,074,850 (1989); H.H.Paradies & U.Schroer, Pharm.Ind., 12, 1387, (1988); S.F.Clancy, D.A.Tanner, M.Thies, H.H.Paradies, Extended Abstracts, Div. of Environ. Chem., S.F.,CA, 203rd Natl.Meet. of the ACS, Vol 32, No1, 907-909, (1992)

[9] S.M.Me'nikov, V.G. Sergeyer, K.Yoshikawa, J.Am.chem.Soc., 117, 2401, (1995)

[10] C.A.Thomas & J.Abelson, in " Procedure in Nucleic Acid Research",eds. G.C.Cantoni, P.R.Davies, Vol.1, 553, Harper, N.Y.(1966)

[11] J.Marmur, Methods Enzymol., 6, 726, (1963)

[12] B.Gerhard & R.A.J.Warren, Biochemistry, 21, 5458, (1982)

[13] C.B.Post & B.H.Zimm, Biopolymers, 21, 2139, (1982); A.Benbasat, Biochemistry, 23, 3609, (1984)

[14] H.H.Paradies, Colloids & Surfaces, 74, 57, (1994)

[15] S.A.Allison, J.C.Herr, D.C.Teller, Biopolymers, 19, 1475, (1980).

SEPARATION OF DNA USING FERROFLUID ARRAY ELECTROPHORESIS

Pascal MAYER[1,2], Jérôme BIBETTE[1] and J.L. VIOVY[3]

1/ Centre de Recherche Paul Pascal, Av. A. Schweitzer, 33600 PESSAC

2/ Present address: GLAXO WELLCOME Research, 14 Ch des Aulx, CP 674, 1228 Plan-les-ouates, GENEVE, SUISSE

3/ Laboratoire de PhysicoChimie Théorique (URA CNRS 1382) ESPCI, 10 Rue Vauquelin, 75005 PARIS and Laboratoire de Physico-Chimie Curie (UMR CNRS 168), Institut Curie Section de Recherche, 11 Rue P.M. Curie, 75005 (corresponding address)

ABSTRACT

We present a new method for electrophoretic separation of DNA, Ferrofluid Array Electrophoresis (FAE). The method uses a stabilized suspension of an hydrophobic ferrofluid in aqueous buffer as the separating medium. When this suspension is placed in a slab cell and submitted to a magnetic field perpendicular to the slab plane, it organizes into a regular array of columns with micron-sized spacing. DNA migrating in this maze leads to size-fractionation. Resolution of lambda phage (48.5 kbp) and T4 (140 kbp) DNA molecules in 30 mn is achieved. The motion of individual DNA molecules during FAE is observed using fluorescence videomicroscopy, and the molecular mechanisms responsible for separation are discussed in the light of recent computer simulations. During migration, large DNA molecules temporarily wrap around the impenetrable ferrofluid columns. They disengage by slippage, like a rope on a pulley, and the dependence of the disengagement time upon DNA size is responsible for the size-fractionation.

1. INTRODUCTION

Electrophoresis, which consists in separating charged object by subjecting them to an electric field in a solution or in a gel, is probably the most widely used separation method for biological molecules. Gel electrophoresis of DNA, in particular, is at the basis of essential projects in molecular genetics, such as physical mapping of genes or sequencing of the genome, and numerous diagnostic applications. More generally, it appears as a necessary step in almost any molecular genetics protocol involving DNA. The separation of duplex DNA is very routinely performed in agarose gels within a few hours, for sizes up to about 20 kilobase-pairs (kbp), and with a resolution of a few percent. For larger sizes, separation is impossible (1). This behavior is now rather well understood, and predicted analytically in the frame of the biased reptation and fluctuations (BRF) model (2-4). Basically, the model represents the motion of the chain as a curvulinear slithering or « reptation » in an effective « tube » representing the sequence of gel pores explored by the chain. Because of electric bias, this tube may orient in the direction of the field, and the orientation (which depends on a rather subtle balance between brownian lateral

Mat. Res. Soc. Symp. Proc. Vol. 463 © 1997 Materials Research Society

and longitudinal fluctuations of the DNA, and electric forces), renders the effective force on the migrating polyelectrolyte chain proportional to its length for chains larger than a critical size N*. N* is proportional to the inverse electrid field, so that the upper limit of separation decreases with increasing field. As a consequence, in a reasonable time, DNA larger than a few tens of kbp cannot be separated by constant field electrophoresis in gels. A more detailed insight of the molecular mechanisms at play in DNA gel electrophoresis was provided by numerical simulations using Langevin (5) or non-local Monte-Carlo dynamics (6,7), and fluorescence videomicroscopy of individual molecules (8,9). It was shown, in particular that DNA chains can undergo motions more complex than simple reptation, involving the creation and annihilation of loops or « hernias », and migration in continuously renewing ramified tubes (6). However, the main predictions of the BRF model, and in particular the saturation of the mobility with increasing molecular size and its connection with the tube orientation, were confirmed by these studies. The limitation of constant-field electropohresis was overcome in 1984, and separation up to several Mbp was made possible, thanks to the invention of pulsed-field gel electrophoresis (10). Several variants of this technique exist, but the general principle is to change the direction or the orientation of the field periodically. This way, since the characteristic time required by a chain to create an oriented tube in the gel (reorientation time) is size-dependent, only chains smaller than an certain size have the possibility to orient in the field. Thanks to the dependence of the mobility upon the orientation of this tube with respect to the field, then, the use of pulsed fiels leads to fractionation in a range of DNA sizes having a reorientation time comparable with the pulse time (11). There are, however, limitations to the application of this nice principle. In particular, it was recognized rather soon that, the larger are the chains one wants to separate, the weaker is the field that can be used: At a given field, bands corresponding to DNA above a critical size are « lost », whatever the pulse frequency, and this critical size decreases rapidly with increasing field. This phenomenon was first reported by Olson (12), and studied in more detail by Slater et al. and by our group (13,14). It was attributed to the irreversible trapping of DNA molecules in the gel. The detailed molecular mechanism of this trapping is not known, although we suggested it could be due to the « knotting » of DNA loops aroung the permanent fibers of the gel (14). Indeed, this trapping, and the breakage of DNA molecules it can induce, are the main reasons for the long duration of pulsed-field runs (typically one day for separations up to 2 kBp, and one week for separation up to about 6 Mbp).

In the last few years, very efficient separations of DNA have been achieved by capillary electrophoresis, using polymer solutions as a sieving medium, and this field is progressing fast (see e.g. 15). Since polymer solutions present no permanent obstacle, one can expect that they would not be prone to DNA trapping as rigid or permanent gels. It has been observed, however, that solutions of large DNA molecules in liquid or viscoelastic medai tend to undergo aggregation when they are submitted to an electric field, which makes the separation of large DNA in solution very difficult (16).

A rather different and innovative method for separating large DNA has been proposed by Volkmuth et al.: these authors have demonstrated that separation can be achieved in a channel containing a micron-sized array of posts, prepared from silicon using microlithography techniques. These planar arrays of obstacles have several advantages as compared with gels: The pore topology is perfectly controlled, and it can be modified at will; pore sizes much larger than the largest pore size achievable in agarose gels, e.g. in the micron range, can be easily obtained. However, this technique also has some drawbacks: because of the limitations of the engraving and bonding technique, the channel must be very thin (typically a few times the width of the posts); the fabrication of these channels require high technology, is expensive and difficult to upgrade to very long channels; electroendosmosis and/or current leakage at the surface are difficult to control.

The present article is the first report of a new separation method, called « ferrofluid array electrophoresis » (FAE), which combines several advantages of capillary electrophoresis in solution and in planar arrays of obstacles. The principle of the method is indeed very simple: it amounts to perform electrophoresis in an array of posts, as in the Wolkmuth-Austin approach (17). The major difference lies in the way the posts are made. In our case, they are obtained by the self-assembling properties of magnetic suspensions (ferrofluids) under the action of an electric field. This assembling is fully reversible, so the separation medium can be switched at will from liquid-like to gellike, making it suitable to protocols issued from CE. The paper is organized as follows. In the following « materials and methods » section, the experimental details are given, together with a brief recall of the self-assembling properties of ferrofluids useful for our present purpose. Examples of actual DNA separations and a discussion of separation mechnaisms are given in section 3 (results and discussion). Conclusions and perspectives are provided in section 4.

2. MATERIALS AND METHODS.

Ferrofluids are made by a suspension of magnetic particles in a fluid (18). Their magnetomechanic properties certainly put them among the most spectacular existing complex fluids, and are at the basis of various applications in fiels as varied as automobiles (variable-viscosity suspension fluids), computers (fluid seals) bioseparations or drug delivery. In the present approach, we use a suspension of a hydrophobic ferrofluid in an aqueous buffer, stabilized with surfactant. The ferrofluid itself is a dispersion of iron oxide particles (of typical size 10 nm) in oil. A monodisperse emulsion of this oxyde-containing oil in aqueous buffer, with droplet size of order 500 nm, is prepared as previously described (18). Each droplet contains numerous monodomain magnetic oxide grains, and behaves as a superparamagnetic object: when the suspension is placed in a slab-cell and submitted to a magnetic field perpendicular to the cell plane, the magnetic droplets develop a strong induced dipole and tend to align in « pearl-necklesses » parallel to the field. These chains of beads can extend on all the thickness of the cell, and form an array of « pillars ». Depending on parameters such as the concentration of the ferrofluid, the magnetic field, and the cell thickness, these chains can later on aggregate laterally to form thicker pillars. The dipole-dipole interaction between columns keep them at regular spacing (the thermodynamically stable state is a hexagonal 2-D crystal, but some hysteresis in the interaction with the wall can lead to defects. and to a glass-like order). This behavior has been studied in detail (19), and it has been demonstrated that the interbead spacing follows a power law with cell thickness:

$$a \propto k\, d^{0.37} \quad (1)$$

The constant k depends on the details of the ferrofluid suspension, but the exponent is universal. This relationship allows to control the pore size.

The ferrofluid used in the present study is prepared from Fe_2O_3 grains with a typical size of 10 nm, with an oil/oxide ratio around 50% (g/g) (Rhone-Poulenc, France). To avoid electroosmosis, the original emulsifying surfactant (SDS) is exchanged with a non-ionic surfactant by repetitive washing with a 0.05% solution of Tergitol NP10 (Sigma, France) in water. The solution is finally supplemented by a stock solution of buffer in order to reach a final O.5X TBE concentration (TRIS 45 mM, boric acid mM, EDTA 1.25 mM, pH 8.3)

The cell was a cross-type slab cell similar in principle with those used for « electropohresis-on-chip » (see e.g. 20), placed in the center of a Helmoltz coil. To improve field homogeneity along the separation channel, the latter was given the shape of a portion of circle, concentric with the Helmoltz coil. (fig. 1).

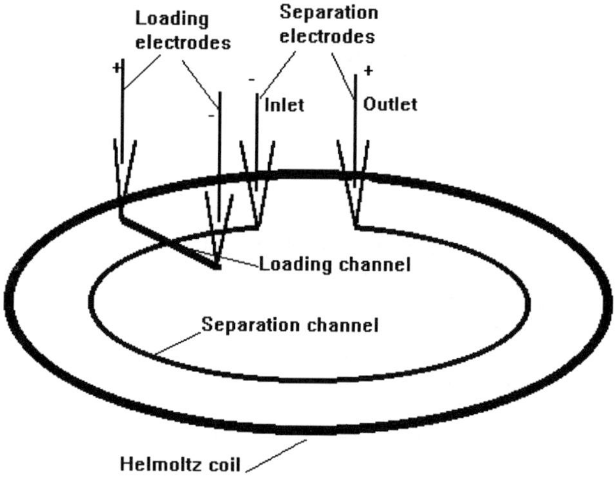

Figure 1: schematic view of the separation cell. The channel is rectangular and enclosed between a thick to glass plate and a thin bottom coverslip, to allow visualization by epifluorescence on an inverted microscope. holes drilled in the top plate at the ends of the circular separation channel and linear injection channel play the role of loading wells for the ferrofluid and for the sample, and of electrode compartments. They are connected with siphons to equilibrate pressures and avoid gravitational flow (not shown)

In the present set of experiments, the cell thickness was of order 50 µm,a nd the intercolumn spacing of order 5 µm. The cell was first entirely filled with the ferrofluid in the absence of any field, and the magnetic field was raised to a value of a few mTesla (this value is not critical provided it is sufficient to « anchor » the ferrofluid colums to the surface). The injection channel was then rinsed with pure buffer using a moderate pressure, and the DNA sample was placed at one extremity of this channel. The injection channel was uniformly filled with DNA by applying a potential drop between electrodes placed at its both ends. The field was then applied along the separation channel, allowing the injection of a finite volume of sample, defined by the width of the injection channel (20). The migration and deformation of individual DNA molecules (Lambda phage, Appligene), T2 (Sigma) or Saccharomyces Cerevisiae (SC) (Bio-Rad) labeled with YOYO-1, Molecular Probes) could be observed and registered using an inverted microsope (NIKON) equipped with a x100 oil-immersion objective, an intensified CCD camera (Hammamatsu) and a VCR. Simultaneous observation of DNA and ferrofluid columns was made possible by a combination of epifluorescence and transmission microscopy. Quantitative detection of electrophoregrams was performed by on-line transfer and integration of the video signal to a workstation.

3 RESULTS AND DISCUSSION

Upon visual observation, the first important point is that DNA molecules migrate in a manner qualitatively very similar to the one reported for microlithographic arrays: they migrate in the space between the ferrofluid colums, without evidence of adsorption or attractive interaction. The smallest DNA essentially avoid the posts, and their eventual collisions with them most often yield little or no observable stretching of the molecules (fig. 2a). The collision of larger DNA, such as T2 (140 kbp) and the chromosomes of SC (ranging from 250 to 2000 kbp), often lead to a very different behavior (2b): different parts of the molecule slide along the post on the two different sides, leading to strongly extended U or J-shapes, with the bend located around the ferrofluid column. The molecule ultimately disengages by slippage around the post, and resumes progression downfield. Molecules of intermediate size, such as T2, often have time to relax partially before encountering a new post. This behavior is consistent with the results of computer simulations by Slater et al. (21), Duke et al. (22) and Yevick et al. (23), who simulated the interaction, entangling and undwinding of long polyelectrolytes with immobile posts. Larger molecules, in the Mb range, stretch on a length of several hundreds of microns, much longer than the distance between posts (fig. 2c). They never relax completely, and tend to hang around several posts simultaneously, and adopt conformations rather similar with the ones observed for constant-field migration in agarose gels, and depicted as a « reptation in a ramified tube » (6). When a polydisperse sample, such as the size standard containing all the chromosomes of SC, is injected, and the cell is observed at a fixed distance (5 cm) from the injection point, it is clear from visual observation that smaller molecules pass first, and that the size of the molecules reaching the observation point increases with time, suggesting a size fractionation effect.

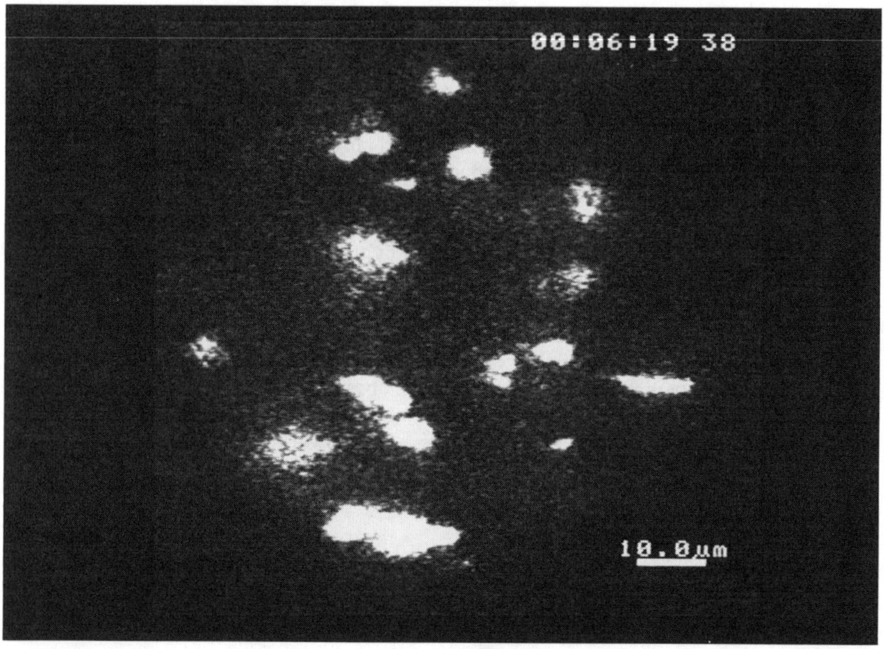
00:06:19 38
10.0µm

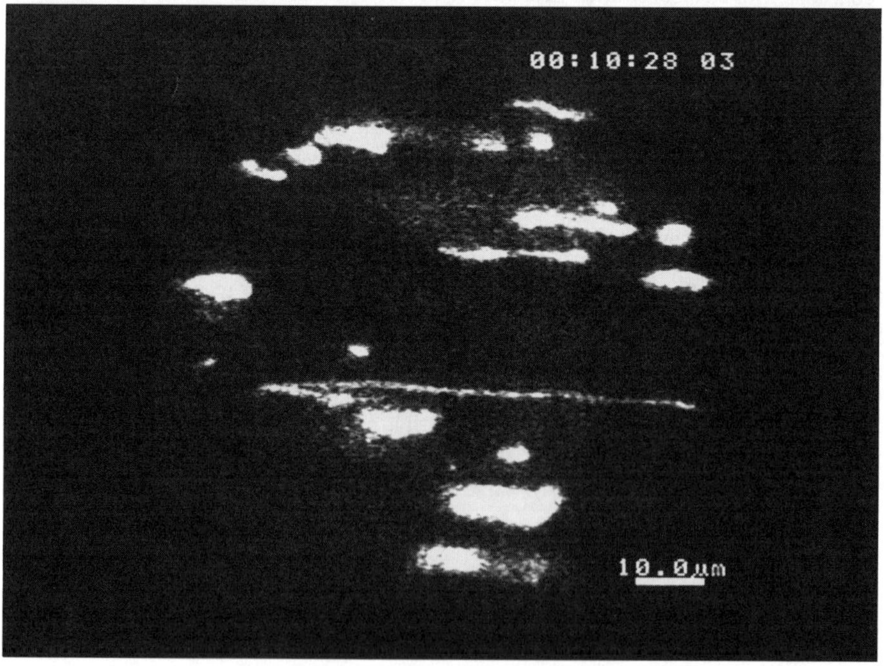
00:10:28 03
10.0µm

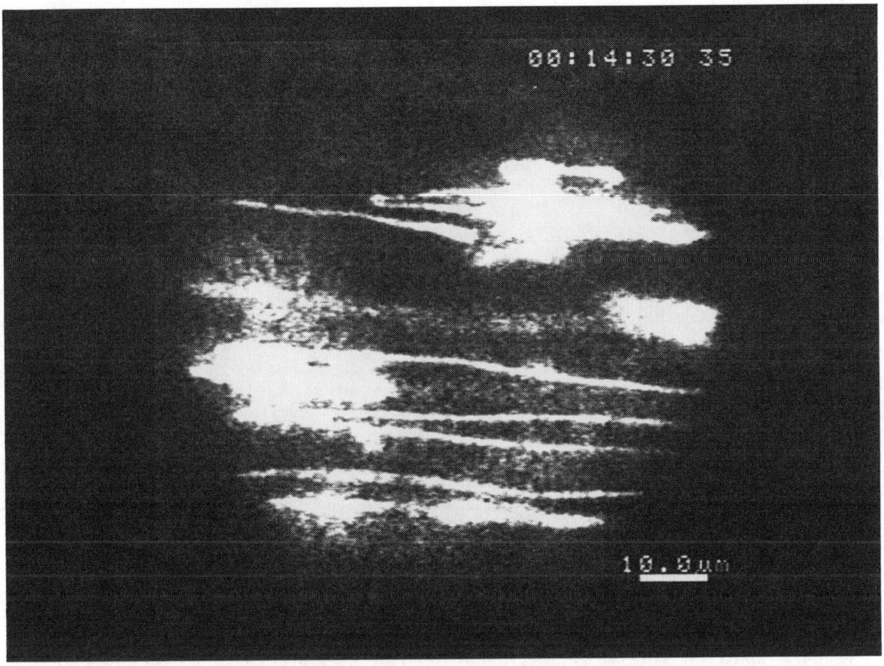

Figure 2: view of the separation channel at 1 cm from the injection cross, at different times following the injection of a polydisperse SC sample (6 mn (2a, top previous page), 10 mn (2b, bottom previous page) and 14 mn (2c, above)) The DNA is prestained with YOYO, and the electric field is 20 V/cm

To be more quantitative, we recorded the integrated full-frame light intensity as a function of time. Within an additive constant, this intensity is proportional to the total quantity of DNA passing in front of the detector at a given time. This mode is equivalent to this used in capillary electrophoresis. Typical results are shown in fig. 3.

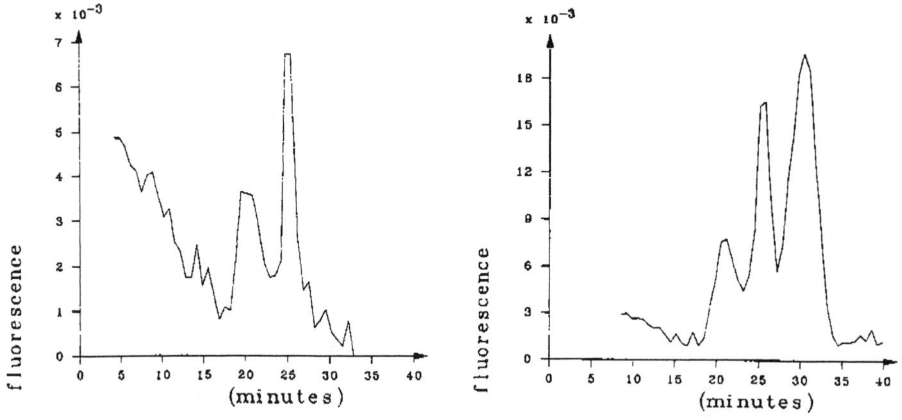

Figure 3: electrophoregrams following the injection of Lambda-phage (48.5 kbp) alone (3a, left) and of a mixture of lambda phage and T4 (140 kbp) (3b, right), obtained 5 cm downfield from the injection cross.

When only lambda-phage DNA is injected(3a), two peaks are observed. We attribute the first peak to degradation products (i.e. smaller DNA): In these experiments, in contrast with e.g. agarose gels, the separating medium is very « open », and the migration of DNA molecules with a radius of gyration smaller than the pore is only weakly slowed down. All small molecules tend to achieve the same velocity, which is the free-liquid mobility. Therefore, all the broken molecules that tend to yield a fast «smear » in agarose gels, because they are far from their limiting mobility and retain a size fractionation, tent to collapse in a « compression band » ahead of the intact molecules, which is the signature of the free mobility (indeed, the position of this peaks corresponds to a mobility of order 2 $m^2/V/s$, which has the right order of magnitude as compared with the commonly accepted free mobility of DNA. When T2 DNA (140 kbp) is co-injected with lambda, a third peak appears (fig. 3b), indicating size-fractionation. This result is consistent with the predictions of Yevick et al. (23). On the basis of semi-empirical arguments, and information drawn from the simulations, the latter authors predicted an expression for the mobility in constant field (relative to the free liquid mobility):

$$\mu/\mu_0 \cong 1 - BN^{3/2} \qquad (2)$$

Where B is a constant depending on the geometric characteristics of the arrangement of posts. This prediction is not quantitatively correct, because the authors neglected important hydrodynamic interactions (24), and in any case it cannot stand for arbitrary values of N (since it leads to negative mobilites for N large). However, the main qualitative prediction, i.e the fact that the interaction of molecules with individual posts can yield size-fractionation in a limited range of sizes, seems in agreement with our observations. Indeed, we observed that the resolution between lambda and T2 is not very good in our experiments. In the present setup, we were also not able to resolve the 13 bands of SC, which tend to provide an unresolved or poorly resolved bunch of peaks piling up to a second « compression band » at the end of the separation domain (we believe this compression is similar to the one encountered in constant-field gel electrophoresis, and related with the independence of DNA alignment in the direction of the field upon the molecular size).

4. CONCLUSIONS AND PERSPECTIVES:

We have demonstrated for the first time the possibility of using ferrofluid arrays self-assembled under a magnetic field, as a separation matrix for electrophoresis. The experiments presented here are still preliminary, and the experimental setup is not fully optimized. In particular, it should be possible to increase significantly the magnetic field, leading to stronger ferrofluid columns, and permitting in turn the use of larger electric field strengths. The system as it is, however, was succesfully applied to the separation of large duplex DNA molecules, and molecules up to 140 kbp were separated in constant field within about 30 mn. In conventional gels, such a separation is impossible in constant field, and even in pulsed fields it requires several hours. This improvement is due mainly to the larger spacing between obstacles, which leads to a larger mobility and shifts the compression band towards larger sizes. The molecular mechanisms of separation, as can be infered from direct observations by videomicroscopy, qualitatively resemble those observed in gels and in microlithographic arrays of posts, although with boundaries between regimes shifted to larger molecular sizes. In particular, one recovers the very general feature, of a strong stretching and aligment of large molecules in the direction of the field. This suggests that much better separation should be obtained with pulsed fields. Work is in progress in this direction. The second important advantage of ferrofluids as compared with gels or lithographic arrays, is that the network of obstacles can be destroyed and reformed at will by switching the magnetic field on and off. In the absence of magnetic field, the suspension has a viscosity of only a few centipoises, and it can be manipulated very easily. Therefore, many advantages of capillary electrophoresis, such as the possibility to replace automatically the separation medium between each run, and to « clean » the separation channel, can be transposed to this new approach with rather simple technical modifications. Also, the ferrofluid droplets can be functionalized, in order to induce tailored interactions of analytes with the obstacles. In this case, the ferrofluid would play the role of a « solid phase », as e.g. in electrochromatography. For these reasons, we believe that the method of ferrofluid array electrophoresis may find various domains of application in the future.

ACKNOWLEDGEMENTS:

This work was supported in part by EC grants CT930018 and CT951158 (Biomed programme).

REFERENCES:

1 D.H. Flint and R.E. Harrington, Biochemistry, **11**,4858 (1972)
2. T. A. J. Duke, A. N. Semenov and J. L. Viovy, Phys. Rev. Lett. **69**, 3260-3263 (1992).
3 T.Duke, J.L. Viovy, A.N. Semenov, Biopolymers **34**, 239-247 (1994)
4 C. Heller, T. Duke, J.L. Viovy, Biopolymers, **34**, 249-259 (1994)
5 J.M. Deutsch, Phys. Rev. Lett. **59**,1255 (1987)
6 T. A. J. Duke and J. L. Viovy, Phys. Rev. Lett. **68**, 542-545 (1992).
7 B.H. Zimm, J. Chem. Phys. **94**, 2187 (1991)
8 S.B. Smith, P.K. Aldridge and J.B. Callis, Science, **243**,203 (1989)
9 D.C. Schwartz and M. Koval, Nature **338**,520 (1989)
10 D.C. Schwartz and C.R. Cantor, Cell, **37**,67 (1984)
11 J.L. Viovy, biopolymers, **26**,1929 (1987)
12 G.F. Carle and M.V. Olson, Trends Biochem. Sci., **12**, 284-287 (1987)
13 C. Turmel, E. Brassard, G.W. Slater and J.Noolandi, Nucleic Acids Res. , 569 (1990)
14 J.L. Viovy, F. Miomandre, M.C. Miquel, F. Caron, F. Sor, Electrophoresis, **13**, 1-6 (1992)
15 <u>Capillary electrophoresis: an analytical tool in Biotechnology</u>, P.G. Righetti Ed., Analytical Biotechnology series, CRC press, N.Y., 1996 (all volume)
16 L. Mitnik, C. Heller, J. Prost, J.L. Viovy, Science, **267**, 219-222 (1995)
17 W.D. Volkmuth and R.H. Austin, Nature, **358**, 600 (1992)
18 J. Bibette, J. Magn. Magn. Mater., **122**, 37 (1993)
19 J. Liu, E.M. Lawrence, A. Wu, M.L. Ivey, G.A. Flores, K. Javier, J. Bibette and J. Richard, Phys. Rev. Lett., **74**, 2828 (1995)
20 D.J. Harrison, K. Fluri, K. Seiler, Z. Fan, C.S. Effenhauser and A. Manz, Science, **261**, 895 (1993)
21 G.I. Nixon and G.W. Slater, Phys. Rev. E, **50**, 5033 (1994)
22 W.D. Volkmuth, T. Duke, M.C. Wu, R.H. Austin and A. Szabo, Phys. Rev. Lett. **72**, 2117 (1994)
23 E.M. Sevick and R.M. Williams, Phys. Rev. Lett., **76**, 2595 (1996)
24 D. Long, J.L. Viovy and A. Ajdari, Phys. Rev. Lett., **76**, 3858 (1996)

DNA END-TO-END DISTANCE CHANGE DUE TO DIVALENT COUNTERION CONDENSATION STUDIED BY PULSE GEL ELECTROPHORESIS

ANZHI Z. LI, HAIYAN HUANG, KENNETH A. MARX*
Department of Chemistry, University of Massachusetts Lowell, Lowell, MA 01854

ABSTRACT

The conformation change of DNA fragments due to divalent counterion condensation onto DNA was investigated by pulse gel electrophoresis, and interpreted by gel models (Reptation and Henry model) and Manning's counterion condensation theory. The measured mobility reductions μ/μ_0 of λ-DNA-Hind III fragments, ranging from 23.13 to 2.027 kilobase pairs, due to interaction with divalent cation Mg^{2+} (1-400 μM), and Ca^{2+}(0-40 μM) in tris-borate buffer were well fit by Manning's Counterion Condensation (CC) theory. We observed the normalized mobility reduction to be shifted by a small amount $\Delta(\mu/\mu_0)$ relative to the CC prediction value. $\Delta(\mu/\mu_0)$ is a function of DNA length, and the ion environment (divalent concentration C_2 and ionic strength). The 'shift' phenomena only occurred close to where C_2 began dominating the counterion binding, a condition described by the monovalent/divalent cation isocompetition point. Combining our observation with theoretical considerations, we conclude that the divalent counterion condensation changes the DNA fragments' conformation, resulting in an end-to-end distance decrease which is molecular weight dependent. The effect was enhanced by an increase of divalent ion concentration and a decrease of the ionic strength.

INTRODUCTION

There exists significant potential for the use of DNA in biomaterial applications. A requirement for its rational use is detailed knowledge of the effect of its environment on its structure and dynamic properties. Our focus in this study is on the interaction of divalent metal cations with DNA in aqueous buffer is to understand the nature of molecular weight effects on the polyion-counterion binding. In other words, how the conformation change of DNA fragments related to the divalent counterion condensation onto DNA?

Pulse gel electrophoresis was employed to measure the mobility reduction of DNA fragments with excess divalent cations Mg^{2+} or Ca^{2+} competing with $tris^+$and Na^+ in TB buffer. When the Henry gel model [1] and a simple ionic strength constant assumption applies [2-3], the mobility reduction μ/μ_0 reflects the effective residual charge fraction Q/Q_0 where μ_0 and Q_0 refer to the controls (without excess multivalent concentration). In our previous studies [3-4], in the presence of relatively small concentrations of trivalent cations $Co(NH_3)_6^{3+}$, spermidine^{3+}, or divalent cation Mg^{2+}, the measured normalized mobility reductions expressed as effective residual charge fractions were found to be well fit with calculated predictions from Manning's Counterion Condensation (CC) theory [5-6]. In fact, the data is well fit by CC theory to a first approximation, when the distribution of μ/μ_0 values with DNA size is ignored. All the experimental data [3-4] showed a distribution of normalized mobility μ/μ_0, or charge neutralization fraction θ related to DNA size from 23.13 to 2.07 kbp while the CC theory, containing no size dependence, predicts only provide a single value. All the measurements showed consistent regularities, the larger the fragment size, the lower the value of mobility reduction, the higher the charge neutralization.

---

* Author for correspondence

Mat. Res. Soc. Symp. Proc. Vol. 463 © 1997 Materials Research Society

We applied the Reptation model [7-8] assuming that ζ, the friction coefficient per unit DNA length, which depends on temperature, gel concentration, and viscosity, remains constant when every experimental condition is the same except for small additions of divalent cations, which does not alter the ionic strength significantly. Then we have:

$$\mu/\mu_0 = Q/Q_0 \left(\langle r_h^2 \rangle / \langle r_{h0}^2 \rangle \right) \qquad (1)$$

where $\langle r_h^2 \rangle$ refers to the end-to-end distance of the DNA chain in excess divalent cation, and $\langle r_{h0}^2 \rangle$ refers to the control parameter without excess divalent cations. The ratio $\langle r_h^2 \rangle / \langle r_{h0}^2 \rangle$ then represents a quantitation of conformation change. In Equation (1), the ratio of effective charge fraction of polyion Q/Q_0 is independent of molecular weight [7] and only depends on binding competition conditions. If we apply Eqn.(1) to two DNA fragments, while keeping the ion environment (divalent concentration C_2 and ionic strength) the same, then we have:

$$(\mu/\mu_0)_{f1} / (\mu/\mu_0)_{f2} = \left(\langle r_h^2 \rangle / \langle r_{h0}^2 \rangle \right)_{f1} / \left(\langle r_h^2 \rangle / \langle r_{h0}^2 \rangle \right)_{f2} \qquad (2)$$

Eqn.(2) creates a connection between mobility reduction measurements and conformation change prediction, which is related to DNA length since $\langle r_h^2 \rangle$ is a function of DNA length. Employing this equation we interpreted the molecular weight effect on counterion condensation binding in terms of $\langle r_h^2 \rangle$. The observed normalized mobility shifts $\Delta(\mu/\mu_0)$ were investigated as a function of DNA length, divalent concentration, and the ionic strength.

EXPERIMENT AND COMPUTATION

Experimental Fragments of λ-DNA-Hind III (23.130, 9.416, 6.557, 4.361, 2.322, and 2.027 kbp) were purchased from New England Biolabs, Inc. A 0.8% mini thin agarose gel (1 mm thick) was used in all our pulse field electrophoresis experiments. Tris-borate-EDTA (TBE) (Bio-Rad, CA) was the electrophoresis buffer. In any single experiment, divalent cations were added during buffer dilution to the designed final ligand concentration before gel electrophoresis. The stock solution (445 mM Tris-borate and 0.1 mM EDTA: 5 X) was diluted to 0.25 X and 0.50X when Mg^{2+} was the divalent cation, and was diluted to 0.15 X, 0.45 X, 0.60 X, and 0.75 X when Ca^{2+} was the divalent cation. The concentration of Mg^{2+} was varied from 0 to 400 μM, while Ca^{2+} was varied from 0 to 40 μM. The pulse field (magnitude 10 v/cm) was set at the ratio of 1 to 3 (2 s on and 6 s off). Total run time was about 4 h and the temperature was controlled at 20 ^{0}C for Ca^{2+}, and 21.5 ^{0}C for Mg^{2+}.

Computational To apply Manning's two variable theory [5-6], we first need to have the ionic strength and monovalent cation concentration C_1 in which DNA was electrophoresed. From known values of pH, buffer concentration, temperature, and pK_a, calculations were carried out to obtain effective concentrations for each species in the buffer based on the Henderson-Hasselbalch equation [9], where the pK_a value was corrected iteratively according to the ionic strength dependence to the pK_a' using the Davies equation [9]. In the case of DNA-Mg^{2+} interaction: (1) 0.25X TB buffer, the ionic strength (I.S.) was calculated to be 8.67 mM and C_1 is 8.65 mM; (2) 0.50X TB buffer, I.S. is 17.70 mM and C_1 is 17.67 mM. In the case of DNA-Ca^{2+} interaction: the I.S. is: 4.84, 14.97, 20.18, and 25.43 mM, and C_1 is 4.83, 14.95, 20.14, and 25.38 mM corresponding to the 0.15 X, 0.45 X, 0.60 X, and 0.75 X buffer concentrations. For competition binding calculations, we first calculate the Debye-Huckel screening parameter κ based on the known ionic strength, then calculate the bound volumes V_{p1} and V_{p2} corresponding to the valences Z=1, 2 [3-4]. We next employ a program written in the mathematica language[10] to solve the simultaneous Eqn.(3) and (4) iteratively, where the charge density parameter $\zeta = 4.2$ and average charge spacing b = 1.7 Å for DNA, and obtain the charge neutralization fractions θ_1, θ_2, and θ.

$$1 + \ln (1000\ \theta_1/C_1 V_{p1}) = -2\xi(1 - \theta_1 - 2\theta_2) \ln (1 - e^{-\kappa b}) \qquad (3)$$

and

$$\ln (\theta_2/ C_2) = \ln (V_{p2} / 1000e) + 2 \ln (1000\ \theta_1 e / C_1 V_{p1}) \qquad (4)$$

The assumption $\mu/\mu_0 = Q/Q_0$ where $Q_0 = 0.24$ [3] and equation $Q = 1- (\theta_1 + 2\ \theta_2)$ were used to convert the mobility reduction to charge binding fraction or vice versa when comparing the measured data with the CC prediction.

When studying the conformation change due to counterion binding for DNA fragments with different size, Eqn.(3) and (4) were employed. The mobility reduction of DNA fragments f_n (n = 1, 2, 3, 4, 5 corresponding to the lengths 23.13, 9.416, 6.557, 4.361, and 2.322 kbp) were normalized by the mobility reduction of fragment f_6 (2.027 kbp): $(\mu/\mu_0)_{fn} / (\mu/\mu_0)_{f6} = (<r_h^2> / <r_{h0}^2>)_{fn} / (<r_h^2> / <r_{h0}^2>)_{f6}$. After processing the measured mobilities in ratio form, we have the ratio of normalized conformation change expressed as normalized $(<r_h^2> / <r_{h0}^2>)_{fn}$, which indicate the end-to-end distance reduced ratio for fragment f_n over fragment f_6.

RESULTS AND DISCUSSION

The counterion binding of divalent cations to DNA is reflected directly in the mobility reduction since each DNA fragment's charge density was lowered. Figure 1 and Figure 2 present normalized mobility vs divalent Mg^{2+} and Ca^{2+} concentration at different ionic strengths. The agreement between experimental data (symbols) and CC prediction (solid line, μ/μ_0 converted from θ) are excellent. Both Figures show that higher divalent cation concentration results in more binding and higher ionic strength results in less binding. Higher ionic strength corresponds to a smaller κ resulting in a lower binding fraction. Also a consistent DNA length effect on μ/μ_0 was observed.

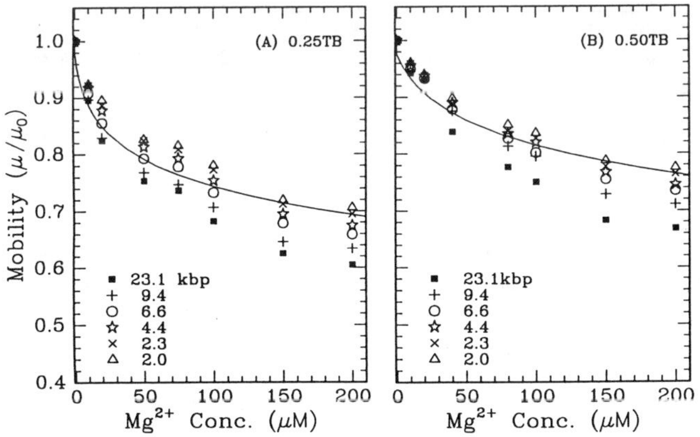

Fig.1 Normalized mobility μ/μ_0 of λ-DNA-Hind III fragments vs excess Mg^{2+} concentration. The experimental data (symbols) were fit by CC prediction (solid line).

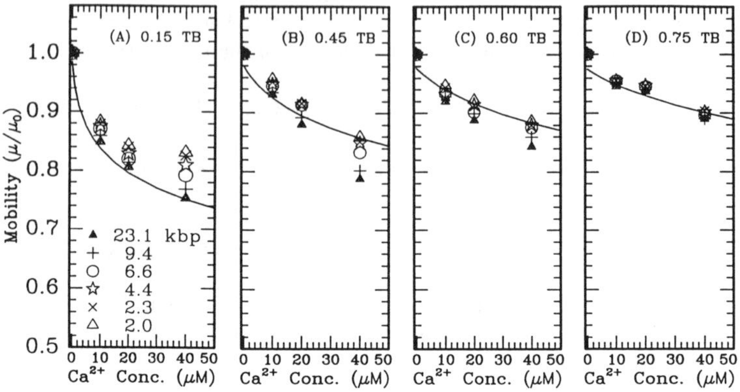

Fig.2 Normalized mobility μ/μ_0 of λ-DNA-Hind III fragments vs excess Ca^{2+} concentration. The experimental data (symbols) were fit by CC theory (solid line).

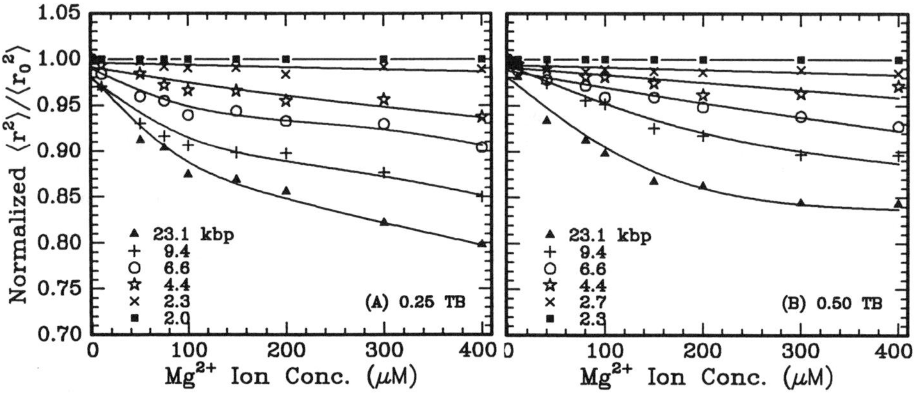

Fig.3 Normalized $<r_h^2> / <r_{h0}^2>$ vs Mg^{2+} concentration. The lines are only for eye guidance

Figure 3 shows the normalized $<r_h^2> / <r_{h0}^2>$ vs Mg^{2+} concentration for different ionic strengths. The DNA length effect on the conformation change is shown clearly. The larger the DNA length, the lower the normalized $<r_h^2>/<r_{h0}^2>$. The above effect was enhanced by increasing divalent Mg^{2+} concentration, and decreasing the ionic strength because the end-to-end distance decrease is due to counterion binding and is eventually governed by ionic strength and C_2.

Figure 4 and Figure 5 show the DNA length effect on the conformation change from a different perspective. The log (normalized $<r_h^2>/<r_{h0}^2>$) vs log(1/N) is displayed under different ionic strength conditions where N is the DNA length. The larger the DNA size, the lower the normalized $<r_h^2> / <r_{h0}^2>$, and the same conclusion applies for the higher C_2. The least DNA size effect, or the least $\Delta(\mu/\mu_0)$ 'shift' was observed in Fig.5 (B) when the ionic strength is relatively high (0.75 TB). Linear fitting resulted in reasonably good fits to all the curves.

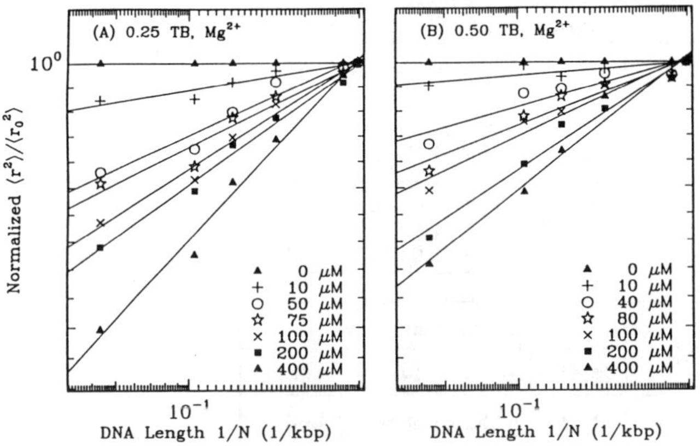

Fig.4 Logarithm of normalized $\langle r_h^2\rangle/\langle r_{h0}^2\rangle$ vs logarithm 1/N show DNA length effects on conformation change. Linear fitting was employed (solid lines).

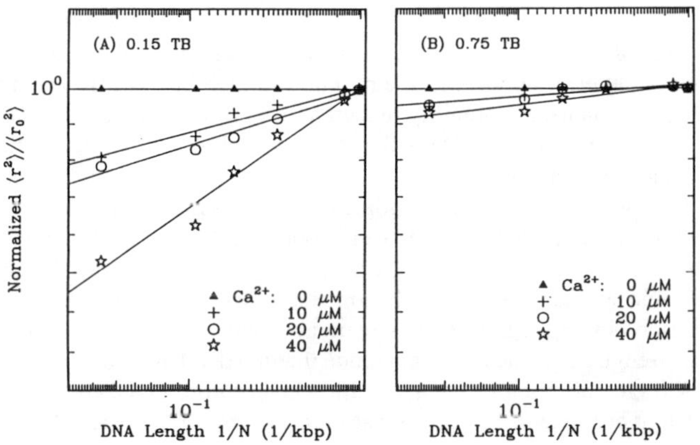

Fig.5 Logarithm of normalized $\langle r_h^2\rangle$ / $\langle r_{h0}^2\rangle$ vs logarithm 1/N. Linear fitting (solid lines).

we next considered how the ionic strength governs the 'shift' phenomena. Table I reveal the 'shift' related to C_2 and ionic strength. The relative shift $\Delta(\mu/\mu_0)/(\mu/\mu_0)$ was calculated by subtracting the μ/μ_0 of the largest fragment (f_1) from that of the smallest (f_6) and normalizing the difference by (μ/μ_0) of f_6. $\Delta(\mu/\mu_0)/(\mu/\mu_0)$ was enhanced by increasing C_2 and decreasing I.S. When I.S. = 25.43 mM (0.75X TB), the relative shift was close to 0.01 and can be ignored. The 'shift' phenomena only occurred when C_2 is dominating the competition binding, which is related to a parameter we term the ***isocompetition point***. The isocompetition point (IP) is a calculated value based on counterion condensation theory where the DNA binding fraction is equal from monovalent as well from divalent cations. Table 1 (bottom row) shows IP corresponding to each ionic strengths. For example, when I.S. = 4.84 mM, which is larger than its IP, a 'shift' could be observed and for I.S. = 25.43 mM, which is much smaller than its IP, the 'shift' can be ignored.

Table I Shift & Isocompetition point
$\Delta(\mu/\mu_0)/(\mu/\mu_0)$

Ca^{2+} Conc. (μM)	Ionic Strength (mM)			
	25.43	20.18	14.97	4.84
10	0.01	0.029	0.029	0.039
20	0.009	0.037	0.039	0.048
40	0.014	0.049	0.084	0.093

I.S. (mM)	Ionic Strength (mM)			
	25.43	20.18	14.97	4.84
*Isocompet. Point (μM)	**111.3**	**69.83**	**38.56**	**4.06**

*Isocompetition point where the DNA binding fraction is equal from monovalent and divalent cations.

CONCLUSION

The conformation change of DNA fragments due to divalent counterion condensation onto DNA was investigated by pulse gel electrophoresis. The normalized mobility was fit well by Manning's CC theory and a 'shift' phenomena, the mobility reduction μ/μ_0 shifted by a small amount $\Delta(\mu/\mu_0)$ relative to the μ/μ_0 of the smallest DNA fragment, was observed. Based on the Reptation model and CC theory, the parameter $<r_h^2> / <r_{h0}^2>)_{fn} / (<r_h^2> / <r_{h0}^2>)_{f0}$, where f_0 is the selected reference DNA fragment, and f_n refers to any length fragment, was introduced to describe the conformation change for different DNA lengths due to counterion binding. We found that both the normalized $(<r_h^2> / <r_{h0}^2>)_{fn}$ and shift $\Delta(\mu/\mu_0)$ are functions of DNA length, divalent cation concentration C_2 and ionic strength, The 'shift' phenomena only occurred close to where C_2 began dominating the counterion binding (described by the monovalent /divalent cation isocompetition point), which is ionic strength dependent. We conclude that the divalent counterion condensation changed the DNA fragments conformation, resulting in an end-to-end distance decrease which is molecular weight dependent. The effect was enhanced by an increase of divalent cation concentration and a decrease of the ionic strength.

REFERENCE

1. S. A. Rice and M. Nagasawa. Polyelectrolyte Solution. Academic Press. New York. (1961)
2. C. Ma and V.A. Bloomfield. Biopolymers. **35**, 211 (1995)
3. A. Z. Li, L. J. Qi, H. H. Shih and K. A. Marx. Biopolymers. **38**, 367 (1996)
4. A. Z. Li, H. Huang, X. Re, L. J. Qi and K. A. Marx. Biophysical J. Submitted, (1996)
5. G. S. Manning. Quart. Rev. Biophys. **11**, 179 (1978)
6. G. S. Manning. Biophys. Chem. **7**, 95 (1977)
7. O. J. Lumpkin and B. H. Zimm. Biopolymers. **21**, 2315 (1982)
8. L. S. Lerman and H. L. Frisch. Biopolymers. **21**, 995 (1982)
9. D. D. Perrin and B. Dempsey. Buffers for PH and Metal Ion Control. Chapman and Hall. London (1979)
10. S. Wolfram. Mathematica: A system for doing mathematics by computer. Addison-Wesley Publishing Company. (1991)

Part III

Biologically Inspired Physics

DNA SEQUENCE ALIGNMENT AND CRITICAL PHENOMENA

DIRK DRASDO(1), TERENCE HWA(2), and MICHAEL LÄSSIG(1)
(1) Max-Planck Institut für Kolloid- und Grenzflächenforschung, 14513 Teltow, Germany
(2) Department of Physics, University of California at San Diego, La Jolla, CA 92093-0319

Abstract

Alignment algorithms are commonly used to detect and quantify similarities between DNA sequences. We study these algorithms in the framework of a recent theory viewing similarity detection as a geometrical critical phenomenon of directed random walks. We show that the *roughness* of these random walks governs the *fidelity* of an alignment, i.e., its ability to capture the correlations between the sequences compared. Criteria for the optimization of alignment algorithms emerge from this theory.

Introduction

The explosion of genetic information has made statistical sequence alignment an indispensable tool in molecular biology [1]. The identity of new genes and relationships about known genes are routinely analyzed by aligning sequences on a computer. This underlies, for example, the retrieval of ancestorial relationships in the history of evolution.

In a typical algorithm [2, 3, 4], each alignment of sequences is assigned a score specified by a set of parameters. Maximization of this score is then used to select the *optimal alignment*, which depends, of course, strongly on the parameters used to define the score. What are then *optimal alignment parameters* making the algorithm most sensitive to the inherited similarities between the sequences? This important problem has so far been solved mostly by trial and error, despite some recent efforts to establish a more solid empirical footing [5, 6].

In this paper, we study the parameter optimization problem using a recent analytical approach to sequence alignment introduced by two of us [7]. The approach is based on a geometrical formulation of sequence alignment [2] and focuses on the morphology of *alignment paths*. This provides a fruitful link (see also Ref. [8]) to various well-studied problems in the statistical physics of critical phenomena.

In a divergent evolution process, similarities between sequences stem from a common ancestor sequence and are gradually destroyed in the course of time. We use a simplified model of evolution: Sequences are altered by a stochastic process of local substitutions, insertions, and deletions. In this model, the mutual similarities between daughter sequences inherited from their common ancestor can be identified uniquely. Hence, we can quantify in an unambiguous way the *fidelity* of an alignment algorithm [7], i.e., its ability to retrieve the inherited similarities from the knowledge of the daughter sequences alone. Then we analyze the dependence of the fidelity on the evolution parameters and the alignment parameters. Maximizing the fidelity defines optimal alignment parameters for given evolution parameters. Conversely, unknown evolution parameters can be reconstructed from alignment data.

In the sequel, we introduce the evolution model and the alignment algorithm used in this work, derive geometrical properties of alignment paths, and discuss how they govern the alignment fidelity and its parameter dependence. In particular, optimal alignment parameters are seen to follow from a simple geometric criterion. Further details of our work are reported in a forthcoming publication [9].

Mat. Res. Soc. Symp. Proc. Vol. 463 © 1997 Materials Research Society

Evolution Model and Alignment Algorithm

The simplified stochastic evolution process used in this paper generates two daughter sequences $\mathcal{D}$ and $\mathcal{D}'$ from a common ancestor sequence $\mathcal{A} = \{\mathcal{A}_k\}$ taken to be a random sequence of length $N \gg 1$. Each element $\mathcal{A}_k$ is with probability 1/4 one of the four different nucleotides A, C, G, T; we neglect any correlations within the ancestor sequence. A daughter sequence $\mathcal{D}$ is generated according to the following rules [7] (see also [10, 11]): (a) Each element $\mathcal{A}_k$ is *deleted* with probability $\tilde{p}/2$. (b) Each element $\mathcal{A}_k$ is *substituted* with probability $(1-\tilde{p}/2)p$ by a randomly chosen nucleotide. (c) If an element $\mathcal{A}_k$ is not deleted, an additional random nucleotide is *inserted* immediately to its right with probability $\tilde{p}/2$. If a random element has been inserted, another random nucleotide is inserted immediately to its right with probability $\tilde{p}/2$, etc. (Notice that this algorithm conserves the average length of the sequence.)

An ancestor element $\mathcal{A}_k$ that is conserved in the evolution process (i.e., not deleted or substituted at any point) gets shifted to a position $i(k)$ due to the insertions and deletions of other elements, and appears as daughter element $\mathcal{D}_{i(k)}$. Since the daughter sequences $\mathcal{D}$ and $\mathcal{D}'$ are generated by independent realizations of the evolution process, a fraction $(1-\tilde{p}/2)^2(1-p)^2$ of the ancestor sequence elements is conserved in $\mathcal{D}$ *and* $\mathcal{D}'$. Each such element $\mathcal{A}_k$ defines a unique pair of daughter elements $(\mathcal{D}_{i(k)} = \mathcal{D}'_{j(k)})$ called a *native pair.*

Alignment algorithms are designed to find the native pairs from the knowledge of the daughter sequences $\mathcal{D}$ and $\mathcal{D}'$ alone. A global alignment of the two sequences is defined as an ordered set of pairings $(\mathcal{D}_i, \mathcal{D}'_j)$ and of gaps $(\mathcal{D}_i, -)$ and $(-, \mathcal{D}'_j)$, each letter $\mathcal{D}_i$ and $\mathcal{D}'_j$ belonging to exactly one pairing or gap (see Fig. 1 (a,b)) [2]. (It is clear that gaps are necessary to account for the shifts due to insertions and deletions and to allow the native pairs to be matched.) We define the *fidelity* of an alignment as the fraction of native pairs $(\mathcal{D}_i, \mathcal{D}'_j)$ that are correctly matched.

In this paper, we use the simplest version of the classic Needleman-Wunsch algorithm [2] to align the sequences $\mathcal{D}$ and $\mathcal{D}'$. An alignment is assigned a score

$$\Sigma = \sqrt{3}\, N_+ - \frac{1}{\sqrt{3}} N_- - \gamma N_g \tag{1}$$

given in terms of its total number N_+ of matches $(\mathcal{D}_i = \mathcal{D}'_j)$, the total number N_- of mismatches $(\mathcal{D}_i \neq \mathcal{D}'_j)$, and the total number N_g of gaps. The scoring function (1) has a single adjustable parameter γ, the effective gap penalty. Without loss of generality, the score contributions of matches and mismatches have been chosen in such a way that a pairing of two independent random elements has the average score 0 and the score variance 1. Maximizing the total score Σ defines the optimal alignment of the sequences $\mathcal{D}$ and $\mathcal{D}'$ for a given value of γ. (From a physicist's point of view, $-\Sigma$ is an energy function that has to be minimized.)

The following geometrical representation of global alignment will prove very useful. Fig. 1(b) shows a two-dimensional grid whose cells are labeled by the index pair (i, j). A given alignment of $\mathcal{D}$ and $\mathcal{D}'$ uniquely defines a *directed path* on the grid [2]: A diagonal bond in cell (i, j) represents the pairing of elements $(\mathcal{D}_i, \mathcal{D}'_j)$. A horizontal bond between cells (i, j) and $(i, j+1)$ represents a gap $(\mathcal{D}_i, -)$ located on sequence $\mathcal{D}'$ between the elements $\mathcal{D}'_j$ and $\mathcal{D}'_{j+1}$. Similarly, a vertical bond between cells (i, j) and $(i+1, j)$ represents a gap located on sequence $\mathcal{D}$ between the elements $\mathcal{D}_i$ and $\mathcal{D}_{i+1}$. Using the rotated coordinates $r \equiv i - j$ and $t \equiv i + j$, this alignment path is described by a single-valued function $r(t)$ measuring its displacement from the diagonal of the alignment grid. The path of the optimal

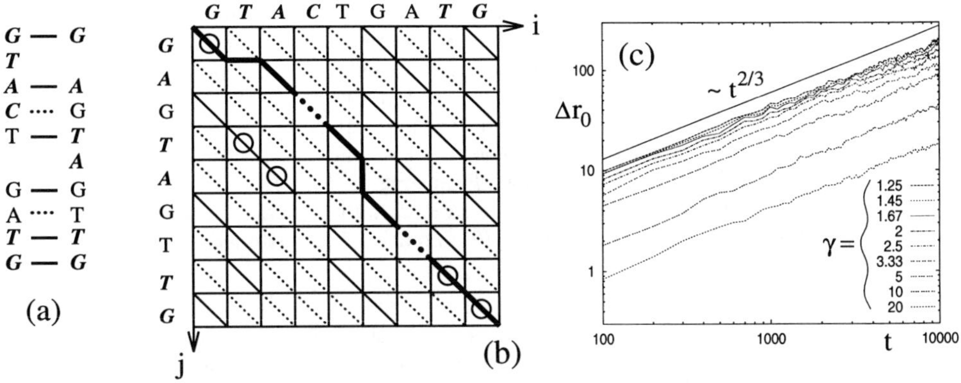

FIGURE 1: (a) One possible alignment of the sequences $\mathcal{D} = \{G, T, A, C, \mathrm{T}, \mathrm{G}, \mathrm{A}, T, G\}$ and $\mathcal{D}' = \{G, A, \mathrm{G}, T, A, \mathrm{G}, \mathrm{T}, T, G\}$; elements conserved from a common ancestor are shown in italics. The alignment has six matches (solid lines), two mismatches (dotted lines), and two gaps. (b) Representation on the alignment grid. Horizontal and vertical bonds represent gaps, solid (dotted) diagonal bonds represent matches (mismatches). Matches corresponding to native pairs are marked with a circle. The directed path $r(t)$ corresponding to the alignment in (a) is shown as a thick line. Its fidelity is 3/5. (c) The mean square displacement $\Delta r_0(t)$ of the optimal alignment path, obtained from a sample of 200 mutually uncorrelated sequence pairs for each value of γ.

alignment is denoted by $r_0(t)$. The set of native pairs resulting from a given evolutionary history corresponds to a set of special diagonal bonds marked by circles in Fig. 1(b). The fidelity of an alignment is the fraction of native bonds that lie on the alignment path $r(t)$. Any shortest trajectory through *all* native bonds defines again a directed path $R(t)$ on the alignment grid called *target path.*

Alignment Statistics and Roughness

The representation on the alignment grid enables us to express the statistics of the evolution process and of sequence alignments in terms of displacement fluctuations measuring the *roughness* of the directed paths $R(t)$ and $r_0(t)$.

Since insertions and deletions are assumed to be independent of each other, the target path $R(t)$ is just a free random walk on the alignment grid; its mean square displacement $(\Delta R(t_1 - t_2))^2 \equiv \overline{(R(t_1) - R(t_2))^2}$ is given by

$$\Delta R(t) = (\tilde{p}\,|t|)^{1/2} \ . \tag{2}$$

Here and throughout the paper, an overbar is used to denote the average over an ensemble of evolution processes for given p and $\tilde{p}$.

At first glance, the optimal alignment path may appear to be a free random walk as well. However, this is generally *not* the case. Consider first the optimal alignment of a pair of random sequences with no mutual correlations (i.e., in the limit $p = 1$). As pointed out in Ref. [7], the mean square displacement $(\Delta r_0(t_1 - t_2))^2 \equiv \overline{(r_0(t_1) - r_0(t_2))^2}$ of the optimal alignment path follows the scaling law

$$\Delta r_0(t) = A(\gamma)\,|t|^{2/3} \tag{3}$$

in the asymptotic regime $|t| \gg A^{-3/2}(\gamma)$, describing a *correlated* random walk.[1] As expected from the theory of critical phenomena, the entire parameter dependence of the displacement $\Delta r_0(t)$ is contained in the amplitude $A(\gamma)$. The exponent 2/3 of the power law is universal (i.e., independent of γ), just as the exponent 1/2 is universal for free random walks. We have verified this numerically (Fig. 1(c)) and have determined the amplitude function $A(\gamma)$ (for details, see [9]). Over the relevant range, $A(\gamma)$ is a monotonically decreasing function of γ, with $A(\gamma) \propto \gamma^{-4/3}$ in the biologically relevant regime $\gamma \gg 1$.

We can compare the roughness of the free random walk $R(t)$ and the correlated random walk $r_0(t)$. Equating the rms. displacements (2) and (3) defines the *roughness matching* scales

$$\tilde{t}(\gamma, \tilde{p}) = \tilde{p}^3 / A^6(\gamma) \, , \qquad \tilde{r}(\gamma, \tilde{p}) = \tilde{p}^2 / A^3(\gamma) \, . \tag{4}$$

For $|t| < \tilde{t}(\gamma, \tilde{p})$, the displacement of $R(t)$ exceeds that of $r_0(t)$. For $|t| > \tilde{t}(\gamma, \tilde{p})$, the displacement of the alignment path becomes dominant since the cost of gaps is outweighed by the gain in score from regions of the grid with an excess number of random matches.

Roughness Matching and Fidelity

For daughter sequences with nonvanishing mutual correlations (i.e., for $p < 1$), the statistics of matches and mismatches on the alignment grid differs from the case of uncorrelated sequences: along the target path $R(t)$, there are $U(p, \tilde{p}) \equiv (1-p)^2(1-\tilde{p}/2)^2$ extra matches per unit of t due to the native pairs. The optimal alignment path contains a finite fraction $\mathcal{F}$ of these extra matches, thereby increasing its score. Hence, it has no longer the displacement (3) and remains confined to the vicinity of the target path $R(t)$. The confinement length r_c is defined by the relative displacement of the two paths, $r_c^2(\gamma, p, \tilde{p}) \equiv \overline{(r_0(t) - R(t))^2}$. Clearly, the average fidelity $\overline{\mathcal{F}}$ decreases with increasing r_c. A rough estimate of the fidelity is just $\overline{\mathcal{F}} \sim 1/r_c$, but there are corrections to this relation [9].

The behavior of the fidelity is well established for an evolution process without insertions and deletions ($\tilde{p} = 0$) [7]. In this limit, $\overline{\mathcal{F}}$ is found to be a monotonically increasing function of γ, reaching its maximum $\mathcal{F}^* = 1$ for $\gamma \to \infty$. This is not surprising since there is no need for any gaps in an alignment if the evolution process has no insertions and deletions.

For small values of $\overline{\mathcal{F}}$, the fidelity has the asymptotic form $\overline{\mathcal{F}} \sim (1+x)\exp(-x)$, where $x \equiv C(\gamma)/U(p,0)$ and $C(\gamma) \sim A^{3/4}$ is another amplitude function. This form is supported by our numerics [9]. For $U \to 0$, the fidelity approaches zero and the confinement length r_c diverges. This singularity marks a continuous phase transition at $U = 0$, which we call the *detectability transition*. Positive correlations between sequences ($U > 0$) are recovered with a finite fidelity $\overline{\mathcal{F}}(\gamma, U)$, while anticorrelations ($U < 0$) cannot be detected (i.e., $\mathcal{F} = 0$). For the alignment path $r_0(t)$, this is a critical (de-)localization transition[2] between the confined regime ($r_c < \infty$) for $U > 0$ and the regime of correlated displacement fluctuations (3) for $U \leq 0$.

[1]This class of random walks is in fact well known to physicists as *directed polymers* in a random medium ([12], see also [13] and references therein). Perhaps the most prominent example occurs in the theory of magnetic superconductors. If these materials are placed into a magnetic field, they develop tubular regions of normal magnetic conductance. These *flux lines* are directed parallel to the applied field (the t direction) and can be described by a displacement vector $r(t)$. In addition, there is often a distribution of point impurities. These act on the flux lines just like a random distribution of matches and mismatches on the alignment grid, causing large displacements of the lowest-energy path $r_0(t)$.

[2]The properties of this phase transition are known from the physics of a magnetic flux line interacting with an attractive linear defect $R(t) = 0$ [14, 15].

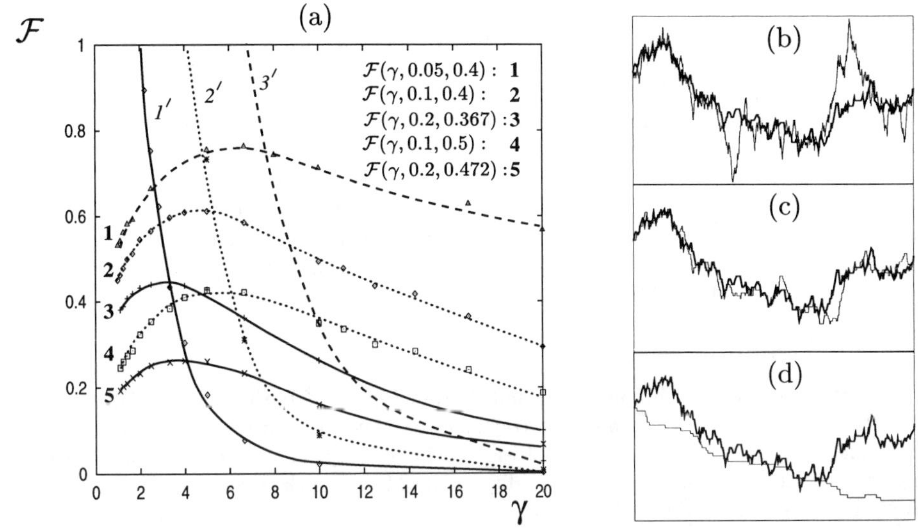

FIGURE 2: (a) The average alignment fidelity $\overline{\mathcal{F}}(\gamma; p, \widetilde{p})$ obtained from a sample of 100 sequence pairs (lines 1-5), and the inverse roughness matching scale $\tilde{r}^{-1}(\gamma; \widetilde{p})$ (lines *1'-3'*) for several values of p and $\widetilde{p}$. The curves belonging to the same value of $\widetilde{p}$ are shown in the same line style (solid: $\widetilde{p} = 0.2$, dotted: $\widetilde{p} = 0.1$, dashed: $\widetilde{p} = 0.05$). The maximum of a fidelity curve is close to its intersection point with the roughness matching curve for the same value of $\widetilde{p}$. (b-d) The optimal alignment path $r_0(t)$ (thin line) for the same sequence pair and the same target path $R(t)$ (thick line) at three different values of γ: (b) in the random fluctuation regime, (c) at the optimal value $\gamma^*(p, \widetilde{p})$, and (d) in the shortcut regime.

For finite insertion/deletion rate $\widetilde{p}$, it is clear that $\mathcal{F}$ should decrease to zero for sufficiently large values of γ, since a high gap penalty prevents the alignment path from following a fluctuating target path $R(t)$. The behavior of the fidelity $\overline{\mathcal{F}}(\gamma; p, \widetilde{p})$ is rather complex. Fig. 2(a) shows the dependence of $\overline{\mathcal{F}}$ on γ for several values of p and $\widetilde{p}$. Unlike for $\widetilde{p} = 0$, these curves have their (single) maximum at a finite value $\gamma^*(p, \widetilde{p})$. Alignment patterns for $\gamma < \gamma^*(p, \widetilde{p})$ and $\gamma > \gamma^*(p, \widetilde{p})$ are clearly distinguished by the roughness of the optimal path $r_0(t)$, as one recognizes from the examples of Fig. 2(b-d): (b) For $\gamma < \gamma^*(p, \widetilde{p})$, the displacement fluctuations of $r_0(t)$ exceed those of the target path $R(t)$. This can be expressed by the relation $r_c > \tilde{r}$ with $\tilde{r}(\gamma, \widetilde{p})$ given by Eq. (4). We call this regime of γ the *random fluctuation regime.* (c) For $\gamma = \gamma^*(p, \widetilde{p})$, the displacement fluctuations of boths paths are seen to be of the same size, i.e., $r_c \sim \tilde{r}$. (d) In the *shortcut regime* for $\gamma > \gamma^*(p, \widetilde{p})$, the dominant fluctuations are those of the target path, while the alignment path $r_0(t)$ has large straight patches with negligible intrinsic roughness.

One can show that in the asymptotic random fluctuation regime and shortcut regime, the fidelity is a monotonically increasing (decreasing) function of γ, respectively. Hence, the fidelity maximum $\mathcal{F}^*$ can be estimated by the *roughness matching condition*

$$\mathcal{F}^* \approx \tilde{r}^{-1}(\gamma^*, \widetilde{p}) \, . \tag{5}$$

As one verifies in Fig. 2(a), the maxima of the curves $\overline{\mathcal{F}}(\gamma; p, \widetilde{p})$ are indeed close to the intersection points with the curves $\tilde{r}^{-1}(\gamma; \widetilde{p})$ given by Eq. (4) with $A(\gamma)$ taken from Fig. 1(b).

This geometric picture of alignment can be extended into a quantitative theory. At any value of γ, the alignment of correlated sequences ($U > 0$) has an average fidelity $\overline{\mathcal{F}} > 0$ and a higher average score than an alignment of uncorrelated sequences ($U = 0$). The score difference comes from the extra matches due to the native pairs, which outweigh the extra gaps imposed on the alignment path by the constraint of following the target path. This score balance determines the alignment fidelity $\overline{\mathcal{F}}$. It turns out to be quite different in the random fluctuation regime and in the shortcut regime, reflecting the different morphologies of the alignment paths in Fig. 2(b-d). The theory predicts fidelity patterns $\mathcal{F}(\gamma, p, \widetilde{p})$ in good agreement with the numerically observed behavior of Fig. 2(a). This will be described in detail in a forthcoming publication [9].

Acknowledgments

The authors are grateful to Stephen Altschul, Steven Benner, Charles Elkan, Walter Fitch, Jeff Thorne, and Michael Waterman for conversations and suggestions. TH acknowledges the financial support of an A. P. Sloan Research Fellowship, an Office of Naval Research Young Investigator Award, and the hospitality of the Max-Planck Institute at Teltow where much of the work was carried out.

References

[1] See review articles in *Met. Enz.* **183**, (1990).
[2] S.B. Needleman and C.D. Wunsch, *J. Mol. Biol.*, **48**, 444 (1970).
[3] T.F. Smith and M.S. Waterman, *Adv. Appl. Math.* **2**, 482 (1981).
[4] For a survey of recent developments, see M.S. Waterman, in *Mathematical Methods for DNA Sequences*, M.S. Waterman ed., CRC Press (1989); and M.S. Waterman, *Introduction to Computational Biology*, Chapman & Hall (1994).
[5] S.A. Benner, M.A. Cohen and G.H. Gonnet, *J. Mol. Biol.* **229**, 1065 (1993).
[6] M. Vingron and M.S. Waterman, *J. Mol. Biol* **235**, 1 (1994).
[7] T. Hwa and M. Lässig, *Phys. Rev. Lett.* **76**, 2591 (1996).
[8] M.Q. Zhang and T.G. Marr, *J. Theo. Biol.* **174**, 119 (1995).
[9] D. Drasdo, T. Hwa, and M. Lässig, Optimal Detection of Similarities in DNA Sequences, preprint (1997).
[10] M.J. Bishop and E.A. Thompson, *J. Mol. Biol.* **190**, 159 (1986).
[11] J.L. Thorne, H. Kishino, and J. Felsenstein, *J. Mol. Evol.* **33**, 114 (1991).
[12] M. Kardar, *Nucl. Phys. B* **290**, 582 (1987).
[13] T. Hwa and D.S. Fisher *Phys. Rev. B* **49**, 3136 (1994).
[14] T. Hwa and T. Nattermann, *Phys. Rev. B* **51**, 455 (1995).
[15] H. Kinzelbach and M. Lässig, *J. Phys. A* **28**, 6535 (1995).

Flexible membranes with anchored polymers

Reinhard Lipowsky, Hans–Günther Döbereiner, Christin Hiergeist, and Vasudevamurthy Indrani

MPI für Kolloid– and Grenzflächenforschung
Kantstr. 55, D–14513 Teltow–Seehof

Abstract

Recent work on flexible membranes such as lipid bilayers with anchored polymers is briefly reviewed. These polymers exert bending moments onto the membranes and, thus, lead to a polymer–induced curvature. From the theoretical point of view, this curvature depends (i) on the membrane–polymer interactions; (ii) on the size of the polymers; and (iii) on the polymer coverage. Experimentally, these curvature effects can be deduced from systematic studies of polymer–decorated vesicles in which one monitors the vesicle shape with varying solvent conditions.

1 Introduction

The systems considered here consist of flexible membranes in aqueous solution such as lipid bilayers in their fluid state and of polymers which are attached to the membranes via anchors. The anchor segments of the polymers consist of hydrophobic blocks which fit into the hydrophobic region of the bilayer as schematically shown in Fig. 1. Several types of polymers with anchors have been synthezised in the chemistry lab; in addition, biology provides a large variety of such systems since most membrane proteins can be considered as anchored polymers.

Apart from the anchor segments, the polymers are taken to be soluble in water and thus behave as self–avoiding chains. When anchored to the membrane, the polymers will experience various interactions with the membrane arising from molecular forces which can be repulsive or attractive. In addition, the chains try to maximize their configurational entropy in front of the membrane surface. If the non–anchored segments of the polymers are effectively repelled from the membrane surface, the anchored polymers form *desorbed mushrooms* or *brushes* depending on the polymer coverage. If this interaction is effectively attractive, the polymers form *adsorbed pancakes*.

Anchored mushrooms and brushes exert bending moments onto the membrane which curve the membrane *away from* the polymers; anchored pancakes, on the other hand, lead to the opposite behavior since the membrane is now curved *towards* the polymers. In this article, we briefly review recent theoretical work on this *polymer–induced curvature* [1, 2, 3] and argue that these curvature effects can be experimentally deduced from a systematic study of vesicle shapes extending the approach in [4].

In what follows, we will first consider the dilute mushroom regime and the semi–dilute brush regime. In these regimes, the polymer–induced curvature depends on the chain length, on the the solvent conditions and on the polymer coverage. We also argue that mushrooms and pancakes, when anchored on both sides of the membrane, have a tendency to form *checkerboard states*, see Fig. 2.

Mat. Res. Soc. Symp. Proc. Vol. 463 © 1997 Materials Research Society

2 Polymer–induced Curvature of Membranes

If both monolayers of the bilayer exhibit a different polymer coverage, the decorated bilayer is asymmetric and will then exhibit a polymer–induced spontaneous curvature. The concept of a spontaneous curvature [5] is appealing from an intuitive point of view but, in general, its magnitude is rather difficult to estimate since it depends on many molecular parameters. In contrast, the spontaneous curvature induced by anchored polymers scales with a few parameters such as the chain length of the polymer which can be experimentally varied in a systematic way.

2.1 Dilute regime and mushroom–induced curvature

First, consider a single mushroom of a linear chain for which the anchor segment is located at one of its ends and for which the non–anchored polymer segments experience effectively repulsive interactions with the membrane surface. The size of the mushroom is comparable to the size of the free polymer, i.e., to $R_{po} \simeq a_{po}N^{\nu}$ with the persistence length a_{po} and the size exponent $\nu \simeq 3/5$ for good solvent conditions.

Such an anchored polymer tends to exert entropic or fluctuation forces onto the bilayer membrane which bend the membrane *away* from the polymer. The resulting polymer–induced mean curvature is given by [1]

$$M_{pi} \sim +T/\kappa R_{po} \sim +T/\kappa a_{po}N^{\nu} \qquad \text{for mushrooms} \tag{1}$$

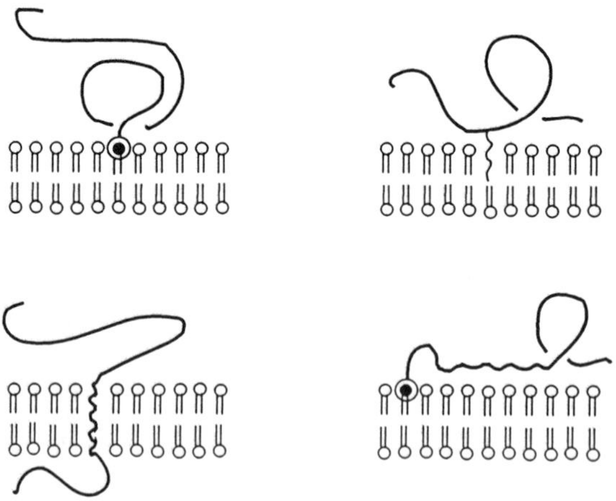

Figure 1: Different ways to anchor polymers into bilayer membranes: (a) Lipid anchor, (b) Hydrophobic side chain; (c) Membrane-spanning polymer segment; and (d) Anchored and adsorbed chain

where T and κ are the temperature and the bending rigidity of the membrane, respectively (here and below, temperature is measured in energy units, i.e., T is a short–hand notation for Boltzmann constant k_B times temperature).

In the above estimate, the effect of the anchored polymer onto the bending rigidity κ and onto the Gaußian bending modulus κ_G of the membrane segment has been ignored. These effects can be determined within a small curvature expansion which gives [2]

$$\kappa_{eff} \approx \kappa + c_1 T \qquad \text{and} \qquad \kappa_{G,eff} = \kappa_G - c_2 T \tag{2}$$

with $c_1 \simeq 0.21$ and $c_2 \simeq 0.17$. Thus, κ and κ_G are *increased* and *decreased* by the polymer mushroom, respectively.

2.2 Overlap coverage and semi–dilute brush regime

It is convenient to introduce the dimensionless coverage $\bar{\Gamma} \equiv (a_{po}/\xi_{an})^2$ where ξ_{an} describes the mean distance between the anchors (i.e., the mean grafting distance). In general, each monolayer of the bilayer will be characterized by a different coverage. We will first focus on the asymmetric case where the polymers are attached only to one side of the bilayer.

When ξ_{an} becomes comparable to the size R_{po} of the mushrooms, i.e., when the coverage is equal to the overlap coverage $\bar{\Gamma}_{ov} = 1/N^{2\nu}$, the polymers start to overlap and to squeeze one another. They then attain conformations which are stretched away from the surface and resemble brush states [6, 7, 8]. In such a situatin, the membrane again bends away from the polymers, and the corresponding polymer–induced curvature M_{pi} increases monotonically with the coverage $\bar{\Gamma}$. For good solvent conditions, scaling arguments lead to $M_{pi} \sim N^2 \bar{\Gamma}^{13/6}$ and $M_{pi} \sim N^{1/7} \bar{\Gamma}^{13/21}$ for small and large $\bar{\Gamma}$, respectively. [2] As in the case of mushrooms, the polymer brush acts to *increase* the bending rigidity κ but to *decrease* the Gaußian bending modulus κ_G. [2, 9, 10].

2.3 Checkerboard states

Two mushrooms which are anchored to two different monolayers, i.e., to two different sides of the bilayer have a tendency to *avoid* each other. Indeed, if we force two such mushrooms to be directly opposite to each other, we loose the free energy $2\Delta\mathcal{F}$ where $\Delta\mathcal{F} \sim \kappa M_{pi}^2 R_{po}^2 \sim T^2/\kappa$ is the free energy gained for each curved membrane segment with an isolated mushroom. A similar situation arises, in general, if one has monolayer domains with area $\mathcal{A}_{do}$ and spontaneous mean curvature M_{do}. In the latter case, $\Delta\mathcal{F} \sim \kappa M_{do}^2 \mathcal{A}_{do}$.

Now assume that the coverage $\bar{\Gamma}$ is of the order of $\bar{\Gamma}_{ov}/2$ on both sides of the bilayer. In such a situation, anchored mushrooms on different sides can still avoid each other if they attain a checkerboard array and the membrane aquires a corresponding curvature modulation as displayed in Fig. 2. In such a situation, the mushroom–covered membrane domains are connected by intermediate membrane segments which consist essentially of saddle points and thus do not contribute to the bending energy of the membrane (in a different context, such a modulation of the membrane curvature, which looks similar to an egg carton, has been obtained from fourth–order curvature terms [11]).

The free energy difference between the checkerboard state and the flat state is $\Delta\mathcal{F} \sim T^2/\kappa$ per domain. Such a situation resembles an antiferromagnet with nearest neighbor

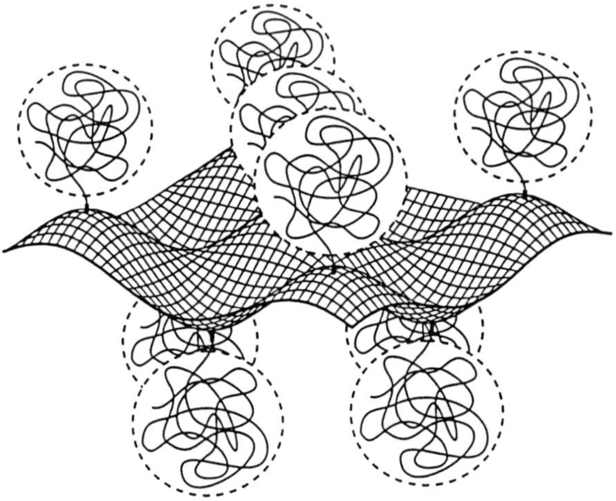

Figure 2: Checkerboard array of mushrooms and checkerboard modulation of membrane curvature, see text.

coupling constant $J \simeq -\Delta\mathcal{F}$ which undergoes a phase transition to an ordered state at $J = J_c$ with $|J_c|/T = \ln(1+\sqrt{2})/2 \simeq 0.44$. For the mushroom domains, one has $|J|/T \sim T/\kappa$ which is presumably too small to stabilize the checkerboard state on large scales. On the other hand, such a state seems to be a good choice for monolayer domains with $\kappa M_{do}^2 \mathcal{A}_{do} \gtrsim T$ which can be fulfilled for anchored pancakes, see below. Furthermore, if the area $\mathcal{A}_{do}$ of these domains grows with time, e.g., because of an underlying phase separation process within the membrane, the checkerboard state should develop as soon as the membrane area has grown up to $\mathcal{A}_{do} = T/\kappa M_{do}^2$.

2.4 Pancake–induced curvature

Finally, let us consider anchored chains for which the non–anchored segments experience effectively *attractive* interactions with the membrane surface. As mentioned, the polymer will then form an adsorbed pancake. It turns out that the membrane now bends *towards* the polymer in order to maximize the number of contact points with the pancake. Therefore, such a pancake induces a spontaneous curvature which has the *opposite* sign compared to the mushroom case and is given by [3]

$$M_{pi} \sim -T/\kappa L_\perp \qquad \text{for pancakes} \tag{3}$$

where the relevant length scale is now given by the pancake thickness $L_\perp$. In [12], a similar parameter dependence for M_{pi} has been proposed but with the plus sign, i.e., with the same sign as in (1). The minus sign as found by us is consistent with related work on polymers adsorbed onto curved surfaces [13, 14, 15]. If the anchored polymers undergo an adsorption–desorption transition, the polymer transforms from a pancake to a mushroom and the curvature of the adjacent membrane segment changes its sign. Therefore, anchored polymers close to such a transition provide a kind of 'curvature switch'. [3, 16]

If the coverage on both sides of the bilayer is of the order of half the overlap concentration, pancakes will also have a tendency to form a checkerboard array. In fact, one now has $\Delta\mathcal{F} \sim \kappa M_{pi}^2 R_{po}^2 \sim (T^2/\kappa)(R_{po}/L_\perp)^2$ which can be large compared to T and, thus, can stabilize the checkerboard state.

3 Experimental perspective

In general, the spontaneous curvature of membranes can be measured by investigating its effect on the morphology of vesicles. From the theoretical point of view, the shape of a vesicle is determined by the minimum of the bending elastic energy of its membrane subject to constraints on vesicle area and volume (for a review, see [17]). Thus, the two principal parameters governing the vesicle shape are the volume-to-area ratio and the spontaneous curvature of the vesicle.

It it important to note that the spontaneous curvature of a vesicle has two rather different contributions. One contribution arises from the spontaneous curvature of the bilayer membrane as determined *locally* by its chemical composition and by the adjacent solution. The second contribution stems from the difference in the number of molecules between the inner and outer monolayer of the closed vesicle. If one can ignore flip–flops between the two monolayers, this difference is fixed and leads to a *global* constraint for the integrated mean curvature of the vesicle. The vesicle shapes which are governed by one of these contributions separately have been calculated some time ago. [18, 19] In general, one has to include both contributions which leads to the so–called area-difference-elasticity model. [20]

Systematic studies, in which experimental observations and theoretical calculations for vesicle shapes have been compared in a quantitative way, have been made only quite recently. [21, 22, 23] In particular, it is now possible to extract the spontaneous curvature of the membrane from the observed vesicle shape. [4] The technique applied in the latter study is quite general; it has also been used, for example, to monitor the effect of small molecules on the spontaneous curvature as will be described elsewhere. Thus, this technique should also be applicable to the polymer–decorated membranes discussed in this article. In order to determine the local contribution to the spontaneous curvature arising from the anchored polymers, one has to perform measurements on the same polymer–decorated vesicle for various solvent conditions.

Another area of membrane physics in which anchored polymers play an important role is the *adhesion* of membranes. Indeed, specific interactions mediated by anchored stickers, i.e., by anchored macromolecules with a sticky end, are responsible for the adhesion of biomembranes. Certain aspects of these systems can be studied in model systems consisting of lipid bilayers with anchored polymers. One example is the recent prediction that the membranes can only adhere by stickers if the sticker concentration exceeds a certain threshold which depends on the lateral tension. [24]

References

[1] R. Lipowsky, Europhys. Lett. **30**, 197 (1995).

[2] C. Hiergeist and R. Lipowsky, J. Phys. France **6**, 1465 (1996).

[3] C. Hiergeist, V. Indrani, and R. Lipowsky, Europhys. Lett. (in press, December 1996).

[4] H.-G. Döbereiner, E. Evans, M. Kraus, U. Seifert and M. Wortis, Phys. Rev. E (submitted).

[5] W. Helfrich, Z. Naturforsch. **28c**, 693 (1973).

[6] S. Alexander, J. Phys. Paris **38**, 983 (1977).

[7] P.-G. de Gennes, Macromolecules **13**, 1069 (1980).

[8] M. Daoud and J. Cotton, J. Phys. (Paris) **43**, 531 (1982).

[9] S. Milner and T. Witten, J. Phys. France **49**, 1951 (1988).

[10] K. Hristova and D. Needham, J. Colloid Interface Sci. **168**, 302 (1994).

[11] R. Goetz and W. Helfrich, J. Phys. II France **6**, 215 (1996).

[12] P.-G. de Gennes, J. Phys. Chem. **94**, 8407 (1990).

[13] P. Pincus, C. Sandroff, and T. Witten, J. Physique **45**, 725 (1984).

[14] J. Brooks, C. Marques, and M. Cates, J. Physique II **1**, 673 (1991).

[15] E. Eisenriegler, A. Hanke, and S. Dietrich, Phys. Rev. E **54**, 1134 (1996).

[16] V. Indrani and R. Lipowsky, in preparation.

[17] U. Seifert and R. Lipowsky, in *Structure and Dynamics of Membranes*, Vol. 1 of *Handbook of Biological Physics*, edited by R. Lipowsky and E. Sackmann (Elsevier, Amsterdam, 1995), pp. 403–463.

[18] S. Svetina and B. Zeks, Eur. Biophys. J. **17**, 101 (1989).

[19] U. Seifert, K. Berndl, and R. Lipowsky, Phys. Rev. A **44**, 1182 (1991).

[20] L. Miao, U. Seifert, M. Wortis, and H.-G. Döbereiner, Phys. Rev. E **49**, 5389 (1994).

[21] K. Berndl, J. Käs, R. Lipowsky, E. Sackmann, and U. Seifert, Europhys. Lett. **13**, 659 (1990).

[22] H.-G. Döbereiner, E. Evans, U. Seifert, and M. Wortis, Phys. Rev. Lett. **75**, 3360 (1995).

[23] H.-G. Döbereiner and U. Seifert, Europhys. Lett. **36**, 245 (1996).

[24] R. Lipowsky, Phys. Rev. Lett. **77**, 1652 (1996).

THE STRUCTURAL CHANGES INDUCED BY ANCHORED POLYMERS ON DROPLET AND BICONTINUOUS MICROEMULSION

JYOTSANA LAL, LOIC AUVRAY
Laboratoire Leon Brillouin, CEN -Saclay, 91191 Gif-Sur-Yvette, Cedex, France.

ABSTRACT

Macroscopically, phase changes (induced due to change of curvature) were observed on the introduction of PEG (polyethyleneglycol) grafted on lipids (anchored polymers) in the Winsor phases of microemulsion. Replacing some of the surfactant in the monolayers of the micro-emulsion with polymer grafted lipids changes the interaction between droplets of micro- emulsion and a greater increase in the percentage of anchored polymer in the monolayer might even bring about structural changes. The results on winsor phases, droplet microemulsions and bicontinous phases with anchored polymers as studied by Small Angle Neutron Scattering (SANS) are reported below, together with the contrast variation techniques used to characterize average curvature in the system.

INTRODUCTION

The microemulsions macroscopically appear as homogeneous stable phases of oil, water and surfactants. Microscopically, the surfactant molecules adsorb to the interface between oil and water domains (~100 Å). since this interface is saturated with surfactants this amounts to vanishing interfacial tension, the energy of such an interface is then dominated by its elastic curvature energy [1,2]. Depending on the curvature of such an interface we can have droplet or bicontinous microemulsions. Here we study the deformation and the change of curvature induced by anchored polymers on the fluid interfaces of the microemulsions. These changes in average curvature and interactions can be characterized by techniques of SANS [3,4].

Theoretically, the question which is less clear is how anchored polymers effect the elastic parameters of the Interface - bending rigidity K, the gaussian rigidity κ and the spontaneous curvature C_0. It has been suggested by deGennes [5] that in case of free polymers that the bending rigidity of the membrane would be enhanced but at the same time he points out the effect of polymer on K and the gaussian rigidity κ is delicate. Brooks et al [6] predict that adsorbed free polymer on surfactant bilayers leads to a decrease in the mean curvature rigidity and an increase in the gaussian rigidity. Leibler [7] has shown that molecules intercalated in lipid membrane (adsorbed drugs, intermembrane proteins etc) could be coupled to the local curvature of the bilayer membrane. They do not seem influence the physical properties in an important way but change the phenomenological elastic parameters K, κ and the spontaneous curvature C_0. In the case of microemulsions it has been known since a long time that the addition of co-surfactants which intercalate into the surfactant built interface diminish the effective rigidity and thus the persistence length [1]. Recently, there has been a series of paper by Helfrich on this subject which say that spontaneous curvature give rise to a reduction in the bending rigidity of mixed surfactant monolayers [8]. Another relevant study is by Lipowsky[9] which describes the bending of membranes by anchored polymers. It points out the relation between curvature and the bending rigidity K in the mushroom and the brush regime of the anchored polymers. We have an ideal model system of fluctuating and flexible surfactant layers of microemulsions with vanishing interfacial tension to test the above proposed hypothesis. These effects have to be probed at very local scales and small angle neutron scattering provides us with a very powerful means to do so.

Mat. Res. Soc. Symp. Proc. Vol. 463 © 1997 Materials Research Society

EXPERIMENTAL TECHNIQUES

Small Angle Neutron Scattering (SANS)

The scattering from the droplets and bicontinous systems both in bulk and film contrast were measured at Saclay (PACE spectrometer). Contrast variation experiments were performed by varying the amount of deuterated water in the bicontinous phases prepared with a known composition with and without the PEG lipids. The cross structure factors were thus deduced. The measurements were done in several different configurations to study different ranges of scattering vector from 5 10^{-3} to 0.12 $Å^{-1}$. Water which scatters isotropically was used to normalize the intensities.

Sample

We have incorporated in the monolayers formed from single tailed-surfactant some lipids with covalently attached polyethylene glycol polymer (PEG-lipid). The lipid molecule abbreviated DSPE is N-(carbamyl-poly(ethylene glycol methyl ether)-1,2 distearyl-sn-glycero-3-phosphoehanoamine, sodium salt (Liposome Technology). The molecular weight of DSPE is 748. The molecular weight of the polymeric moieties was 2000g/mol and 5000g/mol.

Droplet microemulsion-PEG moieties outside water continuum- Oil in water microemulsions (O/S), ϕ=4% (cyclohexane/brine(1%)/hydrogenated SDS, pentanol) were used.The droplets radius R=50 Å. 10% or 5% of the SDS molecules were replaced in mole by 2000PEG-DSPE or 5000PEG-DSPE. The experiments were done in core contrast with the polymer invisible.

Winsor Phases- SDS (0.1g) was added to equal volume of oil i.e. deuterated toluene and water at the salinites of 6.5% and 8%. Butanol was used as a co-surfactant. At the salinity of 6.5% we had a bicontinous phase co-existing with oil and water. At 8% salinity we had O/W microemulsion co-existing with water. The changes in the phase diagram were observed by incorporating various percentages of 2000PEG-DSPE and 5000PEG-DSPE in mole to the above winsor phases.

Bicontinous Phases- The samples studied had composition close to the middle phases. The microemulsions were made of brine with salinty 6.5% by weight, deuterated toluene, sodium dodecylsulphate (SDS) and butanol. The volume of brine or water was the same as toluene , $V_w=V_o$, and the volume fraction $f_w=f_o=V_w/V_w+V_o=0.5$. For a given mass of SDS we adjusted the amount of butanol to obtain a clear, monophasic microemulsion. In calculating the scattering length densities the butanol partition coefficient estimated from the titration curves as 0.06ml of butanol/ml of toluene and 0.02mlof butanol/ml of brine were taken into account. 10% and 15% of the SDS molecules were replaced in mole by 2000PEG-DSPE in the bicontinous microemulsions. The contrast variation experiments were done on a series of 6-8 samples for with and without PEG-DSPE, varying the scattering length of water by using different mixtures of H_2O/D_2O. At 87% D_2O we were at film contrast condition.

RESULTS

Winsor phases

In the winsor phases of water, oil, ionic surfactant and co-surfactant alcohol, as the ionic strength of water is varied, the following phases are observed [3] (Figure 1): (a) an oil in water microemulsion co-existing with an oil excess (b) a three phase equilibria, the middle phase microemulsion containing equal amount of oil and water separating the pure oil and water phases; (c) another two phase equilibria: a water in oil microemulsion coexisting with water excess. At low salinity, the electrostatic interaction between the polar head groups are not fully screened, the surfactant film in this case spontaneously bends towards the oil, resulting

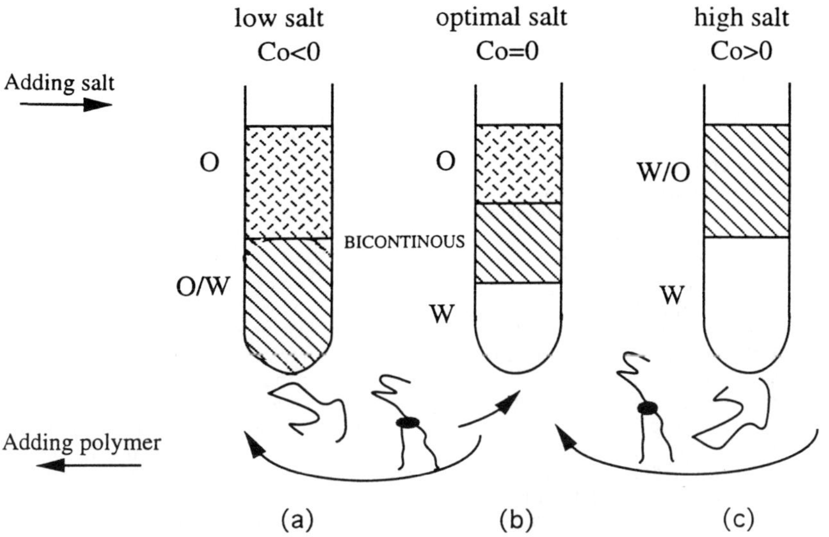

Figure 1 Schematic representation of macroscopic changes which occur in Winsor phases on addition of free and anchored polymers.

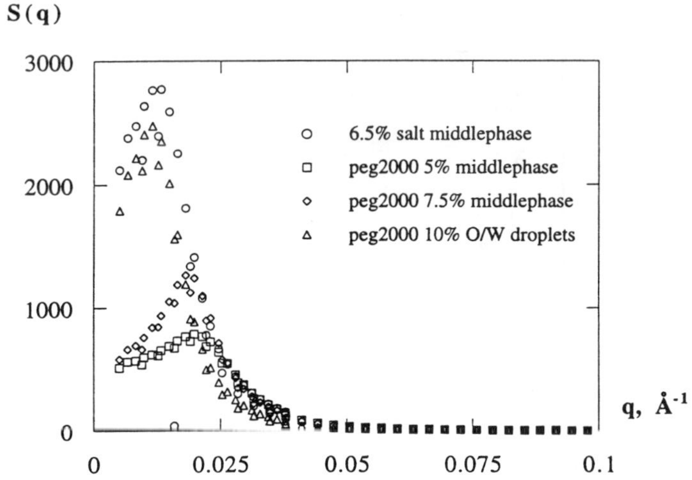

Figure 2 SANS results in representation S(q) versus q showing the microscopic changes induced by anchored polymers when added to the Winsor bicontinous middlephases.

in oil in water microemulsion co-existing with a phase almost pure in oil. Salt addition changes the spontaneous curvature C_0 of the surfactant film between oil and water, by screening the repulsion between the polar head groups. At optimal salt content, the spontaneous curvature of the surfactant film is close to zero, a bicontinous microemulsion phase co-exists with oil and water excess. At very high salinity, the curvature C_0 (>0) inverses completely, the film bends towards the water, a water in oil microemulsion co-exists with pure water.

Polymer grafted on the lipids were added to replace some of the SDS molecules in the winsor phases and the phases to which this system freely equilibriated to were then monitored. These effects were visualized macroscopically in the test tubes and the corresponding microscopic changes in structure have been monitored by SANS. Starting from the middle phase bicontinous microemulsion with 6.5% salt, increasing percentage of 2000PEG-DSPE 5%, 7.5% and 10% were added to it, until at 10% of 2000PEG-DSPE a change in phase from the middlephase bicontinous (Figure 1b) to a two phase: oil in water droplet microemulsion (Figure 1a) co-existing with excess oil was obtained. In the test tubes when a small amount of peg lipids 5% and 7.5% was added, we just achieve a shrinking of the middle bicontinous phase with expulsion of oil and water. The corresponding neutron scattering data S(q) versus q from such middle phases with increasing amount of lipids with grafted polymers is shown in Figure 2. The characteristic scattering peak q* (its origin will be discussed later-it is related to the persistence length of the surfactant film) of the original middle bicontinouse phase (q^*~0.01Å^{-1}, q^{-1}~100Å) shifts first to larger values of scattering vector q (q^*~0.02Å^{-1}, q^{-1}~50Å) when 5% and 7.5% of 2000PEG-DSPE were added. This probably amounts to a decrease in the persistence length of the film in the middle bicontinous phase on addition of polymer grafted lipids. While the characteristic scattering peak q* for oil in water droplet microemulsion formed on addition of 10% PEG lipids in the system corresponds to the typical size of the droplets in the system.

Thus in the above bulk contrast, the concentrated solution of bicontinous middle phase exhibits a correlation peak q^* in the scattering at low q not very different from the case where we have droplets of oil in water. Thus at low spatial resolution (low q) the correlation of the oil and water of a concentrated dispersion of droplets and of a random bicontinous structure with a well defined scale are very similar as seen from Figure 2. The difference between the two will appear on a semi-local scale or in the film geometry. This will be discussed in more detail below.

Droplet Microemulsion

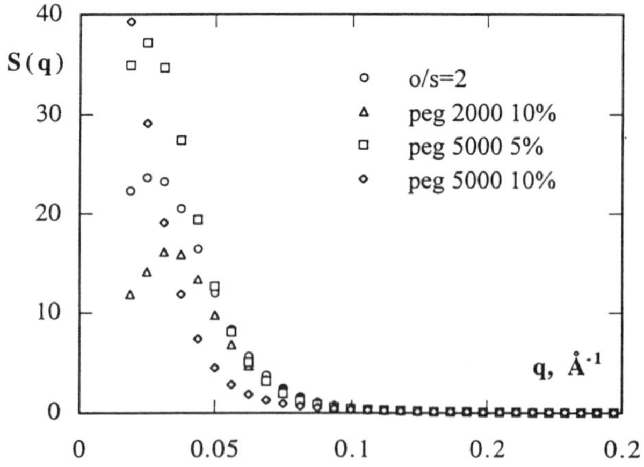

Figure 3 S(q) versus q plots of SANS for oil in water (o/s=2, R~50Å) droplets with varying density and length N of the grafted or anchored polymer chains

In Figure 3, the S(q) versus q plots of SANS are shown for oil in water (o/s=2, R~50Å) droplets with varying density and length N of the grafted polymer chains (PEG-DSPE). The experiments were done in core contrast (deuterated cyclohexane, hydrogenated water 1% salt). The effective interaction between the microemulsion droplets is modified by the grafted polymer which manifests itself by the change of scattering behaviour in the limit q—> 0. Decreasing scattering at q—> 0 implies that the system is not very compressible i.e. the interaction between the droplets is repulsive. The interaction for oil in water droplets is repulsive to start with, on replacing 10% of the SDS molecules in the interfacial film by the 2000PEG-DSPE makes the interaction between the droplets even more repulsive. Decreasing the density and increasing the chain length of the grafted polymer by using 5% of 5000PEG-DSPE makes the interaction between the droplets less repulsive. On increasing the grafting density further by using 10% of 5000PEG-DSPE we change the interaction between droplets from being repulsive to attractive as seen by the great increase of scattering in the limit q—> 0. (One of the reason for this behavior could be physical bridging between droplets or this might indicate a structural change). Thus we can tune the interaction between the droplets by varying the size and the density of the grafted polymer chains.

Bicontinous Microemulsion

The sample made at optimal salinity of 6.5% with equal amounts of oil and water, known quantities of surfactant and co-surfactant results in a bicontinous microemulsion [3]. In the bicontinous phase, interfacial film has almost close to zero spontaneous curvature and is thus very flexible. The characteristic size of this structure is governed by the persistence length ξ of the film [1]. The interfacial film is rigid and flat on scale smaller than ξ and wrinkled on larger scales. The film stiffness forbids the curvature fluctuations at the spatial scales smaller than ξ. In this model ξ=a exp ($2\pi K/K_BT$, thus the persistence length ξ is extremely sensitive to the rigidity constant K.

We have plotted S(q) versus q in bulk contrast. We observe the characteristic scattering peak q* (~ .0186 $Å^{-1}$, q^{*-1}~55.6Å) of the bicontinous structure. This peaks shifts to larger q* values on the addition of 10% 2000PEG-DSPE (~ .022 $Å^{-1}$, q^{*-1}~45.5Å) and 15% 2000PEG-DSPE(~ .025 $Å^{-1}$, q^{*-1}~40Å). We had already made such a macroscopic and microscopic observation in the case middlephase bicontinous structure. Our data implies that introduction of peg lipids in the interfacial film decreases the persistence length ξ of the film i.e. it also diminishes the rigidity K of the film [1].

We have looked just at the interfacial film by doing the experiment in film contrast (87% deuterated water/deuterated toluene) which helps to unravel the structure even more
The apparent average thickness of the film without anchored polymers is d=7 Å. In fact we can estimate the value of d in the case of film with peg lipids by considering the asymptotic behavior of scattering, by assuming that at large q we observe scattering from a plane surface S, of volume V and thickness d, then for small d the scattering is given by $S(q)=2\pi S/V(n_o-n_f)^2(d^2/q^2)(1-d^2q^2/12)$. If we now plot $q^2S(q)$ versus q^2 at asymptotic values of q we get a straight line with a negative slope. The ratio of slope to intercept is equal to $-d^2/12$ and thus we can deduce a value of d from it. From such plots d=15Å for 10% of 2000PEG-DSPE and d=23Å 15% of 2000PEG-DSPE within experimental error.

In order to distinguish between a bicontinous structure or a dense phase of spherical droplets we need to know the sign of the curvature. What is the sign of the curvature induced by the peg-lipids? We already guess the sign of curvature to be negative from the macroscopic middle bicontinous phase experiment which underwent a change to oil in water droplets when the peg-lipids were added. To verify at a microscopic scale, the sign of curvature induced by the peg-lipids we can use equation $S_{fw}(q)=\pi\Sigma c_s\langle C\rangle dq^{-4}-\pi\Sigma c_s d^2q^{-2}$ where c_s is the surfactant concentration and measure $S_{fw}(q)$ [3]. In order to do so we need to carry out a contrast variation experiment varying the contrast length density of water by using different proportions of deuterated water. In Figure 4 we show the results of such a contrast variation experiment. At large angles, the interfacial film should appear flat. S_{fw} (q)=-

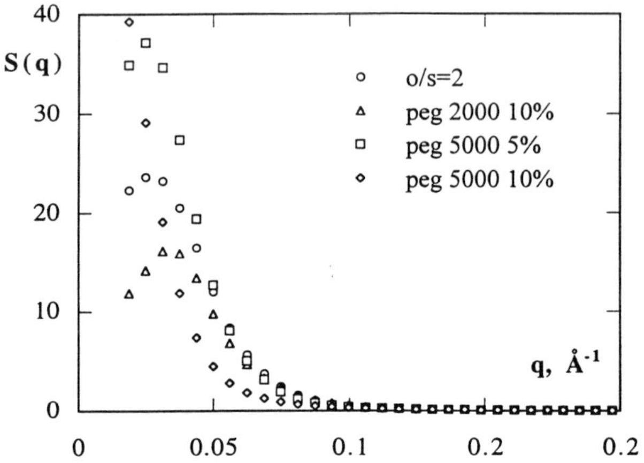

Figure 4 The water film partial structure factor in the representation $q^4S_{fw}(q)$ versus q^2 for different values of peg lipids in monophasic bicontinous phase.

$S_{ff}(q)/2=-\pi\Sigma c_s d^2q^{-2}$. $S_{fw}(q)$ should be negative and proportional to q^{-2}. We also note that there is an another contribution from the curvature $\langle C\rangle$, $S_{fw}(q)=\pi\Sigma c_s\langle C\rangle dq^{-4}$. The samples with a well defined curvature will show a deviation positive or negative from the asymptotic limit q^{-2}. Clearly when we plot $q^4S_{fw}(q)$ versus q^2: for the bicontinous structure we observe a well defined q^{-2} behavior at large q, but with the peg-lipids we see clearly a negative deviation from this behavior. This clearly shows the peg-lipids induce a negative curvature (decrease rigidity [9]) and probably form a dense phase oil in water droplets, the structure perhaps still remains connected. Also, this negative deviation increases with increasing concentration of peg-lipids.

CONCLUSIONS

We have clearly shown that anchored polymers change interaction between microemulsion droplets- it need not be just repulsive but could also be attractive. They induce structural changes or bring about complete phase changes as seen in the Winsor phases when added in sufficiently large amounts. In bicontinous phases the anchored polymers decrease the bending rigidity, the persistence length, increase the apparent average thickness d of the film and in large quantities bring about a negative curvature changes at a semi local scale. Contrary to the naive prediction that anchored polymers should increase membrane rigidity our experiments show a decrease. This is a subtle effect caused by perhaps an indirect coupling between film curvature and concentration fluctuations.

REFERENCES

1. P. G. de Gennes and C. Taupin, J. Phys. Chem. **86**, p. 2294, (1982).

2. W. Helfrich, Z. Naturforsch. Teil C **28**, p.693, (1973).

3. L. Auvray, Ph. D. thesis, Universite de Paris Sud, France, (1985).

4. J. Lal and L. Auvray, J. Phys. II France **4,** p.2119, (1994).

5. P. G. de Gennes, J. Chem. Phys. **94**, 8407, (1990).

6. J. T. Brooks, C. M. Marques and M. E. Cates, Europhys. Lett. **14**(7), p.713, (1991).

7. S. Leibler, J. de Physique **47**, p.507, (1986).

8. W. Helfrich and M. M. Kozlov, J. Phys II, France **3**, p.287, (1993).

9. R. Lipowsky, Europhys. Lett. **30**(4), p.197, (1995).

Part IV

Statistical Mechanics in Biology

Coherent and Incoherent Ripples in Dipalmitoylphosphatidylcholine

P.C. Mason and B.D. Gaulin
Department of Physics and Astronomy
R.M. Epand
Department of Biochemistry
McMaster University,
1280 Main St. West, Hamilton, Ontario, Canada L8S 4M1
and
G.D Wignall and J.S. Lin
Center for Small-Angle Scattering Research
Oak Ridge National Laboratory
Oak Ridge, Tennessee 37831, USA

The rich phase behaviour displayed by phospholipid bilayers and their structural relationship to biological membranes have made them fascinating objects of study. Despite being one of the most often examined of these model membrane systems, dipalmitoylphosphatidylcholine (DPPC) continues to be a source of interest for scientists. In particular, the ripple phase, $P_{\beta'}$, of fully hydrated DPPC has generated a great deal of attention over the years as scientists have tried to understand the structural details of this novel phase[1–5].

An interesting aspect of many phospholipid lamellar systems is the non-equilibrium behaviour they exhibit. Tenchov et al.[6] were the first to report the emergence of non-equilibrium ripple phase in DPPC while recent work by Sun et al.[7] reports non-equilibrium behaviour in the gel ($L_{\beta'}$) phase of longer chain PCs. In this paper, we present neutron and x-ray scattering results examining the history dependent morphology of DPPC, with emphasis on the ripple phase.

We have performed Small Angle Neutron Scattering (SANS) studies on multilamellar vescical (MLV) samples of regular DPPC and samples of DPPC in which the hydrogen atoms in the quatenary ammonium methyl groups have been replaced with deuterium (DPPC-d_9). The SANS experiments were conducted at the W.C. Koehler 30m SANS facility at the Oak Ridge National Laboratory using neutrons with $\lambda = 4.75\AA$ and sample to detector distances of 3.5m and 8.0m.[8] Both samples were placed in a ^2H_2O buffer to increase the contrast between solvent and lipid and to reduce the incoherent scattering from hydrogen.

Figure 1 shows neutron data taken while slowly warming the DPPC-d_9 from 20 °C to 55 °C with a sample to detector distance of 3.5m (q∈[0.02,0.17 $\AA^{-1}$]). The three phases of DPPC are easily identified by the shift in position of the lamellar repeat peak. The inset shows data taken under similar conditions with a sample to detector distance of 8.0m (q ∈ [0.007,0.075 $\AA^{-1}$]). Figure 1 clearly shows that the small angle scattering is sensitive to the phase of DPPC. The small angle scattering (q<0.03 $\AA^{-1}$) doubles upon entering the $P_{\beta'}$ from the $L_{\beta'}$ phase and falls by nearly a factor of four in the L_α phase when compared to the $P_{\beta'}$ phase. The scattering appearing at q∼0.05 $\AA^{-1}$ in the $P_{\beta'}$

Mat. Res. Soc. Symp. Proc. Vol. 463 © 1997 Materials Research Society

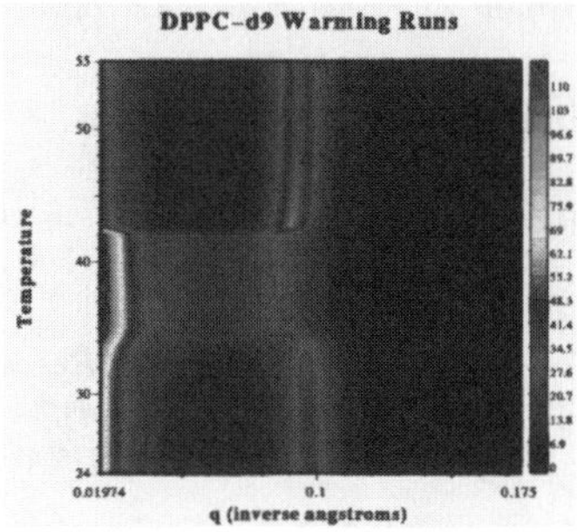

Figure 1: Colour contour plot of DPPC-d_9 warming through its three phases ($L_{\beta'}$,$P_{\beta'}$,L_α) in this temperature range. The plot shows neutron data taken with a sample to detector distance of 3.5m

phase is due to the ripple periodicity. The undeuterated sample have similar scattering profiles but are not shown in this paper.

Figure 2 is similar to figure 1, but shows the neutron profiles of DPPC-d_9 as the sample is slowly cooled. Figure 2 again shows the three distinct phases over this temperature range, but with some noteworthy differences. The scattering due to the ripple periodicity in the $P_{\beta'}$ phase is more prominent and the small angle scattering is markedly different. In the cooling scans, the small angle scattering in the $P_{\beta'}$ phase, while more intense than that of the L_α phase, does not approach the levels attained in the warming scans and reaches its highest intensity in the $L_{\beta'}$ phase.

To compliment the neutron work, x-ray diffraction experiments were conducted using an incident x-ray beam produced with Cu K_α radiation from an 18 kW rotating anode generator which was further monochromated via the (0,0,2) reflection of a pyrolytic graphite monochromator.

Figure 3 shows the diffraction pattern obtained by these two techniques for DPPC warmed into the $P_{\beta'}$ phase. The x-ray data has been scaled so that the data sets overlap. The lamellar repeat peak at q=0.086 $\AA^{-1}$ exhibits the same width in both data sets, indicating that the neutron peak is not resolution limited. The neutron data also show a peak due to the ripple periodicity at q=0.05 $\AA^{-1}$ which does not appear in the x-ray profile.

Comparing these data to the corresponding data taken on cooling the sample into the $P_{\beta'}$ phase from the L_α phase, as shown in figure 4, reveals some interesting contrasts.

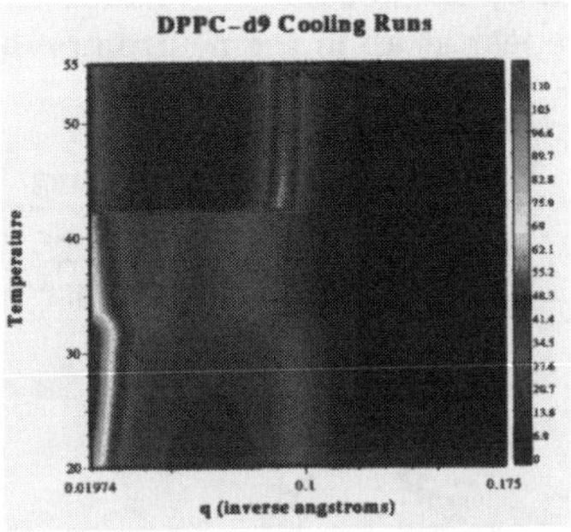

Figure 2: Colour contour plot of DPPC-d_9 cooling through its three phases (L_α,$P_{\beta'}$,$L_{\beta'}$) in this temperature range. The plot shows neutron data taken with a sample to detector distance of 3.5

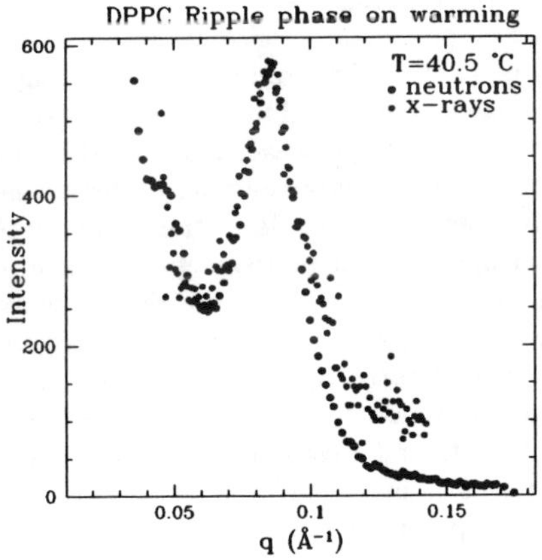

Figure 3: A comparison of neutron and X-ray scattering profiles of DPPC-d_9 in the ripple phase after slow warming from the $L_{\beta'}$ phase.

The neutron peak from ripple periodicity is enhanced in the cooling data and the peak attributed to the lamellar repeat appears broader. The x-ray cooling data shows that this broad peak is, in fact, made up of two closely spaced peaks at q=0.072 $\AA^{-1}$ and q=0.087 $\AA^{-1}$ that are not fully resolved by the neutrons. In addition, the x-rays reveal a peak at q=0.10 $\AA^{-1}$ which does not appear at all in the neutron profile.

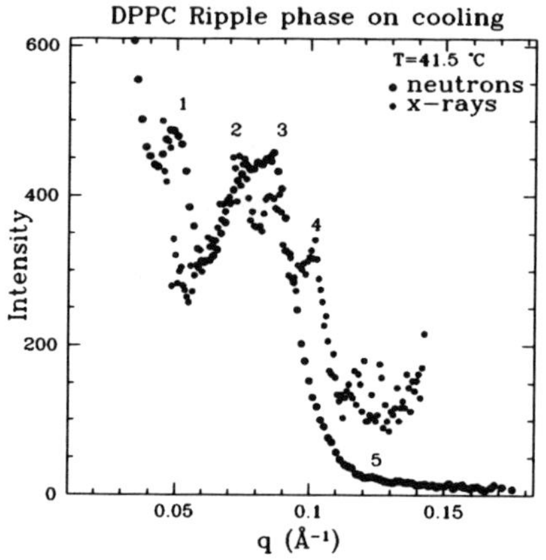

Figure 4: A comparison of neutron and X-ray scattering profiles of DPPC-d_9 in the ripple phase after slow cooling from the L_α phase

We believe the data can be explained as follows. On warming into the $P_{\beta'}$ phase, the formation of ripples in each bilayer is inhibited by the relative "stiffness" of the membrane at these temperatures. The preferred ripple wavelength of ~125 Å is mixed with a variety of longer wavelength ripples leading to a large number of discontinuities and dislocations. This would account for the increased small angle scattering in the the 0.02<q<0.05 region in the ripple phase shown in figure 1 and poorly defined ripple periodicity peak in the neutron data of figure 3. As well, the disorder in the ripples would lead to irregularities in the stacking of the bilayers in the MLV construct, thus accounting for the relatively broad lamellar repeat peak demonstrated by both neutrons and x-rays in figure 3.

On slow cooling into the $P_{\beta'}$ phase from the L_α phase, the fluidity of the bilayers allows them to more easily take their lowest energy arrangement– a single predominant ripple wavelength with each bilayer in the MLV stacked in phase with its neighbouring bilayers. This scenario would explain the more pronounced ripple periodicity peak seen in figures 2 and 4 as well as the relative weakness of the small angle scattering in the $P_{\beta'}$ phase on cooling (cp figs 1 and 2). Furthermore, when cooling into the $L_{\beta'}$ phase this structured $P_{\beta'}$ phase is difficult to erase. Below the pre-transition temperature, the long range order of the ripples is broken but residual ripples remain and are "frozen" into the gel phase, accounting for the *increase* in small angle scattering (figure 2) upon entering the $L_{\beta'}$ phase on cooling. Further evidence for this can be seen by tracing the the path

Peak Number	Peak Label	Calculated($\AA^{-1}$)	Measured($\AA^{-1}$)
1	(01)	—	0.050
2	(10)	—	0.072
3	(11)	0.088	0.087
4	(02)	0.100	0.100
5	(12)	0.123	0.124

Table 1: A comparison of the expected positions of diffraction peaks using our coherent ripple model and the measured positions as shown in figure 4

of the ripple periodicity peak in figure 2 and noting that the peak decreases in q as the pre-transition temperature is approached, and appears to disappear into the small angle scattering at the phase transition.

Finally, one can calculate the expected peak positions for this model of the ripple phase on cooling. Using the labels defined in figure 4, we attribute peak (1) to the ripple periodicity give a ripple wavelength of $\lambda_r = 125\AA$, and peak (2) to lamellar repeat, yielding $d = 87\AA$. The results of these assignments are summarized in table 1. Our value for the lamellar periodicty is significantly higher than the generally accepted value of 72Å obtained when when warming, and the value obtained by Yao et al. on slow cooling.[9]

In conclusion, we believe that, on cooling slowly from the L_α phase, DPPC enters a more ordered $P_{\beta'}$ phase than the $P_{\beta'}$ phase formed on warming. The ordered phase is characterized by a predominant ripple wavelength coupled coherently to ripples throughout the lamelli in the MLV, while in the disordered phase the ripple wavelengths are more loosely defined with no coherence between bilayers even for the majority ripple. The mechanism behind this difference is the fluidity of the L_α phase compared to the relative rigidity of the $L_{\beta'}$ phase as precursors to the $P_{\beta'}$ phase. The softening of the bilayer as it nears the main transition temperature, as well as passage through the anomalous swelling regime could also be important to this effect.[10,11] We have also presented evidence for the non-equilibrium behaviour in the $L_{\beta'}$ phase and the possible existence of ripples frozen into this phase upon cooling.

References

[1] W.-J. Sun, S. Tristram-Nagle, R.M. Suter, and J.F. Nagle, Proc. Natl. Acad. Sci. USA, **93**, 7008 (1996).

[2] H.W. Meyer, Biochim. Biophys. Acta 1302, **138** (1996).

[3] J. Katsaras and V.A. Raghunathan, Phys. Rev. Lett. **74**, 2022 (1995).

[4] H.W. Meyer, B. Dobner, and K. Semmler, Chem. Phys. Lipids **82**, 179 (1996).

[5] T.C. Lubensky and F.C. MacKintosh, Phys. Rev. Lett. **71**, 1565 (1993).

[6] B.G. Tenchov, H. Yao, and I. Hatta, Biophys. J. **56**, 757 (1989).

[7] W.-J. Sun, S. Tristram-Nagle, R.M. Suter, J.F. Nagle, Biochim. Biophys. Acta **1279**, 17 (1996)

[8] W.C. Koehler, Physica (Utrecht), **137B**, 320 (1986).

[9] H. Yao, S. Matuoka, B. Tenchov, and I. Hatta, Biophy. J. **59**, 252 (1991).
[10] E. Evans and R. Kwok, Biochemistry **21**, 4874 (1982).
[11] T. Honger, K. Mortensen, J.H. Ipsen, J. Lemmich, R. Bauer, and O.G. Mouritsen, Phys. Rev. Lett. **72**, 3911 (1994).

SIZE DISTRIBUTIONS OF FLUID MEMBRANE VESICLES FAR FROM EQUILIBRIUM

LEONARDO GOLUBOVIĆ and MIRJANA GOLUBOVIĆ
Department of Physics, West Virginia University, Morgantown, WV 26506

ABSTRACT

We investigate nonequilibrium behavior of a polydisperse ensemble of fluid membrane vesicles by means of a diffusive Boltzmann transport equation which incorporates vesicle diffusion and the reactions between vesicles. This approach is used to study the time evolution of the size distribution of an initially monodisperse vesicle ensemble and its interesting properties such as the internal aqueous, encapsulated volume. We investigate various nonequilibrium paths such ensembles may follow during the equilibration process.

INTRODUCTION

In recent years significant attention has been devoted to the equilibrium statistical mechanics of fluid membranes [1], and their phases [2]-[8]. However, many interesting phenomena involving membranes are non-equilibrium in nature. Technologically important example are liposomes which are potential vehicles for transporting therapeutic and diagnostic agents [9]. They are vesicles formed by bilayers contaning amphiphilic substances like phospholipids dispersed in water. Vesicles such as liposomes have size distributions typically evolving due to reactions between vesicles.

In thermodynamic equilibrium, mono-lamellar vesicles may form isotropic, liquid-like polydisperse droplet phases, which polydispersity properties have been recently investigated experimentally,[10] and theoretically [5]-[8]. These are so-called entropically stabilized vesicles, with the simple membrane curvature free energy, $F = \int dS[\frac{\kappa}{2}H^2 + \overline{\kappa}G]$, where H, and G are, respectively, membrane mean and Gaussian curvature, and κ and $\overline{\kappa}$ are membrane bending and saddle splay rigidity. Here we investigate various nonequilibrium paths to reach the thermodynamic equilibrium in these vesicle phases starting from a nonequilibrium, monodisperse distribution of vesicle sizes.

DIFFUSIVE BOLTZMANN EQUATION FOR REACTING VESICLES

A dilute nonequilibrium polydisperse ensemble of nearly spherical vesicles can be described by a vesicle density $\rho(A, \mathbf{x}, t)$, such that $\rho(A, \mathbf{x}, t)dAd^3x$ is the number of vesicles of the area $A = 4\pi R^2$ in the interval $(A, A + dA)$ contained in a volume element d^3x. A vesicle of area A is free to diffuse and undergo fusions with other vesicles or to split into smaller vesicles. Thus, the vesicle ensemble is a diffusion-reaction system where reactions are vesicle fusion and fission processes preserving total area of vesicles (i.e., total amount of the membrane material). Thus, the time evolution of $\rho(A, \mathbf{x}, t)$ can be described by a diffusive Boltzmann transport equation (TE) of the form

$$\frac{\partial}{\partial t}\rho(A, \mathbf{x}, t) = D(A)\Delta_{\mathbf{x}}\rho(A, \mathbf{x}, t) + R_a + R_b + R_c + R_d, \tag{1}$$

Mat. Res. Soc. Symp. Proc. Vol. 463 © 1997 Materials Research Society

where the first term is ordinary diffusion with the size dependent diffusion constant $D(A) = k_B T/6\pi\eta(A/4\pi)^{1/2}$, according to the Einstein-Stokes law, with η, the viscosity of the solvent. R_a to R_d in (1) are reaction rates associated with vesicle fusions [R_a and R_b], and fissions [R_c and R_d]. If $\rho(A, \mathbf{x}, t)$ is slowly varying in space, these rates can be generally written in the form

$$R_a = -2\int_0^\infty dA_1\ \Gamma(A_1, A)\ \rho(A_1, \mathbf{x}, t)\ \rho(A, \mathbf{x}, t), \tag{2a}$$

$$R_b = \int_0^A dA_1\ \Gamma(A_1, A - A_1)\ \rho(A_1, \mathbf{x}, t)\ \rho(A - A_1, \mathbf{x}, t), \tag{2b}$$

$$R_c = 2\int_0^\infty dA_1\ \Pi(A_1, A)\ \rho(A + A_1, \mathbf{x}, t), \tag{2c}$$

$$R_d = -\int_0^A dA_1\ \Pi(A_1, A - A_1)\ \rho(A, \mathbf{x}, t), \tag{2d}$$

with

$$\Gamma(A_1, A) \approx p_{fus}\frac{k_B T}{\eta}\left[2 + \left(\frac{A}{A_1}\right)^{1/2} + \left(\frac{A_1}{A}\right)^{1/2}\right],$$

where p_{fus} is the probability that an encounter between the vesicles leads to a fusion [11]. Reaction kernels Γ and Π in Eqs. (2) are related by the detailled balance between fusions and fissions in the thermodynamic equilibrium. This gives the condition

$$\Pi(A_1, A_2) = \frac{\rho_{eq}(A_1)\ \rho_{eq}(A_2)}{\rho_{eq}(A_1 + A_2)}\Gamma(A_1, A_2), \tag{3}$$

where $\rho_{eq}(A)$ is the equilibrium vesicle density. For example, for the *entropically stabilized* vesicles [5][6]

$$\rho_{eq}(A) = \frac{C}{A^{5/2}}\left(\frac{A}{A_{min}}\right)^{4/3} e^{-A/A_{max}}, \tag{4}$$

with A_{min}, the minimum area of a vesicle; $C = const.(\kappa/k_B T)^2 exp(-E_b(A_{min})/k_B T)$, and $E_b(A_{min}) = 8\pi\kappa + 4\pi\bar{\kappa}$ the curvature energy of the smallest vesicles (with the area= A_{min}). The power law prefactor $(A/A_{min})^{4/3}$ in ρ_{eq} arisis from the dependence of the membrane curvature rigidities κ and $\bar{\kappa}$ on vesicle size,[12] whereas the prefactor $C/A^{5/2}$ originates from the entropy of vesicle's collective modes [5][6]. A_{max} in (4) is, effectively, the area of the largest vesicles present in the system. It is determined by the total amount of the membrane material, i.e, total membrane area A_{tot} present in the system

$$A_{tot} = \int d^3x\ dA\ A\ \rho(A, \mathbf{x}, t). \tag{5}$$

Vesicle fusions and fissions processes included in our TE (1) preserve A_{tot} and drive the density $\rho(\mathbf{x}, A, t)$ towards the equilibrium density (4) at long times. We remark that the actual value of A_{max} in (4) does not explicitely enter the TE (1) [see Eqs. (3) and (4)].

EVOLUTION OF INITIALLY MONODISPERSE VESICLES

Here we discuss evolution of a spatially uniform, initially monodisperse ensemble of vesicles. Let n_o be the density of these vesicles all initially, at $t = 0$, having the same area A_o. Thus $\rho(A, \mathbf{x}, t = 0) = n_o\delta(A - A_o)$. The TE (1) will drive ρ to an equilibrium distribution of the form (4) with A_{max} determined by the conservation of A_{tot}, Eq. (5). This conservation law gives A_{max} through the equation $\Phi_A = n_o A_o = \int dA\ A\ \rho_{eq}(A)$, (Φ_A, membrane area per unit volume) yielding $A_{max} = \left(\frac{\Phi_A A_{min}^{4/3}}{C}\right)^{6/5} = A_o\left(\frac{n_0}{n^*}\right)^{6/5}$, with

$$n^* = \frac{C}{(A_o)^{1/6} A_{min}^{4/3}}. \tag{6}$$

The volume fraction encapsulated by vesicles, $\Phi_V(t) = \frac{1}{3(4\pi)^{1/2}} \int dA\ A^{3/2}\ \rho(A,t)$, and the number of vesicles per unit volume $n_v(t) = \int dA\ \rho(A,t)$, evolve from their initial values $\Phi_V(0) = \frac{1}{3(4\pi)^{1/2}} n_o A_o^{3/2}$, and $n_v(0) = n_o$, to their equilibrium values

$$\Phi_V(\infty) = \Phi_{V,eq} = \frac{A_{min}^{4/5}\Phi_A^{8/5}}{C^{3/5}} = \frac{A_{min}^{4/5}(n_o A_o)^{8/5}}{C^{3/5}} = \Phi_V(0)\left(\frac{n_0}{n^*}\right)^{3/5}, \tag{7}$$

and

$$n_v(\infty) = n_{v,eq} = \frac{6C}{A_{min}^{3/2}}\left[1 - \left(\frac{A_{min}}{A_{max}}\right)^{1/6}\right] = \frac{6C}{A_{min}^{3/2}}\left[1 - \left(\frac{C}{n_o A_o A_{min}^{1/2}}\right)^{1/6}\right]. \tag{8}$$

Above and hereafter, we assume that membrane area density $\Phi_A = n_o A_o$ is well above the critical vesicle concentration, $\Phi_{A,cvc} = n_{o,cvc} A_o = \frac{C}{A_{min}^{1/2}}$. This ensures that $A_{max} \gg A_{min}$. We remark that $\Phi_{V,eq} < 1$ throughout the fluid-like vesicle phases, whereas $\Phi_{V,eq} \approx 1$ at the first order transition from the vesicle to the lamellar phase [6],[7],[8]. So, at the transition, $\Phi_{A,crit} = n_{o,crit} A_o \approx \frac{C^{3/8}}{A_{min}^{1/2}}$. Thus, for a given initial vesicle size A_o, previous equations define several characteristic vesicle densities: $n_{o,cvc}$, n^*, $n_{o,crit}$, and $n_{v,eq}$. Finally, as it must be the case that $\Phi_V(0) < 1$, it follows that $n_o < n_{o,max} \approx A_o^{-3/2}$. As $A_o > A_{min}$, one has $n_{v,eq} > n^* > n_{o,cvc}$. The existence of several characteristic scales for n_o implies the existence of several types of the equilibration behavior discussed in the following.

Type I equilibration. Initial vesicle density is above its equilibrium value, $n_o \gg n_{v,eq}$. Then, also, $n_o \gg n^*$, and, thus, $A_o \ll A_{max}$, and $\Phi_V(0) \ll \Phi_{V,eq}$. Thus, the equilibration must be dominated by vesicle fusion processes [R_a and R_b terms of the TE (1)], which decrease the number density of vesicles $n_v(t)$, and, as $A_1^{3/2} + A_2^{3/2} < (A_1 + A_2)^{3/2}$, increase the encapsulated volume fraction $\Phi_V(t)$. This is illustrated in Fig. 1, obtained by numerically solving the TE (1). We see that $\Phi_V(t)$ grows as $t^{1/2}$ until it saturates to its equilibrium value $\Phi_{V,eq}$ at times longer than some equilibration time-scale t_{eq}. This growth can be understood analytically, as for $t \ll t_{eq}$, one can ignore the fissional terms in the TE, i.e., $\frac{\partial}{\partial t}\rho(A,t) \approx R_a + R_b$. This equation has an automodel solution of the form $\rho(A,t) = \Phi_A A(t)^{-2} f(A/A(t))$ [with $\int dx\, x f(x) = 1$], characterized by a growing vesicle area scale $A(t) = A(0) + \Phi_A \Gamma_o t$, with $\Gamma_o = p_{fus} k_B T/\eta$. Thus, for $t \ll t_{eq}$, the typical vesicle size, $R \sim A(t)^{1/2}$, grows as $t^{1/2}$, whereas $\Phi_V(t) = const.\Phi_A[A(t)]^{1/2} \sim t^{1/2}$, and $n_v(t) = const.\Phi_A/A(t) \sim t^{-1}$. For $t \approx t_{eq}$, the area scale $A(t)$ reaches A_{max}, $A(t_{eq}) \approx A_{max}$, or, equivalently, $\Phi_V(t_{eq}) \approx \Phi_{V,eq}$. This yields

$$t_{eq} = \frac{A_{min}^{8/5}}{C^{6/5}\Gamma_o}(\Phi_A)^{1/5}, \tag{9}$$

for $n_o < n_{o,crit}$, i.e., $\Phi_A < \Phi_{A,crit}$, when the equilibrium state is the dilute vesicle phase. On the other side, if $n_o > n_{o,crit}$, the equilibrium state is the lamellar phase and vesicles will evolve via fusions until the encapsulated volume fraction becomes $O(1)$. At that time scale the monolamellar vesicles will transform into a multilamellar longlived vesicles (corresponding to confocal defects of a Smectic-A phase). Thus, the equilibration time scale (from mono- to multilamellar vesicle state) can be estimated from $\Phi_V(t_{eq}) \approx 1$, yielding

$$t_{eq} = \frac{1}{\Gamma_o}(\Phi_A)^{-3}, \tag{10}$$

for $\Phi_A > \Phi_{A,crit}$. By Eqs. (9) and (10), t_{eq} has a maximum for $\Phi_A \approx \Phi_{A,crit}$, i.e., at the first order transition from the droplet to the lamellar phase at which $t_{eq} \approx A_{min}^{3/2}/C^{9/8}\Gamma_o$.

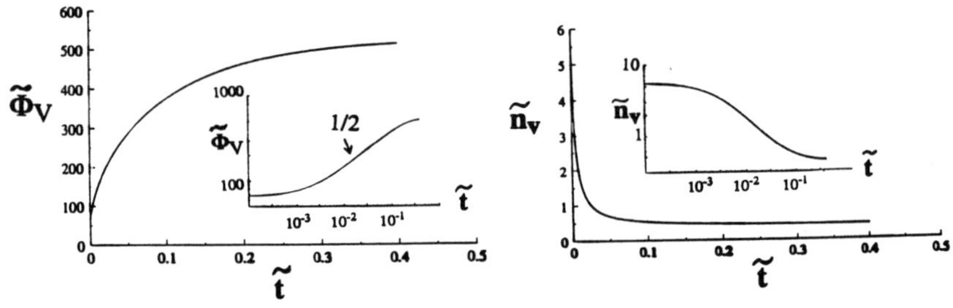

FIG. 1: Example of the Type I equilibration. Here, and in the figures 2 and 3, $\tilde{t} = t/[t]$, $\tilde{\Phi}_V(\tilde{t}) = \Phi_V(t)/[\Phi_V]$, and $\tilde{n}_v(\tilde{t}) = n_v(t)/[n_v]$, with $[t] = A_{min}^{3/2}/C\Gamma_o$, $[\Phi_V] = C$, and $[n_v] = C/A_{min}^{3/2}$. In this example $\tilde{A}_o = A_o/A_{min} = 5$, and $\tilde{n}_v(0) = 60$.

Type II equilibration. Initial vesicle density n_o is in the range $n_{o,cvc} \ll n_o \ll n^*$ (then, also, $n_o \ll n_{v,eq}$). Then $A_o \gg A_{max}$, and $\Phi_V(0) \gg \Phi_{V,eq}$. Thus, the equilibration must be dominated by vesicle fission processes [R_c and R_d terms of the TE (1)], which increase the number density of vesicles $n_v(t)$, and, as $A_1^{3/2} + A_2^{3/2} < (A_1 + A_2)^{3/2}$, decrease the encapsulated volume fraction $\Phi_V(t)$. This is illustrated in Fig. 2, obtained by numerically solving the TE (1).

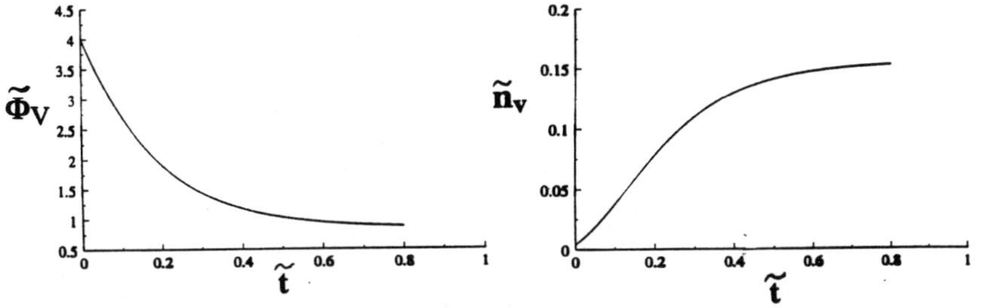

FIG. 2: Example of the Type II equilibration. Here $\tilde{A}_o = 100$, and $\tilde{n}_v(0) = 0.042$.

We see that $\Phi_V(t)$ decreases, whereas the vesicle density $n_v(t)$ increases, until they saturates to their equilibrium values at times longer than some equilibration time-scale t_{eq}. This growth can be understood analytically, as for $t \ll t_{eq}$, one can ignore the fusional terms in the TE, i.e., $\frac{\partial}{\partial t}\rho(A,t) \approx R_c + R_d$. By analysing this equation, we find that initially monodisperse size distribution gets replaced by a strongly polydisperse distribution of the form $\rho \sim A^{-11/6}$, for $A < A(t)$, and $\rho \approx 0$, for $A > A(t)$. Here, $A(t) = A_o\left[1 - \frac{t}{t_{eq}}\right]^6$, with

$$t_{eq} = \frac{A_{min}^{4/3}}{C\Gamma_o} A_o^{1/6} \sim R_o^{1/3}. \tag{11}$$

Thus, the equilibration is dominated by the fissional decay of vesicles which occurs in a *finite time* proportional to $R_o^{1/3}$, where R_o is the initial vesicle size. The above results for $\rho(A,t)$ and t_{eq}, Eq. (11), are consequences of the particular form of the equilibrium vesicle distribution, Eq.(4) [which enters only the fissional part of the TE]. Thus, they reflect the lengthscale dependence of membrane bending and saddle splay rigidity on the vesicle size [12], as well as entropy of vesicle collective degrees of freedom incorporated in ρ_{eq}, Eq. (4).

Type I/II equilibration. Initial vesicle density n_o is between n^* and $n_{v,eq}$ ($n^* \ll n_o \ll n_{v,eq}$). By numerically solving the TE (1), we obtain $\Phi_V(t)$ and $n_v(t)$ in Fig. 3.

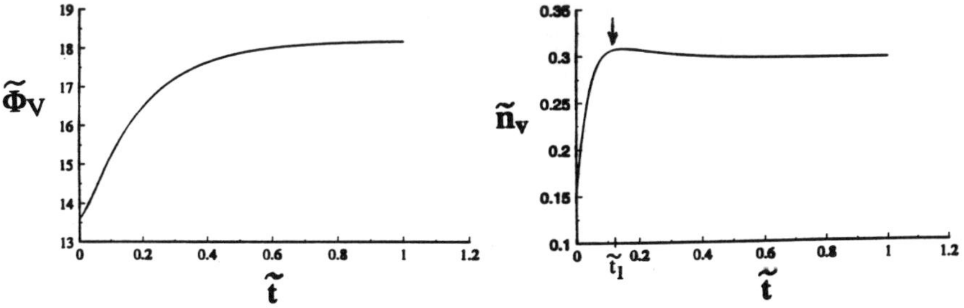

FIG. 3: Example of the Type I/II equilibration. Here $\tilde{A}_o = 20$ and $\tilde{n}_v(0) = 1.6$.

As $n_o < n_{v,eq}$, the vesicle density $n_v(t)$ increases untill it saturates to $n_{v,eq}$. In this respect, this equilibration is simillar to the type II equilibration. As there, this increase of $n_v(t)$ is produced here by the fissional terms of the TE. However, as $n^* < n_o$ here, one has $A_o < A_{max}$, and $\Phi_V(0) < \Phi_{V,eq}$. Thus, the encapsulated volume fraction $\Phi_V(t)$ increases until it saturates to $\Phi_{V,eq}$. In this respect, this equilibration is similar to the type I equilibration. As there, this increase of $\Phi_V(t)$ is produced here by the fusional terms of the TE. From Fig. 3, we see that the fissional increase of $n_v(t)$ produces an ***over-shoot*** above the equilibrium value, yielding a maximum of $n_v(t)$ at some characteristic time scale t_1. For $t < t_1$, the fission produced vesicles are small, and do not significantly affect the encapsulated volume fraction which continuously increases (see Fig. 3) due to fusions of the largest vesicles which contribute most to Φ_V. Fusions, however, significantly less affect the number density of vesicles n_v. For example, after the over-shoot in n_v at $t = t_1$, Fig. 3, there is only a small decrease of $n_v(t)$, which is due to fusions of the largest vesicles in the tail of the distribution. Thus, most of the production of n_v is due to fissions and it is practically over already at the fissional time-scale $t \sim t_1 \approx \frac{6A_{min}^{4/3}}{C\Gamma_o} A_o^{1/6}$, as in the type II equilibration, Eq. (11). At this time-scale, fusions still go on and increase Φ_V, see Fig. 3. Thus, the ultimate equilibration time t_{eq} is determined by fusions. t_{eq} is thus given by the fusional time scale $t_{eq} = \frac{A_{min}^{8/5}}{C^{6/5}\Gamma_o}(\Phi_A)^{1/5}$, as in the type I equilibration, Eq. (9). This t_{eq} is larger than the fissional time-scale t_1 as $t_1/t_{eq} \approx (n^*/n_o)^{1/5} < 1$ here.

ACKNOWLEDGMENTS

This work is supported by a Mylan Laboratories research grant.

REFERENCES

1. Statistical Mechanics of Membranes and Surfaces, edited by D. R. Nelson, T. Piran, and S. Weinberg (World Scientific, Singapore, 1989).
2. S. A. Safran et al., Phys. Rev. Lett. **57**, 491 (1986).
3. L. Golubović and T. Lubensky, Phys. Rev. B**39**, 12110 (1989).
4. L. Golubović and T. Lubensky, Phys. Rev. A**41**, 4343 (1990).
5. D. A. Huse and S. Leibler, J. Phys. (France) **49**, 605 (1988).
6. D. Morse and S. Milner, Europhys. Lett **26**, 565 (1994), and Phys. Rev. E**52**, 5918 (1995).
7. L. Golubović, Phys. Rev. E**50**, R2419 (1994).
8. D. Morse, Phys. Rev. E**50**, R2423 (1994)
9. Liposome technology, edited by G. Gregoriades (CRC Press, Boca Raton, 1984).
10. P. Hervé et al., J. Phys. II (France) **3**, 1255 (1993).
11. L. Golubović, in Disordered Materials and Interfaces, edited by H. Z. Cummins et al., MRS Symposium Proceedings, Vol. **407**, 275 (1996).
12. See, F. David, in Ref. 1.

Conformational and Dynamic Properties of Polymer Loops and Their Mixtures at an Impenetrable Interface

H.S. Gulati†, D.C. Driscoll ‡, R.L. Jones‡, R.J. Spontak‡ and C.K. Hall†
Departments of †Chemical Engineering and ‡Materials Science and Engineering
North Carolina State University, Raleigh, NC 27695

ABSTRACT

In this study we investigate the equilibrium conformational properties and dynamic relaxation behavior of polymer loops grafted at an interface using the discontinuous molecular dynamics simulation technique. Differences and similarities between the structural and dynamic properties of polymer loops and tails are identified. The conformational properties of mixtures of polymers loops and tails are also studied using the bond-fluctuation method. The effect of mixture composition on the conformational properties of the individual components in the mixture is discussed.

INTRODUCTION

Polymers adsorbed or grafted to an interface are of critical importance in technologies as varied as adhesive/protective coatings, antithrombogenic biomaterials, colloidal stabilization and polymer blend compatibilizers.[1] The dynamic behavior of grafted polymer layers has also become a subject of great interest due to its relevance in the processing of grafted layers. Most studies of grafted polymer layers have concentrated on elucidating the properties of layers of chains grafted at one end (polymer "tails"). Extensive theoretical formalisms[2–4] have been proposed, and simulation studies[5–7] conducted, to address the structural and dynamic properties of polymer tails. While single-tethered polymer chains have received considerable attention, other tethered conformations are also of practical and fundamental interest. In particular, double-tethered chains (or polymer "loops") provide an additional opportunity to tailor the properties of thin polymer coatings for specific applications.

In this work, we analyze the equilibrium conformational and dynamic relaxation properties of looped chain molecules grafted at an impenetrable interface by applying the discontinuous molecular dynamics (DMD) simulation technique. By comparing our simulation results for loops to previously available data for tails, we are able to discern the similarities and differences between polymer loops and tails, and identify the role of a second tethered site on conformational and dynamic properties. We have also studied the structural properties of grafted layers consisting of mixtures of polymer tails and loops using the bond-fluctuation (BF) simulation technique. The effect of mixture composition on grafted layer properties, as well as on the structural properties of the individual component has been ascertained.

SIMULATION OVERVIEW

The conformational and dynamic properties of looped chain molecules are obtained using the DMD simulation technique. We limit this study to athermal cases and assume that there are no attractive monomer-monomer or monomer-wall interactions. In DMD simulations, looped polymers are modeled as chains of N tangent hard spheres, tethered at both ends to the surface. The chains are simulated us-

Mat. Res. Soc. Symp. Proc. Vol. 463 © 1997 Materials Research Society

ing the Rapaport[8] algorithm. Figure 1 depicts a typical conformational ensemble containing 20 looped chains of 40 spheres each.

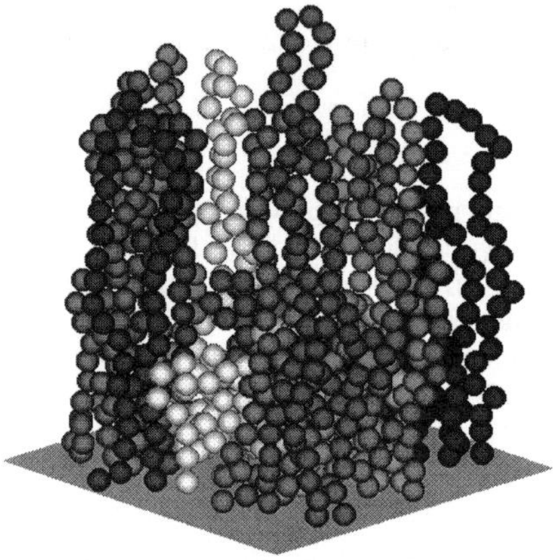

Fig. 1 Typical initial configuration of 20 hard sphere double-tethered chains of length N=40 at an anchor density σ=0.25.

The molecular dynamics algorithm involves calculating the order in which particles collide with each other, advancing the system to the impending collision, and calculating the post-event velocities of the participating segments using Newton's equations of motion. Precise details of the DMD technique are discussed elsewhere.[9,10] The DMD simulations are performed for chains of length $N = 20$, 40 and 80. Surface densities σ range from 0.01 to 0.48, where σ is defined as the fractional surface coverage:

$$\sigma = \frac{\pi}{4}\frac{N_a d^2}{A}. \tag{1}$$

Here, N_a is the number of anchor sites, d is the diameter of the particles, and A denotes the area of the impenetrable surface (at $z = 0$) in the primary simulation cell.

The conformational properties of mixtures of loops and tails are studied using the BF simulation method. Unlike the DMD approach, BF simulations are conducted on a 3-D cubic lattice (a general description of the BF algorithm is provided by Deutsch and Dickman[11]). The BF algorithm used in this study is identical to the one used earlier by Jones and Spontak[12] to study the conformational properties of looped chain layers.

RESULTS

Conformational Properties of loops

The equilibrium conformational properties of grafted chains yield important information about the structure of the chains in the layer, as well as layer characteristics such as the effective surface coverage and layer height. The average height h of a grafted layer can be quantified in terms of the normal component of the radius of gyration, $\langle R_{gz}^2\rangle^{1/2}$. Classical SCF theory for polymer tails[4] predicts that the height of a grafted polymer tail layer scales as $h \sim N\sigma^{1/3}$. In Fig. 2, $\langle R_{gz}^2\rangle^{1/2}/\sigma^{1/3}$ from both DMD simulations of polymer loops and SCF predictions for polymer tails of $N/2$ repeat units are plotted as a function of anchor density.

As evident in this figure, two scaling regimes are clearly discernible. At intermediate to high densities, $\langle R_{gz}^2\rangle^{1/2}$ scales as $\sigma^{1/3}$. This region is termed the "scaling" regime, where the observed scaling behavior of polymer loops is consistent with SCF predictions for polymer tails of $N/2$ repeat units. At low σ, however, $\langle R_{gz}^2\rangle^{1/2}/\sigma^{1/3}$ scales as $\sigma^{-1/3}$; i.e., $\langle R_{gz}^2\rangle^{1/2}$ does not change with increasing σ. This regime is termed the "mushroom" regime, where chains are isolated and assume an inverted mushroom shape, thus covering the surface in patches. The layer height in this regime scales only with chain length ($h \sim N^{3/5}$).

This density-induced transition in the scaling behavior of grafted layer height from the mushroom to scaling regime is referred to as "chain impingement."[12] The chain impingement density (σ_c) can be gleaned from Fig. 3, which shows the radius of gyration of looped chains of various lengths as a function of anchor density. Initially, as σ is increased, R_g decreases, indicating that the chains assume more spherical conformations. As σ increases further, neighboring chains eventually begin to touch each other at the so-called impingement density.

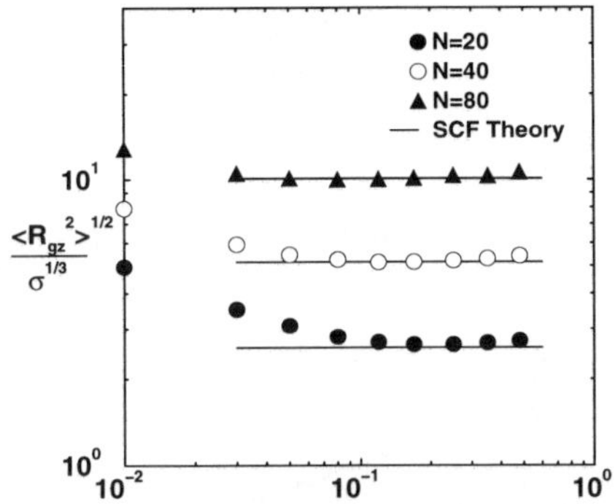

Fig. 2. Density scaling behavior of reduced layer heights, expressed in terms of $<R_{gz}^2>^{1/2}/\sigma^{1/3}$. Predictions from SCF theory for tails with $N/2$ segments are shown as solid lines.

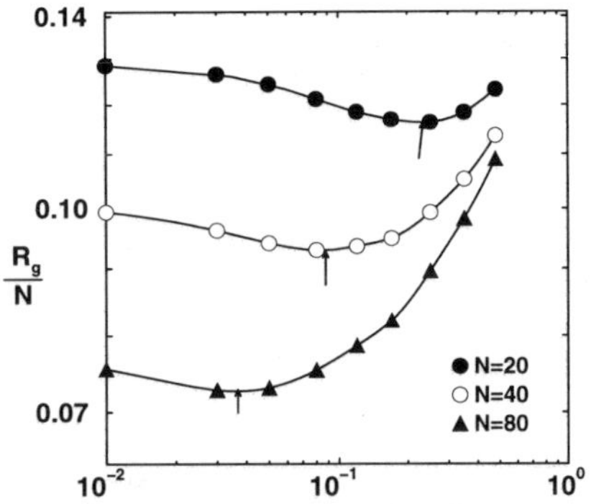

Fig. 3. Normalized radius of gyration, R_g/N, of looped chains as a function of σ. Arrows indicate the minima in the curves which correspond to the impingement density for a given N.

Mixtures of Loops and Tails

A mixture of double-tethered polymer loops and single-tethered tails represents the case of dual-end-functionalized chains in which only a fraction of the chains manage to attach to the interface at both ends. We investigate the effect of composition on the height of the loop layer in a tail/loop mixture where both chain species possess an identical number of repeat units. In Fig. 4, $< R_{gz}^2 >_{loop}^{1/2} /\sigma^{1/3}$ is presented as a function of anchor density for several loop compositions, X_l.

As seen in Fig. 4, $< R_{gz}^2 >_{loop}^{1/2}$ increases with increasing X_l. Tails tend to suppress the loop layer below, forcing the loops closer to the surface. As a result, $< R_{gz}^2 >_{loop}^{1/2}$ decreases with increasing tail composition. We also note that the high-density SCF scaling behavior observed in Fig. 2 for pure loops is also valid for loops in tail/loop mixtures; i.e., $<R_{gz}^2>_{loop}^{1/2} \sim \sigma^{1/3}$ at intermediate to high densities. One interesting difference, however, is that the onset of the scaling regime for loops occurs at lower density as X_l is decreased. Since loops have two tethered sites, the effective density per chain for loops is higher than that of tails. Therefore, with increasing loop composition, the effective density at the surface increases, resulting in the onset of the scaling regime at lower anchor density.

Dynamic Properties of Loops

The dynamic behavior of grafted polymer layers is of considerable interest as it provides a microscopic understanding of the internal stress relaxation process, as well as of the dynamic response of such layers to external shear deformation. The relaxation behavior of polymer loops is evaluated by monitoring the time correlation functions of the radius of gyration (C_{R_g}). This time autocorrelation function is given by

$$C_{R_g}(t) = \frac{\langle R_g(t)R_g(0)\rangle - \langle R_g\rangle^2}{\langle R_g^2\rangle - \langle R_g\rangle^2}. \tag{2}$$

The dependence of C_{R_g} on time for looped chains at anchor densities ranging from 0.05 to 0.48 is presented in Fig. 5. Simulation times are presented in reduced units defined as

$$t^* \equiv t(k_BT/md^2)^{1/2}, \tag{3}$$

where t is the simulation time, k_B is the Boltzmann constant, T denotes absolute temperature, and m and d are the mass and diameter, respectively, of each segment.

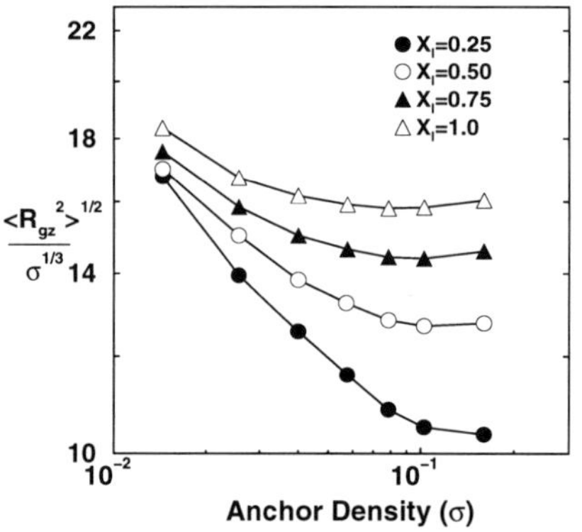

Fig. 4. Effect of composition on the height of loop layer ($< R_{gz}^2 >_{loop}^{1/2}$) in a tail/loop mixture.

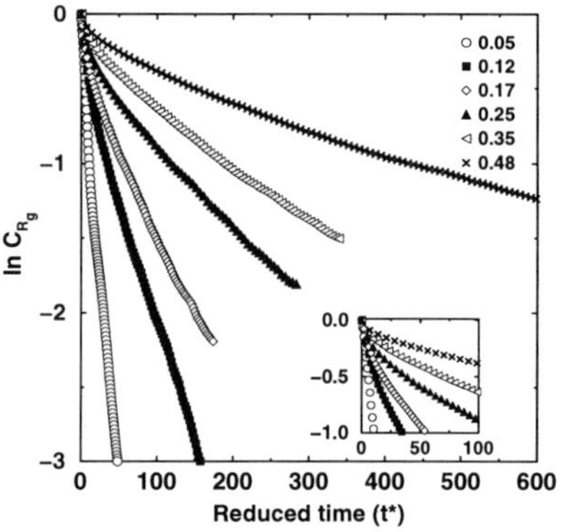

Fig. 5. Radius of gyration time autocorrelation function C_{R_g} for loops with N=40. The inset shows the non-exponential decay of C_{R_g} at small t^*.

The decay of the autocorrelation function displays two regimes. At very short times, interactions are predominantly *intra*chain as chains do not feel the presence of other chains in the vicinity and are highly mobile in their localized pocket. In this regime, C_{R_g} exhibits a rapid non-exponential decay as shown in the inset of Fig. 5. At longer times, as chains diffuse out of their local pockets, *inter*chain interactions slow down the motion of the chain, hence slowing the decay of C_{R_g}. As might be expected, the region of initial non-exponential decay becomes smaller as σ increases, since the local pocket of a chain shrinks in size and interchain interactions occur at earlier times.

From the autocorrelation function data presented in Fig. 5 we can calculate the average relaxation time τ for looped chains using the techniques described by Murat and Grest.[13,9] Relaxation times are obtained for looped chain layers composed of N=20, 40, and 80 repeat segments at various densities. Figure 6 presents a double-logarithmic plot of $\tau/\sigma^{4/3}$ as a function of σ. As indicated in the figure, the relaxation time τ scales as $\tau \sim \sigma^{4/3}$. The relaxation times are fitted to an expression of the form $\tau \sim N^a \sigma^b$ to obtain

$$\tau_{loop} \sim N^3 \sigma^{4/3} \tag{4}$$

The relaxation time of polymer tails obtained from previous BF simulations[14] and theoretical frameworks[15] is found to scale as $\tau_{tail} \sim N^3 \sigma^{2/3}$. Thus, we find that the scaling dependence of loop relaxation time on chain length ($\tau \sim N^3$) is identical to the scaling behavior of polymer tails. The density dependence of loop scaling behavior, however, is much stronger (squared) compared to that of tails. We are presently unable to provide a theoretical explanation for this behavior. While an explanation for this difference in the density scaling behavior of loops and tails will have to await a more detailed theoretical analysis, the doubled exponent in density scaling of loops versus tails does offer some interesting speculations about the scaling behavior of chains with multiple grafted sites (i.e., polymer "trains").

The lateral self-diffusion coefficients are obtained from the average mean square displacement of the centers of mass of the looped chains.[16] Our results are presented in terms of the dimensionless reduced lateral diffusivity (D^*), defined as $D^*=D/[k_BTd^2/m]^{1/2}$. Figure 7 shows the dependence of D^* on σ for looped chains with 20, 40, and 80 repeat units. In order to highlight the scaling behavior of D^* at low density, $D^*\sigma^{4/3}$ is plotted as a function of σ.

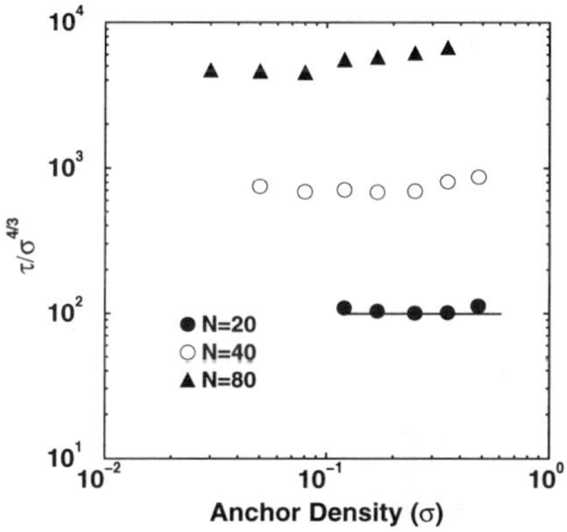

Fig. 6. Variation of relaxation time τ with anchor density. We plot $\tau/\sigma^{4/3}$ as function of σ to show the scaling behavior of τ at low density.

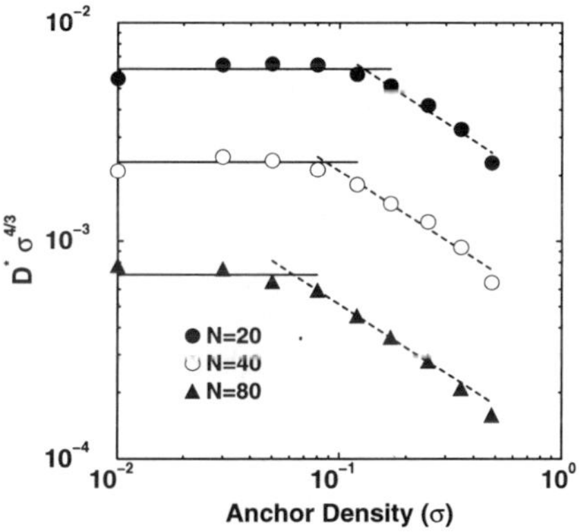

Fig. 7. Variation of reduced lateral diffusivity D^* with density. We plot $D^*\sigma^{4/3}$ as function of σ to show the scaling behavior of D^* at low density.

As in Fig. 2, two different scaling regimes are clearly discernible in Fig. 7. At low densities, lateral diffusivity scales with density as $D^* \sim \sigma^{-4/3}$. As density increases beyond the point where chain impingement occurs, the scaling behavior undergoes a gradual transition to a region where $D^* \sim \sigma^{-2}$. The scaling dependence of D^* on chain length also goes through a smooth transition from $D^* \sim N^{-4/3}$ at low densities to $D^* \sim N^{-2}$. As was the case for scaling behavior of τ, D^* for looped chains exhibits a much stronger σ dependence ($D^* \sim \sigma^{-4/3}$ to σ^{-2}) than that exhibited by polymer tails ($D^* \sim \sigma^{-2/3}$).[14]

CONCLUSIONS

We have studied the structural and dynamic properties of grafted polymer loops and mixtures of loops and tails. Comparing the equilibrium conformational properties of polymer loops and tails, we find that polymer loops of N repeat units are structurally similar to polymer tails of $N/2$ repeat units. However, comparing the scaling relationships of the dynamic properties of looped chains to those of polymer tails, we observe that despite the presence of the second anchor, the chain-length dependence of relaxation time and lateral diffusivity of polymer loops and polymer tails is found to be similar. In contrast, the dependence of relaxation time and lateral diffusivity of polymer loops on density is much stronger (squared) than that of polymer tails.

ACKNOWLEDGMENT

This work was supported by the Director, Office of Energy Research, Office of Basic Sciences, Chemical Sciences Division of the U.S. Department of Energy under Contract No. DE-FG05-91ER1481. Partial support is also provided by the Donors of the Petroleum Research Fund, administered by the American Chemical Society.

REFERENCES

1. S. Wu, Polymer Interfaces and Adhesion. (Dekker, New York, 1982).
2. S. Alexander, J. Phys. (Paris), **38**, 983 (1977).
3. P. G. de Gennes, Macromolecules, **13**, 1070 (1980).
4. S. T. Milner, T. A. Witten, and M. E. Cates, Macromolecules, **21**, 2610 (1988).
5. M. Murat and G. S. Grest, Macromolecules, **24**, 704 (1991).
6. P.-Y. Lai and E. B. Zhulina, J. Phys. II (France), **2**, 547 (1992).
7. J. F. Marko and A. Chakrabarti, Phys. Rev. Lett., **48**, 2739 (1993).
8. D. C. Rapaport, J. Chem. Phys., **71**, 3299 (1979).
9. H. S. Gulati, C. K. Hall, R. L. Jones, and R. J. Spontak, J. Chem. Phys., **105**, 7712 (1996).
10. S. W. Smith, B. D. Freeman, and C. K. Hall, J. Comput. Phys. (in press).
11. H.-P. Deutsch and R. Dickman, J. Chem. Phys., **93**, 8983 (1990).
12. R. L. Jones and R. J. Spontak, J. Chem. Phys., **103**, 5137 (1995).
13. M. Murat and G. S. Grest, Macromolecules, **22**, 4054 (1989).
14. P.-Y. Lai and K. Binder, J. Chem. Phys., **95**, 9288 (1991).
15. L. I. Klushin and A. M. Skvortsov, Macromolecules, **24**, 1549 (1991).
16. D. A. McQuarrie, Statistical Mechanics. (Harper and Row, New York, 1973).

DYNAMIC ASPECTS OF FORMATION OF SYNTHETIC MEMBRANES

M.THIES*, H.H.PARADIES*, S.F.CLANCY**
*Biotechnology & Physical Chemistry, Märkische Fachhochschule, Iserlohn, D-58644, Frauenstuhlweg 31, P.O.BOX 2061, Germany,HParadies@aol.com
**Witco Corporation, Research & Development, P.O.Box 646, Dublin, OH 43017, & Safety,Health, & Environmental Affairs, One American Line, Greenwich, CT 068131-2559,USA

ABSTRACT

A detailed physical analysis of data obtained from static light and dynamic light scattering experiments with polymer-like distearyldimethylammonium OH (DSDMA X^-) micelles has been undertaken at dilute and semidilute concentrations, and compared with those in the presence of racemic and enantiomeric anions, e.g. S-(+)-lactate or the R-(-)-enantiomer.

INTRODUCTION

A range of amphiphilic molecules forms lamella phases that can swell in a solvent. For ionic systems the swelling in water and in oil-like environments is caused by electrostatic double layer forces [1-3]. For liquid bilayers the swelling in water or saline solutions have traditionally been interpreted as due to a hydration force whose molecular origin is connected to a structure formed in the saline system, including counterions. In light of the challenges of this view [4-7], particularly not considering configurational arrangements of the head groups or the protrusions of the entire molecules, or a combination of both effects, and as a function of chiral, counterions, we investigated the dynamic aspects of the formation of synthetic membranes [8]. The synthetic membranes are built from distearyldimethylammonium (DSDMA) X^- compounds in the presence of inorganic anions with X^- = OH, Cl or chiral counterions e.g. R-(-)-lactate or the S-(+)-enantiomer. In an effort to delineate the importance of the above mentioned contributions but also the influence of the counterions X^- on these factors, we have performed a series of static and dynamic light scattering experiments to address the dynamic aspects of the formation of these fully synthetic membranes with respect to simple vs. chiral anions.– By adding achiral anionic compounds, e.g. propionic acid, myristic acid (MA) or palmitic acid (PA) or OH^- and most unexpectedly S-(+)-lactate or the R-(-)-enantiomers in a certain ratio to the double-chained cationic N-surfactants, one can induce the formation of large entanglements. Also the addition of MA can induce the formation of non-lamella inverted hexagonal (H_{11}) and /or inverted bicontinuous cubic phases as observed for dimyristoylphosphatidylcholine (DMPC) and dipalmitoylphosphatidylcholine (DPPC), respectively [9]. Didodecyldimethylammonium X^-salts which are much more water soluble than the DSDMA X^- compounds have been studied extensively by Brady et al. [10] but not noticing the remarkable effects of the solution properties of DDMA X^- in the presence of enantiomeric lactates or racemic lactate in the semidilute regime e.g. increase of entanglement, swelling, changing of the persistence length and viscosity effects due to the formation of wormlike micelles, which follows in fact the same pattern as found for DSDMA X^- [11].- Furthermore, there are no experimental data available on the DSDMA X^- with X = R-(-)-lactate, the S-(+)-enantiomer or on the racemic α-hydroxy-propionic acid with respect to formation of vesicles or micelles including their solution structures. However, it has been shown that DSDMA OH reveals a fairly unusual pattern in aqueous solution : at low molarity of salt it forms vesicles, at high salt as well as high surfactant concentrations it forms threadlike micelles. Since the theory of the vesicle or spherical-to-wormlike micelle transition including the theory of the rheological behavior of wormlike micellar solution presently lack predictive power laws, particularly for these environmental and biological active materials [12], we believe it is important to establish convincing physical data of this system undergoing spherical (vesicular) to

Mat. Res. Soc. Symp. Proc. Vol. 463 © 1997 Materials Research Society

wormlike micelle transition for DSDMA OH in comparison to DSDMA X^- with X = S-(+)-lactate, the R-(-)-enantiomer and the racemic R,S-lactate, particularly for the threadlike micelles.

EXPERIMENTAL

Materials

The preparation of DSDMA or DDMA X^-, X = CL, Br, OH including purification and characterisation has been reported [11]. The preparation of DSDMA- or DDMA ≈ S-(+)- or R-(-)-lactate has been performed by ion-exchange chromatography of the appropriate chloride on Amberlite IRA-400 resin loaded with the enantiomeric or racemic α-hydroxy-propionic acid or sodium salt, respectively, at pH 7.2 (20^0C), or by direct ion exchange of the chloride in the presence of 0.1 M sodium α-hydroxy-propionate in water-ethanol mixtures (35% v/v) [20]. The completeness of exchange was determined by analysis of chloride content or FAB-MS Spectroscopy (Finnigan 8200).

Methods.

Static and Dynamic Light Scattering (QELS) experiments were performed on an ALV-LSE Goniometer (ALV Langen, Germany), equipped with an He-Ne-laser (NEC, 50 mW). An ALV-5000 multibit, multi-τ-correlator and a computer controlled stepping motor driven variable angle detector system was applied [11]. A thermostat permitted control of temperature of $\pm 0.1^0$ C (Lauda 15) for all experiments performed.

RESULTS AND DISCUSSION

The strongest evidence in form of flexible micelles and observation of their changes into other hydrodynamic structures in aqueous solutions in the presence of different counterions i.e. OH vs. lactates, was obtained in the semidilute regime at surfactant concentrations large enough that the elongated micelles can overlap by forming a transient network. Also in this regime the system exhibits a viscoelastic behavior which is reminiscent of that of entangled polymer solutions. Since the concentration range studied is very similar to that used in the preparation of synthetic membrane systems for measuring proton transfer rates of intercalated biological active molecules [8] or in the biotransformation through active intermediates e.g. the corresponding N-oxides and Cope-arrangement [13]. The physical results obtained are fundamental in respect to the structures dealt with in the biochemical reactions (ATP, ADP) or bacterial cell structures.

Static Light Scattering.

In the semidilute regime, i.e. concentrations of 1 mM DSDMA in the presence of R-(-)- or S-(+)-lactate, the flexible chains overlap at concentrations still low enough for the system to be described by one interaction parameter, the screening length ζ. This can be interpreted as the mesh size of the transient network formed by entangled chains. Fig.1 shows that $1/I_S$ increases linearly with q^2, which is also observed for the most dilute solution of DSDMA– R-(-)- or S-(+)-lactate of 0.5-1.0 mM, whereas for DSDMA OH the same behavior has been observed at concentrations of 5 mM. The crossover between dilute and semidilute solutions of DSDMA and enantiomeric lactates as seen by plots of log I_S vs. log M (data not shown) can be related to the strength of the maximum intensity for long DSDMA-chains in the presence of the chiral counterions (1mM) in reasonable good solvents. For long thin chains $qR_G \ll 1$ is not obeyed over the entire scattering range of q-values, which are not accessible experimentally. Therefore, for a fixed value the maximum I_S was extrapolated at the crossover concentration where $q\zeta=1$. To overcome this difficulty I_S was extrapolated to q=0 for determining the overlapping of chains. The q values have been introduced in Fig.1 as data points, and also for obtaining $I_S(q=0)$ at different surfactant concentrations at $c_2 \geq 1$ mM up to 5 mM.

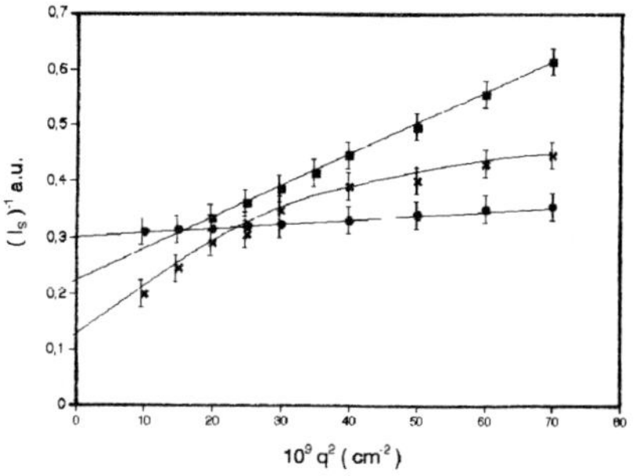

Fig.1 Variation of $1/I_s$ vs. g^2 for DSDM~ S-(+)-lactate solutions in the presence of 5 mM S-(+)-lactate solutions in the presence of 0.5 mM S-(+)-Lactate with C_{DSDMA} = 0.5 mM (x—x); 1.0 mM (•—•); 1.5 mM (■—■); q in cm $^{-1}$.

The slope was determined to be -0.39 vs -0.31, as predicted by Des Cloizeaux according to $\zeta = \varnothing^{-0.77}$ [14]. The decrease of the scattered intensity upon increasing DSDMA-lactate concentration can be attributed to the decrease in correlation range which is also supported by the q^2 dependence of I_S. Since no curvature is seen in Fig.1 we estimate the radius of gyration of R_G = 839 Å. Table I lists all the hydrodynamic parameters for the DSDMA-lactate system and for DSDMA OH for comparison.

Table I

Molecular Parameters of DSDMA X^- in Water and in the Presence of 10 mM $X^{\ominus}$ (20°C)$^{a.)}$.

Parameter	OH^-	S-(+)-Lactate	R-(-)-Lactate	R,S-Lactate
R_g, Å	815	839	840	835
M_W, x 10^{-3}	3.500	3.700	3.780	3.790
A_2, x 10^{-5} mL·mol·g^{-2}	4.55	2.95	2.98	3.05
L_c, Å	6.234	6.419	6.420	6.430
l, Å	330	340	345	349
R, Å	1969	2.000	2.090	2.050
R_g, Å calc.	805	820	825	825
$<D>_Z$ x 10^{-6} cm^{-2}· s^{-1}	4.09	3.75	3.70	3.73
Relaxation times				
τ_1, exp, µs	324	321	319	325
τ_1, Zimm, free draining, µs	324	326	328	330
τ_1, Fujima, coil, µs	596	610	625	620
τ_2, Fujima coil, µs	163	175	180	179
c^*	1.85	1.15	1.10	1.09

a.) L_c " M/M_L, with M_L = 565 $Å^{-1}$, the molecular weight per unit length; L is the computed from $3R_g^3/l^2 = (L_c / l)- [1 + \exp(-L_c / l)]$. $R^2 = [L_c^2/ \gamma L]$ with $1/ \gamma = 2l$; $R_{g\,calc}^2 = R_g^2/ 6$. $c^* = 3M_r / (4\pi R_g^2 N_A)$; $\tau_1 = 2R_g^2/(D\pi^2 n^2)$

That we do not observe any significant curvature in the plot of $1/I_S$ vs.q^2 as for DSDMA-lactate concentrations increase (1mM-5mM) above 1 mM, and not below 5 mM, is an indication that the DSDMA-lactate micelles do not show any significant polydispersity, and there are no changes of the characteristic length ζ of this system as long as the narrow concentration range is fixed. Another explanation can well be an interpenetration of DSDMA-lactate micelles resulting in a new characteristic length, ξ, which is independent of the micelle length rather than polydispersity. So ξ is independent of the DSDMA-lactate micelle length with no overlapping of micelle size with surfactant concentration, and the correlation length can be identified with R_G.

Dynamic Light Scattering.

The values of the average diffusion coefficients obtained by the cumulants method as a function of angle increase due to contributions of the internal modes for DSDMA-lactate concentration of c=1.5mM (Fig.2).

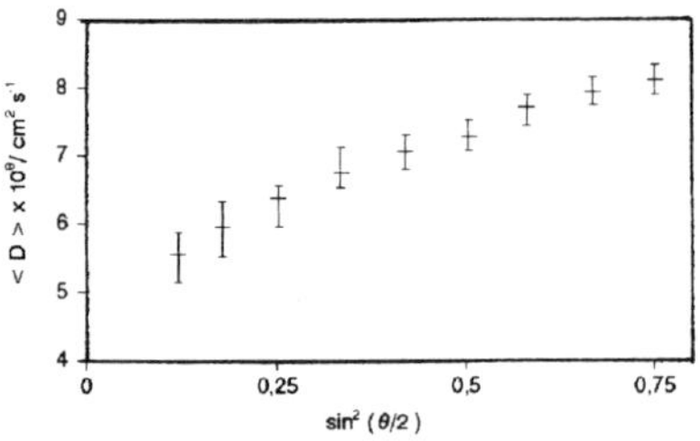

Fig. 2 Dynamic light scattering of DSDMA~R-(-)-lactate for C_{DSDMA} = 1.5 mM as a function of scattering angle (θ) (20°C).

The values of the self-diffusion coefficient were obtained from plots of Γ vs q^2, where Γ is the average decay rate for different surfactant concentrations in the presence of racemic or enantiomeric lactates, respectively. No significant differences in the self diffusion coefficient between the DSDMA-R-(-)-lactate and the DSDMA-S-(+)-lactate as well a for the racemic lactate have been found, although there are differences in the CMC between the DSDMA-R,S-lactate and the enantiomeric DSDMA-lactates having a smaller CMC than the DSDMA-R,S-lactate. The corresponding correlation functions were analysed by the CONTIN program [15], as shown in Fig.3, wheras Fig. 4 shows the relative amplitudes of the slow and fast component of DSDMA-S-(+)-lactate. The fast component is q^2 dependent, showing diffusive behavior. The slow component is independent of q, which can be reconciled of a structural relaxation process of a stabilized remark, resulting also in viscoelastic properties as noticed by rheological experiments, too. The slow component may derive from the disruption/coalescense kinetics of the DSDMA·X aggregates, which is in accord with rheological experiments (Paradies, Clancy unpubl.)

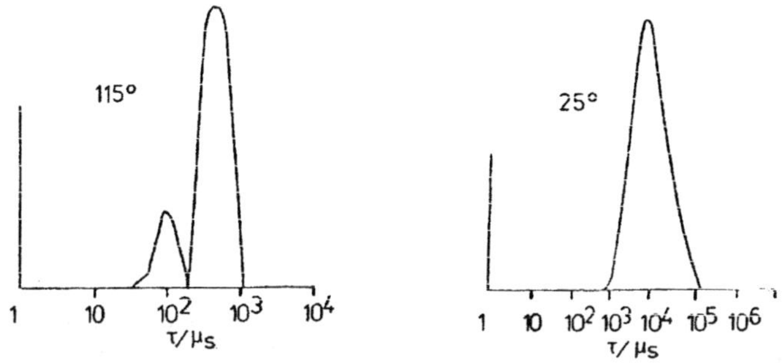

Fig.3 Decay time distribution function at different angles obtained throgh Laplace inversion with CONTIN for C_{DSDMA} = 1.5 mM and 0.5 mM S-(+)-lactate.

At lower angles only one peak is to be seen, representing the translational diffusion coefficient terms. By increasing the magnitude of the scattering vector length, a second peak appears in Fig.3 at a faster decay time and hence smaller in apparent size. However, the proximity of the image peak to the minor peak causes a perturbation of the peak position of the minor component of the time scale. In order to eliminate this drawback the correlation function was divided by exp(-qDt) to yield a "normalised" correlation function. By applying the correct limiting value for D the contribution from the internal modes can be retained only by considering a single coil. The position of the fast-moving peak on the time scale corresponds then to both the diffusional and internal mode contributions of $g^{(1)}(t)$ At low scattering angles, e.g. $qR_G << 1$, the translational diffusion coefficient is the only one which contributes to $g^{(1)}(t)$, so the correlation function decays approximately as a single exponential. But, when qR_G becomes larger with increasing angle the internal modes will contribute to the correlation function, so it becomes multiexponentional (qR_G = 1.95-2.20).The increase in the average diffusion coefficient (Fig.2) with scattering angle , also shows an increase in the relative variance from 0.65 to 1.04, in the angular range of 25-110^0. Furthermore, when the DSDMA-lactate concentration increases, the variance of the correlation function decreases. This is in accord with the view that a progressive interpenetration of the flexible chains occurs which indicates that the diffusion coefficient becomes independent of the internal characteristics of the DSDMA-lactate micelle, but only dependent on the average distances between closest chain contact.

The experimental data have been prosecuted by applying a double exponential expression with a fixed D-value which is related to the slow relaxation rate, Γ_S, according to eqn. (1):

$$(1) \qquad G^{(2)}(t) = 1 + \beta[A_S\exp(-\Gamma_S \cdot t) + A_F\exp(-\Gamma_F \cdot t) + C]^2$$

with $G^{(2)}(t)$ the measured correlation function, β is a constant which accounts for the deviations from ideal correlation, and C is the base line. The variables Γ_S and Γ_F are the relaxation rates of the slow and fast components, respectively, with Γ_S reflecting the translational mode, and Γ_F combines the fast internal modes and the diffusive component. This satisfies equation 11 so that $\Gamma_F - \Gamma_S = 2/\tau_1$. The diffusion coefficient determined was fixed to be 4.09 x 10^{-6} $cm^{-2}s^{-1}$ as measured at low angles. Fig. 4 shows the amplitude of the fast mode dependence on angle for c = 0.005 M DSDMA-lactate and the slow mode dependency, which is an indication for viscoelasticity.

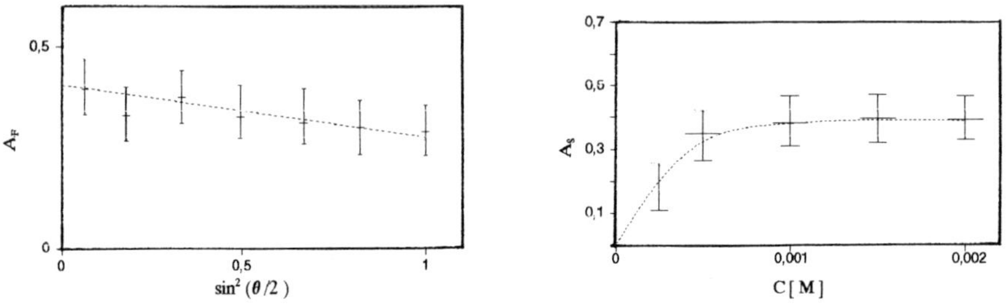

Fig. 4 Relative amplitudes of the slow (A_S) and fast component (A_F) as a function of scatteringangle (left)
at C_{DSDMA} = 1.0mM in the presence of R-(-)-lactate (1.5 mM), and of DSDMA-R-(-)-lactate (-1.5 mM) (right).

The average τ_1 value calculated according to eqn. (1) fit the $G^{(2)}(t)$ measured correlation function, and was found to be 309 μs. This obtained value can be compared with the one computed from the first normal mode τ_1 for a free-draining coil according to Rouse [16] and Zimm [17]. Normally, the value is determined by extrapolation to infinite dilution, but using the low concentration applied here, the experimental value obtained in this study seems to be reliable. Now considering different interpretations of $<D>_Z$, τ_1, τ_2, and R_H, e.g. if a Gaussian coil model can be physically applied, the size of this model can be estimated. A Gaussian coil has to be long and very flexible. Following the Fujima notation for flexible macromolecules [18], $\gamma \cdot L >> 1$ with varying degrees of flexibility, the flexibility can be expressed in terms of the product of the particle length, $\gamma \cdot L$, where γ is the inverse Kuhn segment length, $1/\gamma = 2l$ with l the persistence length, and L is the length of the macromolecule in solution. Taking n=6020 as the apparent aggregation number for DSDMA-lactate, for a spherical micelle the radius of the hydrocarbon core, R_{HC} would be:

$$(2) \qquad R_{HC} = (3V_{HC} / 4\pi)^{1/3} = 12.77 \text{ Å}$$

with:

$$(3) \qquad V_C = n[27.4 + 26.9(n_c - 1)] = 485 \cdot n$$

where n_c = 18, the number of carbons in the chains. For a cylinder of the same radius we calculate:

$$(4) \qquad \pi\gamma_{HC}^2 \cdot L(n) = 485 \cdot n$$

which furnishes us with a value of L(6020) = 2498.9 Å. If $\gamma \cdot L$ =10, for a Gaussian coil where minimum flexibility is assumed, then the statistical segment length of $l' = L / \gamma L$ = 249.9 Å is obtained with $n' = L/l'$ =10. The radius of gyration of such a Gaussian coil is calculated according to:

$$(5) \qquad R_g = n'l^2 / 6$$

and would be 322.6 Å. This is a much smaller value than the 840 Å obtained for DSDMA-lactate, indicating that a Gaussian coil is not consistent with the experimental data obtained. The limit $\gamma \cdot$

L>>l corresponds to a completely flexible coil, and would agree with the Rouse-Zimm model. We find a γL value of 9.9 (10) which indicates that the coil is not in the extreme limit of flexiblity. If we introduce these values into the Fujima model [18] we calculate τ_1 = 596 μs. This is much lower than the value obtained experimentally of 311 μs. One explanation for this discrepancy is that we are dealing with a semidilute solution so the overall rotational motion, which is related to τ_1, is suppressed. The τ_1 value observed can also be related to a combination of τ_1 and τ_2 which for is flexible rod is of the order of τ_2 = 100 μs. In addition, the τ_1 measured can be a combination of τ_1 *and* higher terms on the time scale which was not measured experimentally.

However, the best qualitative description of the findings in this particular salt region (1.0 mM - 5.0 mM R-(-)-/or S-(+)-lactate) in the concentration range of DSDMA - lactate (0.8 x 10^{-3} M to 1,5 x 10^{-3} M, 1.15 c* - 2.12 c*) is that of a semidilute polymer solution, comprised of polymeric chains which are not at the extreme limit of flexibility, but revealing some stiffness of the hydrocarbon chains.

Furthermore, the fastest decrease of D levelling off D_{self} follows a power law $C^{-\delta}$ with δ=0.7 in 1.0 - 5.0 mM lactate solutions. These values are closed to but significantly smaller than those previously observed for other systems. Note, Ott et. al. [19] observed a value of 1.35 ± 0,05 for lecithin micelles which seem to agree with the description of tracer diffusion in living polymers as stated by Cates [20]. This is not the case for our DSDMA-lactate system studied here where δ is larger than 1.35 for R-(-)- or S-(+)-lactate concentrations smaller than 5.0 mM, and where, at larger concentrations for R-(-)- or S-(+)-lactate of 10.0 mM a sort of anomalous diffusion takes place at T=20 - 25°C [21].

CONCLUSIONS

An important contribution in understanding the DSDMA-lactate system is the concentration dependence of the chiral lactates surfactant upon the dynamically time scale; that is, at higher surfactant concentration the distribution of decay times becomes bimodal. The fast mode reflects the co-operative motions of the transient network formed throughout the chain entanglements, which has a positive exponent ($\tau \propto c^{0.68}$) similar to the previous study of the DSDMA OH system. In fact the DSDMA OH system parallels very closely the system studied here, however with the exception that the concentrations used are a magnitude lower than for the DSDMA OH hydroxide. It should be remarked that the application of the theory applied here is only valid for the description of the short time-scale dynamics, since the effect of topological constraints which dominates the long time-scale dynamics was not taken into account.

REFERENCES

[1] V.A.Parsegian, Trans. Faraday Soc.62, 848, (1966)
[2] B. Jönsson, H. Wennerström, J. Colloid Interface Sci., 80, 482, (1981)
[3] H.Wennerström, Prog. Colloid Polymer. Sci., 74, 31, (1987)
[4] B.Jönsson, H.Wennerström, J.chem.Soc., Faraday Trans. 2, 79, 19, (1983)
[5] J.N.Israelachvili, H.Wennerström, Langmuir, 6, 873, (1990)
[6] U.Nilsson, B.Jönsson, H.Wennerström, Faraday Discus. 90, (1990)
[7] B.Jönsson, H.Wennerström, Chem. Scr., 25, 117, (1985)

[8] H.H.Paradies, J.Phys.Chem.,90, 5956,(1986); H.H.Paradies, Medicinale , 14, 1-13,(1984); H.H.Paradies, Colloids. Surf., 6, 405, (1983); H.H.Paradies, Angew.Chem., 94, 793, (1982); H.H.Paradies, Angew.Chem.Supplement, 1970-1981, (1982)

[9] A.Ben-Shaul, W.M.Gelbart, in *Micelles, Membranes, Microemulsion & Monolayers*. Edited by W.M.Gelbart,A. Ben-Shaul, D.Roux; New York: Springer pp 1-104, (1994).

[10] J.F.Brady, D.F.Evans, G.G.Warr, F.Grieser, B.W.Ninham, J.Phys.Chem.,90, 1853, (1986)

[11] S.F.Clancy, P.H.Steiger, D.A.Tanner, M.Thies, H.H.Paradies, J.Phys.Chem., 98, 11143, (1994)

[12] S.F.Clancy, D.A.Tanner, M.Thies, H.H.Paradies, *Extended Abstracts*, Division of Environmental Chemistry, San Francisco, CA, 203rd National Meeting of the American Chemical Society, Vol.32, No 1, 907-908, (1992)

[13] H.H.Paradies, M.Thies, S.F.Clancy, J.Org.Chem., submitted (1996)

[14] J.Des Clizeaux, J.Phys.Chem.,36, 281, (1975)

[15] S.W.Provencher, Makromol.Chem.,180, 201, (1979)

[16] P.E.Rouse, J.chem. Phys., 21, 1272, (1955)

[17] B.H.Zimm, J.Chem.Phys., 24, 269, (1956)

[18] S.Fijima & M. Maruyama, Macromolecules, 6, 237, (1973)

[19] A.Ott, W.Urbach, D.Langevin, P.Schurtenberger, R. Scartazzini, P.L.Luisi, J.Phys. : Condens. Matter, 2, 5907,(1990)

[20] M.E.Cates, Macromolecules, 20, 2289, (1987); R.Granek & M.E.Cates, J.chem. Phys.,96, 4756, (1992)

[21] M.Thies, S.F.Clancy, H.H.Paradies, J.Phys.Chem., 100, 9881, (1996); H.H.Paradies, D.W.Wagner,W.R.Fischer,Ber.Bunsen.Ges.Phys.Chem.,100, 1299, (1996); H.H.Paradies, M.Thies, S.F.Clancy, J.Phys.Chem., submitted (1996)

VECTOR LATTICE MODEL OF AMPHIPHILIC MIXTURES: INTERFACIAL BEHAVIOR

C. BUZANO[1], L. R. EVANGELISTA[1,2] and A. PELIZZOLA[1]
[1]Dipartimento di Fisica, Politecnico di Torino, c. Duca degli Abruzzi 24, 10129 Torino, Italy
[2]Departamento de Fìsica, FUEM, Av. Colombo, 5790, 87020-900, Maringà, Paranà, Brazil

ABSTRACT

The interfaces between the water-rich, oil-rich and disordered phases of a balanced ternary mixture of oil, water and amphiphile are studied, on the basis of a vector lattice model, by means of an exact partial integration and a local mean field approximation. The phase diagram exhibits a water/oil-rich phase, structured (ordered bicontinuous and lamellar) phases and a disordered phase with a microemulsion region. The interfacial tension and profiles at oil/water/disordered phase coexistence are calculated and the possibility of a wetting transition is discussed.

INTRODUCTION

In this paper we consider a mixture of water, oil and amphiphilic molecules [1], which are characterized by the presence of a polar head (which tends to attract water) and a non-polar tail (which tends to attract oil). When amphiphilic molecules are added to a mixture of oil and water, they tend to form layers at the interface to oil and water, thereby strongly reducing the interfacial tension. This feature is of fundamental importance in many biological and industrial processes.

Amphiphilic mixtures have also attracted a great deal of interest from the theoretical point of view because they are complex fluids exhibiting several distinct thermodynamic phases, like water/oil-rich phases, liquid-crystalline phases, and a disordered phase in which a microemulsion region can be identified. From the point of view of thermodynamic phases a microemulsion is still a disordered fluid, but it is characterized by peculiar short-range correlations, related to the presence of fluctuating layers of amphiphile separating oil and water. Such short-range correlations are of course due to the amphiphilic interactions, and can be regarded as a reminiscence of the long-range ordered structures appearing at lower temperatures. From the theoretical point of view the boundary between the microemulsion and the ordinary disordered fluid is usually defined by means of the Lifshitz line or the disorder line [1].

It is both experimentally observed and theoretically predicted that under suitable thermodynamic conditions the disordered fluid coexists with the water-rich and oil-rich phases. The purpose of the present study is then to calculate interfacial profiles and tensions at such a coexistence and to discuss the possibility of a wetting transition, for a balanced system (we assume that oil and water behave symmetrically) and for varying amphiphilic interaction strength.

THEORY

The model on which our investigation is based is a vector lattice model proposed by Ciach, Høye and Stell [2, 3]. The model is most easily written in terms of two sets of variables. For each lattice site, the z-component $s(\mathbf{r})$ of a spin-1 operator, taking values $\pm 1, 0$, identifies

Mat. Res. Soc. Symp. Proc. Vol. 463 © 1997 Materials Research Society

the chemical species ($+1$ corresponding to water, 0 to amphiphile and -1 to oil), while a vector variable $\mathbf{n}(\mathbf{r})$, meaningful only for $s(\mathbf{r}) = 0$, specifies the amphiphile orientation. Since we are going to consider the model on the simple cubic lattice, the possible values of the orientational variables will be restricted to those pointing towards the six nearest-neighbors of a given site, for the sake of simplicity: it seems reasonable to believe that such an assumption will not alter the basic features of the model. Assuming oil/water symmetry, the model Hamiltonian (divided by $k_B T$, where as customary k_B is the Boltzmann's constant and T the absolute temperature) can be written as

$$\begin{aligned}\frac{\mathcal{H}}{k_B T} = & - \frac{J}{2}\sum_{\mathbf{r},\delta} s(\mathbf{r})s(\mathbf{r}+\delta) - \frac{K}{2}\sum_{\mathbf{r},\delta} s^2(\mathbf{r})s^2(\mathbf{r}+\delta) + D\sum_{\mathbf{r}} s^2(\mathbf{r}) \\ & - A\sum_{\mathbf{r},\delta}\left[1 - s^2(\mathbf{r})\right] s(\mathbf{r}+\delta)\, \mathbf{n}(\mathbf{r})\cdot\delta, \qquad (1)\end{aligned}$$

where J and K are related to the chemical interaction strengths between different species by $J = (\epsilon_{ww}+\epsilon_{oo}-2\epsilon_{wo})/4$ and $K = (\epsilon_{ww}+\epsilon_{oo}+2\epsilon_{wo})/4-(\epsilon_{wa}+\epsilon_{oa})+\epsilon_{aa}$ ($\epsilon_{\alpha\beta}$ being the interaction energy between molecules of species α and β), D is related to a difference of chemical potentials by $D = \mu_a - (\mu_w + \mu_o)/2 + 3(\epsilon_{wa} + \epsilon_{oa} - 2\epsilon_{aa})$ (μ_α being the chemical potential for species α) and can be regarded as a rescaled chemical potential for the amphiphile, A is the strength of the amphiphilic orientation-dependent interaction, which characterizes the strength of the amphiphile and δ is a vector of length equal to one lattice spacing pointing towards nearest neighbors.

In previous papers [4, 5], we studied the phase diagram of this model doing first an exact integration of the orientational degrees of freedom, which is possible since these interact only with the spin ones, and not among themselves. This leads to a model in the spin variables only, with many multispin interactions (up to 7 spins for the simple cubic lattice), which is then studied by a local mean-field approach. Details of our approach have been given in Ref. [5]. As a result we found the water-rich, the oil-rich, the bicontinuous ordered and lamellar phases with different periodicities, and a disordered phase with a microemulsion region, which we identified by means of the Lifshitz line. In the temperature vs. chemical potential plane, we have found the three-phase coexistence line between water, oil and disordered phases, which emerges from the multiphase point at which these phases coexist also with a structured phase (bicontinuous or lamellar, depending on the amphiphile strength). The lamellar phases (the so-called L_2 and L_3 phases) are found only at non-zero temperature and for strong enough amphiphile, and occupy always tiny regions close to the multiphase point.

In the framework of the local mean-field approach to the model which results from the integration of orientational degrees of freedom, interfacial properties like order parameter profiles and interfacial tensions can be studied simply by considering a one-dimensional sample of $N \times 1 \times 1$ sites, with periodic boundary conditions in the "short" directions and fixed boundary conditions in the "long" one (if we are going to study the interface between phases A and B, the order parameters at sites 1 and N of our sample will be constrained to the corresponding bulk values). The interfacial tension σ_{AB} at coexistence is then given by the excess free energy with respect to the bulk value, while interfacial profiles can be deduced from the couples (m_n, q_n), $n = 1, 2, \ldots N$, defined by

$$\begin{aligned} m_n &= \langle s_n \rangle = \rho_{w,n} - \rho_{o,n} \\ q_n &= \langle s_n^2 \rangle = 1 - \rho_{a,n}, \qquad (2)\end{aligned}$$

where s_n is the spin-1 operator at the n-th site of our sample and $\rho_{\alpha,n}$ ($\alpha = w, o, a$ for water, oil and amphiphile respectively) is the concentration of the species α at the same site.

Among the possible interfacial phenomena, wetting of the water/oil interface by the disordered fluid is of particular interest. In order to investigate this possibility we shall determine the interfacial tensions σ_{wo}, σ_{wd} and σ_{do}, where d denotes the disordered phase. If the water/oil interface is dry the condition $\sigma_{wo} < \sigma_{wd} + \sigma_{do}$ holds [1], while if the interface is wet (which means that a macrosopic region occupied by the disordered phase appears between the water-rich and oil-rich bulk phases) the condition $\sigma_{wo} = \sigma_{wd} + \sigma_{do}$ holds [1]. In the case of a first order wetting transition, a metastable interface with $\sigma_{wo} > \sigma_{wd} + \sigma_{do}$ is expected at temperatures higher than the transition temperature [6]. Notice that, since we are dealing with a balanced system in which oil and water behave symmetrically, we have $\sigma_{wd} = \sigma_{do}$.

RESULTS

We have studied the interfacial properties of our model for $K/J = 1$ (which amounts to assume $\epsilon_{wo} + \epsilon_{aa} = \epsilon_{wa} + \epsilon_{oa}$) and varying A/J, and report here the results for two typical cases of weak and strong amphiphile, corresponding to $A/J = 1.1$ and 3.0 respectively. We consider first the case $A/J = 1.1$, for which the portion of the phase diagram which is relevant for the present study is reported in Fig. 1.

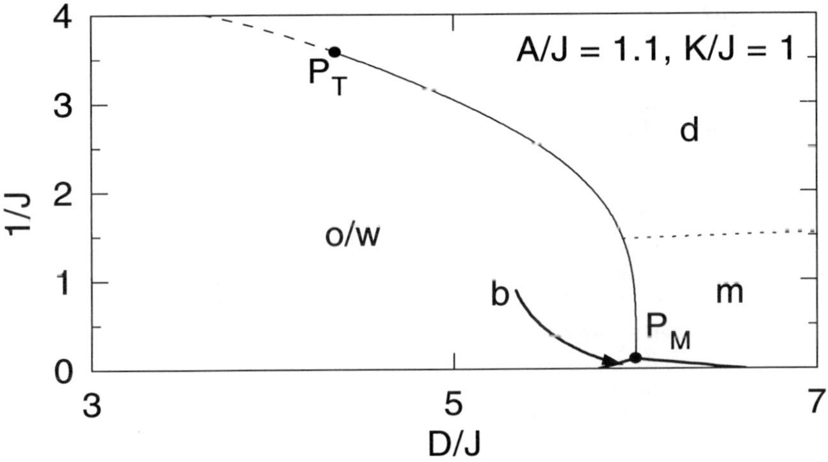

Figure 1: Phase diagram for $K/J = 1$ and $A/J = 1.1$ in the temperature ($1/J$) vs. chemical potential (D/J) plane. Solid, dashed and dotted lines denote first order, second order and Lifshitz lines, respectively, while P_T and P_M are the tricritical and multiphase point and w/o, b, d and m stand for water/oil-rich, bicontinuous ordered, disordered and microemulsion.

Here the bicontinuous ordered phase occupies a rather small region of the temperature ($1/J$) vs. chemical potential (D/J) plane (the multiphase point P_M is at $1/J = 0.11$, while the tricritical point P_T is at $1/J = 3.58$ and the Lifshitz line meets the first-order, or three-phase coexistence, line at $1/J = 1.46$), since A/J is very close to the value 1, at which this phase disappears at all. However, such weak amphiphile (i.e. small A/J) cases are very important to study since, as we shall see in the following, the possibility of wetting is enhanced for weak amphiphile.

In Fig. 2 we have plotted the interface tensions vs. temperature along the coexistence line.

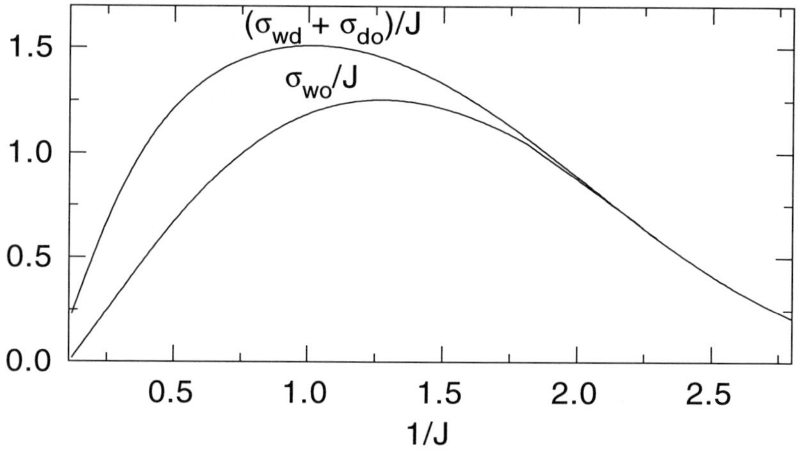

Figure 2: Interfacial tensions vs. temperature for $K/J = 1$ and $A/J = 1.1$.

In the vicinity of the multiphase point, we have a water/oil interface tension $\sigma_{wo} \sim 10^{-2}$, that is two orders of magnitude smaller than the corresponding tension without amphiphile. The behavior of the tensions σ_{wo} and $\sigma_{wd} + \sigma_{do}$ shows that increasing the temperature one passes from a region in which the interface is evidently dry to a region in which the two curves are indistinguishable on the figure scale, but this does not imply rigorously a wetting transition. Numerical results indicate that, for any temperature up to $1/J = 2.8$ (higher temperatures are very computationally demanding), among the many metastable solutions (the presence of many metastable solutions in a lattice model is a consequence of the fact that in real systems the interfacial profile varies continuously as the temperature increases) which are found for the w/o interface, taking a sample with N large enough (we used $N \leq 100$) it is always possible to find one with an interface tension σ_{wo} smaller than $\sigma_{wd} + \sigma_{do} = 2\sigma_{wd}$. As a consequence, a thermodynamic wetting transition does not take place for $1/J \leq 2.8$ (but we cannot definitely rule out such a possibility at higher temperatures), although a substantial, even if not macroscopic, amount of amphiphile is found at the interface.

A situation which resembles a very weak first order prewetting transition is instead observed at $1/J = 1.8$, but this is clearly a spurious effect induced by the lattice.

In order to examine more deeply the possibility of a wetting transition we have plotted the interface profiles at $1/J = 2.0$ (Fig. 3) and 2.8 (Fig. 4).

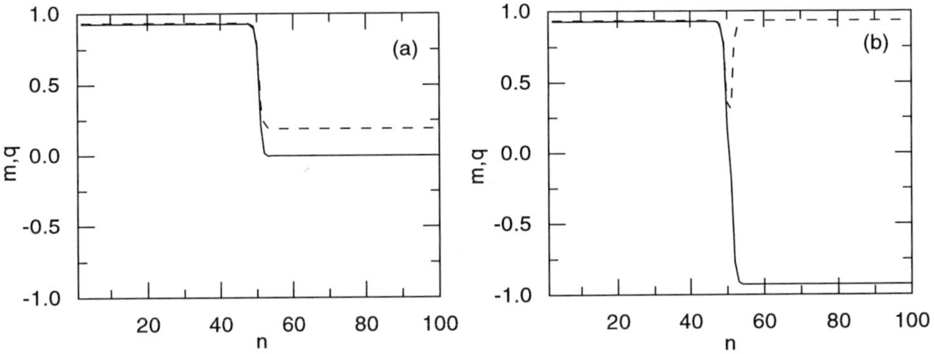

Figure 3: Interfacial profiles of the order parameters m (solid lines) and q (dashed lines) at the water/disordered (a) and water/oil (b) interfaces for $K/J = 1$, $A/J = 1.1$ and $1/J = 2.0$.

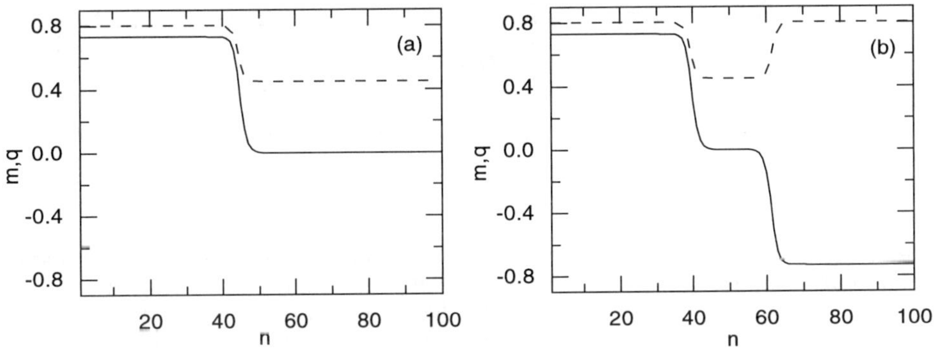

Figure 4: The same as Fig. 3 for $1/J = 2.8$.

It is readily seen that the interfacial regions are spread for the higher temperature: in particular, at the water/oil interface the amphiphile excess with respect to a completely dry interface (which would diverge at a second order wetting transition) is much higher at the higher temperature. Nevertheless, it has to be noticed that in the central part of Fig. 4(b) the values of m_n and q_n never stabilize on the bulk values for a disordered fluid (although again this is not visible on the figure scale), as one would expect for a truly wet interface.

Let us now consider the strong amphiphile case $A/J = 3.0$, for which the phase diagram was reported previously [5]. The multiphase point P_M was found at $1/J = 4.76$, while the Lifshitz line met the coexistence line at $1/J = 3.10$ and the tricritical point P_T was at $1/J = 3.75$. For $A/J = 3.0$ the two interface tension curves (Fig. 5) are much more separated than in the previous case, and hence a wetting transition is much less probable here. Furthermore, the interfacial profiles for $A/J = 3.0$ in the range of temperatures that we have been able to consider resemble very closely those depicted in Fig. 3. Finally, as far as the water/oil interface tension near the multiphase point is concerned, this stays of order 10^{-2} for $A/J = 3.0$, but drops to the order 10^{-3} for $A/J = 5.0$.

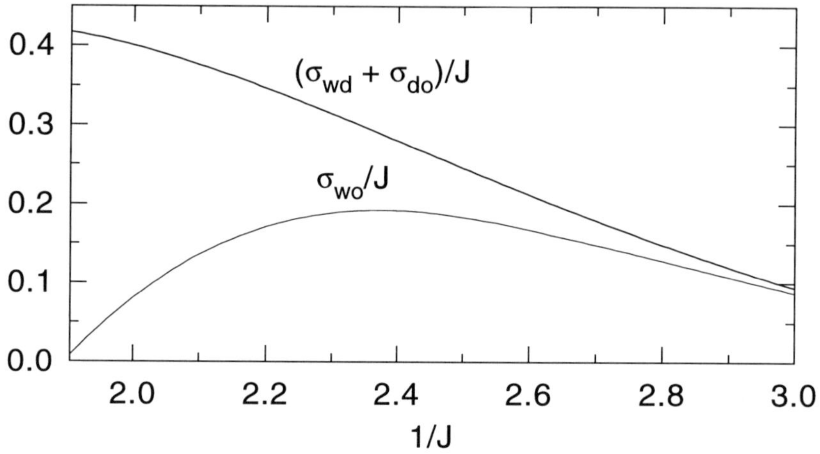

Figure 5: The same as Fig. 2 for $A/J = 3.0$.

CONCLUSIONS

We have studied the interfacial behavior at three-phase coexistence of a model for amphiphilic mixtures proposed by Ciach, Høye and Stell [2, 3], by means of a partial integration followed by a local mean-field approach [4, 5]. The analysis of the interfacial tensions confirms, as already found in the context of other lattice models [1, 6], that the presence of the amphihile strongly reduces the water/oil interface tension. A significant tendency to wetting of the water/oil interface by the disordered phase is observed, especially for weak amphiphile, but no thermodynamic wetting transition has been found in the range of temperature that could be examined with a reasonable computational effort.

REFERENCES

[1] G. Gompper and M. Schick, in *Phase Transitions and Critical Phenomena*, vol. 16, edited by C. Domb and J. L. Lebowitz (Academic Press, London, 1994) and references therein.

[2] A. Ciach, J. S. Hoye and G. Stell, J. Phys. A **21**, L777 (1988).

[3] A. Ciach, J. Chem. Phys. **96**, 1399 (1992).

[4] C. Buzano, L. R. Evangelista and A. Pelizzola, Mol. Cryst. Liq. Cryst., in press.

[5] C. Buzano, L. R. Evangelista and A. Pelizzola, submitted to Phys. Rev. E.

[6] M. W. Matsen and D. E. Sullivan, J. Phys. II **2**, 93 (1992).

VARIATION OF THE RATE OF EXTENSION OF ACTIN NETWORKS

DONALD J. OLBRIS and JUDITH HERZFELD*
Dept. of Chemistry, Brandeis University, Waltham, MA, 02254-9110
* Corresponding author

ABSTRACT

Self-assembling protein filaments are important components of a cell's superstructure. Among these, actin filaments form the backbone of protrusions and extensions such as pseudopodia. The rates at which these structures extend cover a startlingly wide range: the acrosomal process of the sea cucumber may extend 90 μm in 10 seconds, which is more than 20 times the speed at which an epithelial goldfish keratocyte crawls. We seek to explain this range by examining the delivery of actin monomers to the growing filament ends. We show that the diffusive flux of actin monomers is adequate for fueling the slower movement of crawling cells, but is insufficient to propel the quicker acrosomal process of the sea cucumber. By introducing bulk fluid flow in response to the diffusive movement of water through the cell membrane, actin delivery can be enhanced. We compare the calculated speeds to experimental observations and discuss future refinements to the model.

INTRODUCTION

Cells reach and crawl, yet they do so without specialized muscle tissue. Instead, they actively rearrange their cytoskeletons. By selectively polymerizing and depolymerizing actin, cells create a variety of extensions and protrusions that serve their motile needs [1-3]. For example, epithelial goldfish keratocytes extend lamellipodia—wide, shallow appendages that span the front of a moving cell. As the cell moves, the lamellipodium is continuously assembled at the front and absorbed at the rear, each portion staying stationary with respect to the surface while the cell slides around it [4, 5]. In a different context, the sperm of the sea cucumber *Thyone briareus* penetrates the egg by shooting forth a very thin projection, the acrosomal process [6-8].

Although the cell extensions are all dependent on actin filament assembly, their velocities differ dramatically [3]. In particular, the acrosomal process of *Thyone* extends twenty fold faster than the goldfish keratocyte crawls. One key determinant of velocity must be the rate of delivery of actin monomers to the front. This is non-trivial, since diffusion delivers actin too slowly to maintain the observed extension speed of the acrosomal process of *Thyone* [9]. Here we will focus on the epithelial goldfish keratocyte and the *Thyone* acrosomal process, seeking particularly to account for the impressive speed of extension of the latter.

THEORY

The problem of actin-based cell motility may be crudely broken into two pieces, protrusive force generation and actin delivery. As our focus is on actin delivery, we take an existing model for protrusive force generation and incorporate it into our model for monomer delivery. For monomer delivery, we first consider actin diffusion alone. Then we introduce osmotically driven bulk flow.

The Brownian Ratchet

The "Brownian Ratchet" model of force generation takes advantage of fluctuations in the cell's membrane [10]. Actin filaments abutting the cell membrane are unable to lengthen due to obstruction of the ends by the membrane. When thermal fluctuations move membrane segments away from the filament, gaps larger than an actin monomer may be created, allowing polymerization to occur. The now-longer filament prevents the membrane from returning to its original position—this is the "ratchet." The speed of this extension is [10]

$$v_{ext} = \delta\left[k_+ c_L e^{-\omega} - k_-\right], \tag{1}$$

Mat. Res. Soc. Symp. Proc. Vol. 463 © 1997 Materials Research Society

where k_+ (k_-) is the on (off) rate constant for actin filament assembly, c_L is the actin monomer concentration at the front of an extension of length L, and $\omega = f\delta/k_BT$ is the dimensionless work done against the load force f in adding a monomer of length δ to the filament. This approximation is valid for $c_L k_+ \delta << 2D/\delta$ and $k_- \delta << 2D/\delta$, which are satisfied by the systems of interest in this paper. The flux of actin monomers onto the filaments for the Brownian Ratchet is

$$j_{BR} = n\left(k_+ c_L e^{-\omega} - k_-\right), \tag{2}$$

where n is the number of filament ends involved per unit cross-sectional area.

Actin Diffusion

As the Brownian Ratchet draws down the concentration of actin monomers in the fluid at the front of the extension, diffusion tends to dissipate the gradient, bringing actin forward from the cell body. In the goldfish keratocyte, this is a steady state process. As the shape of the lamellipodium and the cell's speed both remain approximately constant, the flux of actin forward from the rear of the cell must balance the flux of monomers being added to the actin network.

The diffusive flux of monomers to the tip can be easily estimated by Fick's law. If c_0 is the concentration in the cell body (i.e., at $x = 0$), and c_L is the concentration at the front (i.e., at $x = L$), then the flux of actin monomers (in particles per unit time per unit cross-sectional area) moving to the front of the extension is

$$j_D = (D_a/L)(c_0 - c_L), \tag{3}$$

where D_a is the diffusion coefficient of actin. The resulting concentration profile of actin monomers is linear.

Equating the fluxes of Eqs. (2) and (3) gives the steady state actin concentration at the front of the extension

$$c_L = \frac{c_0 + (nLk_-/D_a)}{1 + (nLk_+/D_a)e^{-\omega}}. \tag{4}$$

The steady state extension speed is then obtained by substituting c_L from Eq. (4) into Eq. (1).

The Actin Strainer

The Actin Strainer Model takes into account the possibility of osmotic effects. When polymerization draws down the actin concentration, water will tend to move to neighboring regions of higher solute concentration. This results in net transport of water through the permeable cell membrane. For a solute concentration difference of Δc, there will be a water flux

$$j_w = P_f \Delta c \tag{5}$$

through a membrane with a hydraulic permeability coefficient of P_f [11]. This flux will clearly be largest where the concentration difference is largest. Assuming that the cell body, with actin concentration c_0, is in osmotic equilibrium, we consider only the concentration difference elsewhere due to actin polymerization and obtain the driving force

$$\Delta c = (c_0 - c). \tag{6}$$

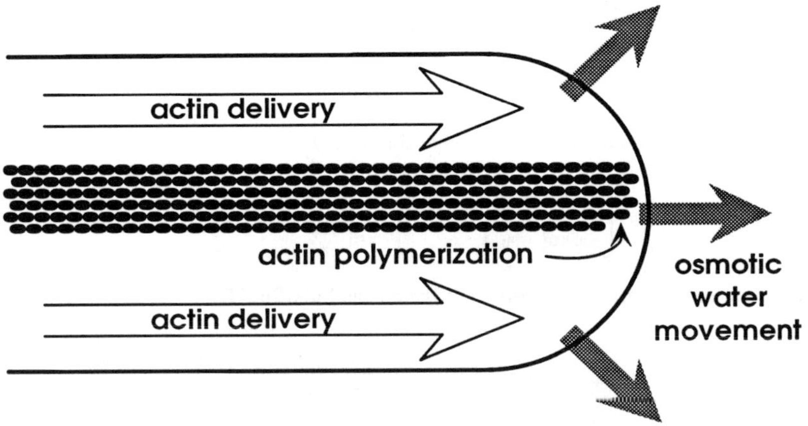

FIGURE 1. A schematic of the Actin Strainer Model for the acrosomal process. Polymerization takes place at the front. Actin monomers are carried forward both by diffusion and by bulk fluid flow in response to the escape of water through the cell membrane.

Since the volume of a cellular extension is stabilized by the presence of the actin network, we consider the volume of the extension to be constant. Neglecting the small compressibility of water, constant volume implies that water lost through the membrane around the extension must be replaced from the cell body. This fluid will carry a comparatively high concentration of actin. Thus actin delivery is enhanced by bulk flow when water leaving the membrane at the front causes actin-rich fluid to enter from the rear. Fig. 1 shows a schematic of this effect for the acrosomal process.

Due to diffusion driven by assembly at the front, the concentration of actin will be depressed relative to its value in the cell body virtually everywhere along the extension. As a result, an osmotic pressure difference will develop everywhere along the extension (acrosomal process or cell lamellipodium), leading to water exit through every element of the membrane around it. However, bulk flow will cause the concentration profile to be relatively flat near the cell body and drop sharply near the front. Therefore, in steady state, most of the water discharge will occur near the front. This situation can be approximated in a relatively tractable way by assuming that essentially all the osmotic flux occurs through the small area A_{mem} of membrane near the front that surrounds the region in which the actin monomer concentration is depleted at least halfway from c_0 to c_L. Under this assumption, water and actin monomers can only leave or enter the system at the two ends. With this simplification, the equations for the concentration of actin and the extension speed can be written explicitly and solved numerically.

Again we seek the steady state in which the fluxes of actin monomers in and out are in balance. In addition to the flux of actin by diffusion, there is a flux of actin carried by the fluid. This flux may be written $j_f = c_0 v_f$, where fluid moving with velocity v_f carries a concentration c_0 of actin from the cell body. The magnitude of the bulk flow is determined by how much water leaves the extension per unit time. Multiplying the water flux given by Eqs. (5) and (6) by the membrane area A_{mem} and the volume of a water molecule v_w gives the volume of water escaping from the front per unit time. The fluid speed is then

$$v_f = v_w P_f (c_0 - c_L)(A_{mem}/A_x), \tag{7}$$

where A_x is the cross sectional area of the extension.

The one-dimensional diffusion-convection equation at steady state is

$$D_a \frac{d^2 c}{dx^2} - v_f \frac{dc}{dx} = 0. \tag{8}$$

In the geometry described above, the solution to Eq. (8) is

$$c(x) = \left[\frac{c_0 \exp\left(v_f L / D_a\right) - c_L}{\exp\left(v_f L / D_a\right) - 1}\right] - \left[\frac{c_0 - c_L}{\exp\left(v_f L / D_a\right) - 1}\right] \exp\left(v_f x / D_a\right). \tag{9}$$

The total monomer flux carried forward by diffusion and bulk fluid flow is then

$$j_{tot} = -D_a \frac{dc}{dx} + c v_f = \frac{c_0 \exp\left(v_f L / D_a\right) - c_L}{\exp\left(v_f L / D_a\right) - 1} v_f. \tag{10}$$

At steady state, this flux is equal to the flux of monomers onto the filaments in Eq. (2). The steady state value for c_L is obtained numerically and the corresponding extension speed is then given by Eq. (1).

RESULTS

As we are interested in matching the experimentally observed speeds, it is important that we use realistic values for the parameters in the model. The values used are given in Table I, along with their sources.

For the goldfish keratocyte system, the diffusion-only model presented above, with the parameters listed in Table I, predicts a speed of 0.31 μm/s, in good agreement with the observed value of about 0.5 μm/s [4, 5]. Given that the concentration of actin, the load force, and other parameters may vary in living cells, we may consider the diffusion-only model to describe the movement of the goldfish keratocyte rather well.

The extension of the acrosomal process is a different story. Even when the diffusion-only model is applied to the acrosomal process at a short length (minimizing the distance that actin monomers need to travel), the model predicts an extension speed of only 17 nm/s. This is almost a thousand fold slower than the 6-12 μm/s observed experimentally [12]. It is also slow compared to the fish keratocyte. This is because the growing ends of the actin filaments in the acrosomal process are much more densely packed, requiring more monomers per unit extension.

To apply the actin strainer model to the acrosomal system, the determination of the membrane permeability and driving force for water movement through the membrane must be done with some care. The extension of the acrosomal process is preceded by other stages of the acrosomal reaction. In particular, there is an initial dramatic swelling of the periacrosomal compartment due to rapid influx of water [8]. This event suggests that water movement is functionally significant in this system. From the overall change in volume we can estimate an "effective osmotic activity coefficient" of γ_{eff}=160 for the actin monomers that are being released from storage aggregates at this stage. This large number actually reflects concomitant ion movements. However, in the absence of a more detailed model, we do not consider ion movements explicitly and lump their effects into the effective osmotic activity coefficient for actin which we insert in Eq. (7). From the rate with which the periacrosomal compartment approaches its expanded volume, we can obtain an estimate for the membrane permeability. This gives $P_f = 7.2 \times 10^{-2}$ cm/s, a value that falls in the general range for biological membranes. By comparison, red blood cells have a hydraulic permeability of about $P_f \approx 10^{-2}$ cm/s [11], while some renal membranes have permeabilities as high as $P_f \approx 3 \times 10^{-1}$ cm/s [13].

TABLE I. Parameter values used in calculating extension speeds.

Parameter (source)	Value
ratchet step size δ (1/2 monomer) [14]	2.7 nm
actin on rate constant k_+ [15]	11.3 $(\mu M\ s)^{-1}$
actin off rate k_- [15]	1.4 s^{-1}
actin diffusion coeff D_a [16]	10^{-8} $cm^2\ s^{-1}$
temperature T (assumed)	27°C
Lamellipodium of epithelial fish keratocyte:	
length L [4, 5]	3.0 μm
height h [17]	0.2 μm
end density n [18, 19]	4.0×10^8 cm^{-2}
reservoir concentration of actin c_0 [20, 21]	10 mg/ml
load force per fiber f [a]	0.875 pN
Acrosomal process of *Thyone briareus*:	
length L [12]	0-90 μm
radius r [12]	0.025 μm
end density n [b]	3.1×10^{10} cm^{-2}
reservoir concentration of actin c_0 [c]	55 mg/ml
load force per fiber f [a]	0.1 pN
membrane permeability P_f [d]	7.2×10^{-2} cm/s
effective osmotic activity coefficient γ_{eff} [d]	160

a. Calculated for each system assuming a membrane tension of 0.035 dyne/cm [10] distributed over the stated density of growing filament ends.
b. Electron micrograph show 60 filaments in the cross section of a 0.05 μm diameter process [12].
c. A 90 μm process of 60 filaments contains 2×10^6 monomers, originally stored in the periacrosomal compartment whose volume (after doubling) is 10^{-12} cm^3 [12].
d. See text.

Using these parameters and the others listed in Table I, the Actin Strainer Model predicts a velocity of 4.3 μm/s for the extension of the acrosomal process. Although falling a bit short of the 6-12 μm/s observed in the *Thyone* system, it is more than a hundred-fold improvement over the diffusion-only results (17 nm/s) in this system. The calculated velocity is very sensitive to some of the parameters we have estimated, particularly the membrane permeability. If the permeability of the membrane were double what we have used (still well within the known range of permeabilities of biological membranes), the calculated speed would climb to 6.6 μm/s. Changes in other parameters such as actin polymerization rate and the concentration of actin monomers in the cell body could similarly affect the calculated extension rate.

CONCLUSIONS

We have presented a model for the extension of actin filament networks incorporating both diffusion and bulk flow delivery of actin monomers to the growing ends. The bulk flow is driven by the escape of water through the cell membrane in response to osmotic pressure changes. Addition of actin monomers to the filament network is described by the Brownian Ratchet Model.

Speeds of extension are calculated with realistic parameters for two systems, epithelial goldfish keratocytes and the acrosomal process of *Thyone*. We find that the observed velocity of goldfish keratocyte locomotion is adequately supported by actin diffusion alone. The rapid extension of the acrosomal process of *Thyone* requires bulk flow and is adequately modeled by the Actin Strainer Model introduced here.

Some of the assumptions made in this model will be relaxed in future work. In particular, water should be allowed to escape through any region of the membrane, not just at the front. In addition, the steady state restriction should be relaxed for the extending acrosomal process. This will enable calculation of the time course of the extension of the acrosomal process and comparison of the length dependence of the velocity with experimental observations. A more explicit treatment of ionic movement in the acrosomal system would also be desirable.

ACKNOWLEDGMENTS

Supported by NIH grant HL-36546.

REFERENCES

1. T. J. Mitchison and L. P. Cramer, Cell **84**, 371 (1996).
2. J. A. Cooper, Annu. Rev. Physiol. **53**, 585 (1991).
3. J. Condeelis, Annu. Rev. Cell Biol. **9**, 411 (1993).
4. J. Lee, A. Ishihara, J. A. Theriot, and K. Jacobson, Nature **362**, 167 (1993).
5. J. A. Theriot and T. J. Mitchison, Nature **352**, 126 (1991).
6. L. H. Colwin and A. L. Colwin, Biol. Bull. **109**, 357 (1955).
7. J. C. Dan, in *Fertilization*, edited by C. B. Metz and A. Monroy (Academic Press, New York, 1967), Vol. I, p. 237.
8. S. Inoué and L. G. Tilney, J. Cell Biol. **93**, 812 (1982).
9. A. S. Perelson and E. A. Coutsias, J. Math Biol. **23**, 361 (1986).
10. C. S. Peskin, G. M. Odell, and G. F. Oster, Biophys. J. **65**, 316 (1993).
11. R. I. Macey and J. Brahm, in *Water Transport in Biological Membranes*, edited by G. Benga (CRC Press, Boca Raton, 1989), Vol. II, p. 26.
12. L. G. Tilney and S. Inoué, J. Cell Biol. **93**, 820 (1982).
13. P. Agre, G. M. Preston, B. L. Smith, J. S. Jung, S. Raina, C. Moon, W. B. Guggino, and S. Nielsen, Am. J. Physiol. **265**, F463 (1993).
14. D. J. DeRosier and L. G. Tilney, in *Cell and Muscle Motility*, edited by J. W. Shay (Plenum Press, New York, 1984), Vol. 5, p. 139.
15. T. D. Pollard, J. Cell Biol. **103**, 2747 (1986).
16. S. Felder, PhD thesis, Washington University, 1984.
17. D. Bray and J. G. White, Science **239**, 885 (1988).
18. K. Luby-Phelps, D. L. Taylor, and F. Lanni, J. Cell Biol. **102**, 2015 (1986).
19. J. H. Hartwig and P. Shevlin, J. Cell Biol. **103** (1986).
20. T. P. Stossel, in *Inflammation: Basic Principles and Clinical Correlates*, edited by J. I. Gallin, I. M. Goldstein and P. Snyderman (Raven Press, New York, 1988).
21. T. D. Pollard, K. Fujiwara, R. Handin, and G. Wiess, Ann. NY Acad. Sci. **283**, 218 (1977).

AN AVOIDANCE MODEL FOR SHORT-RANGE ORDER INDUCED BY SOFT REPULSIONS IN SYSTEMS OF RIGID RODS

JINING HAN and JUDITH HERZFELD*
Dept. of Chemistry, Brandeis University, Waltham, MA, 02254-9110
* Corresponding author

ABSTRACT

The effects of soft repulsions on hard particle systems are calculated using an avoidance model which improves upon the simple mean field approximation. The method not only yields a better free energy, but also gives an estimate for the short-range positional order induced by soft repulsions. The results indicate little short-range order for isotropically oriented rods. However, for parallel rods short-range order increases to significant levels as the particle axial ratio increases.

INTRODUCTION

Colloidal suspensions are often described as hard particle gases. However, hard particle systems are scarce in reality. In particular, suspended particles are often charged and electrostatic forces must be taken into account. In a seminal paper [1], Onsager suggested absorbing the effects of soft repulsions in an enlarged effective size for the particles. This approach has been elaborated by Stroobants et al.[2]. In particular, they found that for very long, rigid, charged rods in solution, the effects of the soft repulsions can be characterized by two parameters: the effective diameter and the so-called "twisting" parameter which represents the torque between two rods interacting with soft repulsions. These workers used a second order virial expansion in their derivation which gives asymptotically correct results as the concentration goes to zero. A similar effective hard particle model (using virial expansion to higher orders) has been used by Minton and Edelhoch to fit light scattering [3] and osmotic pressure [4] data for a charged globular protein (bovine serum albumin) at volume fractions less than 10%. Unfortunately, these methods of treating soft repulsions cannot be applied at higher concentrations. If the effective diameter obtained in the low concentration limit is assumed naively to be independent of the concentration, the osmotic pressure is overestimated at higher concentrations. Clearly, as the concentration increases, the particles must make greater use of configurations that they might populate infrequently at low concentrations. The effective radius must therefore decrease with increasing concentration, and a single value valid for low concentrations is not appropriate for higher concentrations.

Historically, there is another way of treating soft repulsions. The method, introduced by van Laar [5], and later used extensively by Flory [6], employs a mean field approximation. Whereas Onsager introduced an enlarged effective radius, which represents a kind of short-range order, the mean field approximation ignores any short-range order. In other words, the mean field approximation assumes that the positions and orientations of any two particles are completely uncorrelated except as necessary to avoid interpenetration of their hard cores. With this assumption, the average potential energy can be calculated easily. Combined with an estimate of the configurational entropy, the theory provides the free energy of the system and thus a full thermodynamic description.

The basic mean field approximation is simple, but does not consider the possible short-range order induced by the soft repulsions. With soft repulsions, it is energetically unfavorable for particles to approach one another. Approximating this correlation in an all-or-none fashion, it is as if particles with a larger effective size interact with weaker soft repulsions. Since this effective size is different from Onsager's, subsuming only part of the soft repulsions, we will refer it as the "avoidance size." A larger avoidance size reduces (but does not eliminate) the energy due to soft repulsions. On the other hand, it also reduces the configurational entropy. Under suitable conditions, one can find an optimal avoidance size that minimizes the free energy. This

Mat. Res. Soc. Symp. Proc. Vol. 463 © 1997 Materials Research Society

variational method has two advantages over the usual mean field approximation: it provides a better estimate of the free energy and also gives a semi-quantitative picture of the short-range order induced by soft repulsions.

The soft interactions between mesoscopic particles in solution can be complicated, including, e.g., electrostatic interactions, van der Waals interactions, and interactions due to perturbations of water structure. To illustrate the method here, we will use a screened electrostatic force. The shapes and orientations of particles can also vary widely. For mathematical simplicity, we will consider only monodisperse spherocylinders with isotropic or parallel orientations. For these systems, we use scaled particle theory [7] to estimate the configurational entropy as a function of avoidance size. The residual soft repulsion energy for any avoidance size is calculated readily in the mean field approximation. Then the minimization procedure outlined above can be carried out to obtain a better estimate of the free energy and avoidance size.

THEORY

Consider a system of isotropic or parallel spherocylinders with hard core diameter D_0 and cylinder length L. As a measure of particle asymmetry, it is convenient to imagine that each spherocylinder is a linear aggregate of n spheres with the same diameter and conserved volume, in which case $L=(2/3)(n-1)D_0$. The free energy per particle (in units of $k_B T$) includes three contributions:

$$f = f_{\text{ideal}} + f_{\text{conf}} + f_{\text{rep}}. \tag{1}$$

The ideal part of the free energy per particle is given by

$$f_{\text{ideal}} = \ln\left(c_p \Lambda^3\right) - 1, \tag{2}$$

where c_p is the particle number density and $\Lambda = h/(2\pi m k_B T)^{1/2}$ is the thermal wavelength of a particle of mass m.

f_{conf} is the configurational free energy per particle. Scaled particle theory [7] for hard spherocylinders gives:

$$f_{\text{conf}} = -\ln(1-v_a) + B\left(\frac{v_a}{1-v_a}\right) + C\left(\frac{v_a}{1-v_a}\right)^2, \tag{3}$$

where $v_a = c_p\left(\pi D^3/6 + \pi D^2 L/4\right)$ is the apparent volume fraction for the avoidance diameter D, and the coefficients are given by

$$B = \begin{cases} 3+(3/2)(L/D)^2/[1+(3/2)(L/D)] & \text{for isotropic rods} \\ 3 & \text{for parallel rods} \end{cases} \tag{4}$$

and

$$C = \begin{cases} (3/2)[1+(L/D)]^3/[1+(3/2)(L/D)]^2 & \text{for isotropic rods} \\ (3/2)[1+3(L/D)+2(L/D)^2]/[1+(3/2)(L/D)]^2 & \text{for parallel rods} \end{cases}. \tag{5}$$

Defining $v_s = c_p(\pi D_0^3/6 + \pi D_0^2 L/4) = nc_p(\pi D_0^3/6)$ to be the hard core volume fraction of solute and $\alpha = D/D_0$, it follows that

$$v_a = v_s \alpha^2(\alpha + n - 1)/n = \begin{cases} \alpha^3 v_s & \text{for } n = 1 \\ \alpha^2 v_s & \text{for } n >> 1 \end{cases}. \tag{6}$$

To calculate f_{rep}, the contribution of soft repulsions to the free energy per particle, we approximate the soft repulsion energy between two cylinders as the sum of that between pairs of constituent spheres with the pair radial distribution function

$$g(r) = \begin{cases} 0 & \text{for } r \le D \text{ (the avoidance region)} \\ 1 & \text{for } r > D \text{ (the mean field approximation)} \end{cases}. \tag{7}$$

There are n^2 such pairs. Suppose the soft repulsion potential energy between two spheres is given by

$$u(r) = \frac{J}{r}\exp(-r/\lambda), \tag{8}$$

where J and λ are the strength and the decay length of the interaction, respectively. Then,

$$f_{rep} = \frac{1}{2} n^2 c_p \int u(r) g(r) 4\pi r^2 dr. \tag{9}$$

Substitution of Eqs. (7) and (8) into Eq. (9) gives

$$f_{rep} = 12 v_s (nJ\lambda^2/D_0^3)(1 + D/\lambda)\exp(-D/\lambda). \tag{10}$$

This calculation of the dependence of f_{rep} on the avoidance diameter is obviously exact for $n = 1$. For $n > 1$ it is accurate within a geometric factor of order of magnitude one.

The optimal avoidance diameter D is that which minimizes the free energy per particle given by Eq. (1). For given values of n and v_s or c_p, the minimization can be accomplished numerically to give α. If $\alpha > 1$, then there is short-range order with an avoidance diameter $D = \alpha D_0$. Otherwise we simply have $D = D_0$, indicating that in this case it is not advantageous for a particle to have an avoidance diameter larger than the hard core diameter. The value of α represents a trade-off between f_{conf} and f_{rep}, since f_{ideal} has no D dependence. The solution value of α obviously depends on the hard core volume fraction v_s. In general, at lower concentrations the configurational free energy is small compared to the repulsive free energy and there is a solution with $\alpha > 1$. As the concentration increases, the configurational free energy increases faster than the soft repulsion free energy (i.e., the soft repulsions become relatively less important) and, as a result, α approaches 1. Above some concentration, $D = D_0$ (and $v_a = v_s$). Once α is obtained, the free energy [Eq. (1)] and the osmotic pressure ($\Pi = c_p^2 \, \partial f / \partial c_p$) can also be calculated.

RESULTS

We have used the avoidance model to study both isotropic and parallel spherocylinders. For the isotropic case, we do not find $\alpha > 1$ unless the solution is very dilute or repulsions are long range (λ is large) and strong (say $J > 10$, using energy units of $k_B T$ and distance based on volume units of $\pi D_0^3/6$). Even then, the free energy density and osmotic pressure calculated using our method are not substantially different from that calculated using Flory's simple mean field approximation (i.e., $\alpha = 1$).

For parallel spherocylinders, the situation changes drastically: soft repulsions with moderate J values can have an appreciable effect for large n because $\left(f_{\text{conf}}/f_{\text{rep}}\right)$ for parallel particles is reduced with increasing axial ratio. The range of the soft repulsion λ is also important. Figure 1 shows three examples with different combinations of λ and n. In all cases, $J = 2$. We do not show the effect of varying J since Eqs. (2)-(6) show that, for parallel particles, when n is large, f_{conf} loses its dependence on n. In this case, $(\partial f/\partial D)$ depends on J and n only as the product Jn from Eq. (10).

An assessment of the degree of short-range order is provided by α, the ratio of the avoidance diameter to the hard core diameter. As can be seen from Figure 1A, α starts with a finite maximum value in the limit of zero volume fraction and decreases with increasing volume fraction. As discussed in the introduction, this is the qualitatively expected behavior. As the system becomes more and more crowded, avoidance becomes more difficult and eventually ceases altogether ($\alpha = 1$). Figure 1A also shows that the short-range order decreases when either the range of the force (λ/D_0) or the length of the rods (n) is reduced.

Another measure of short-range order is given by the behavior of the apparent volume fraction as a function of the hard core volume fraction. Figure 1B shows that, as the hard-core volume fraction increases from zero, the apparent volume fraction initially increases very sharply and then saturates. The effect is more pronounced for the longer rods and longer interaction ranges. In the cases shown, it is only at very high (probably physically irrelevant) volume fractions that the apparent volume fraction becomes the same as the hard core volume fraction.

It is also interesting to compare the osmotic pressures obtained using the avoidance model and the simple mean field approximation. We found that in most cases the two are not substantially different. The largest effect is seen when the soft repulsion free energy and the configurational free energy are comparable in magnitude. In this range, the avoidance model gives a lower osmotic pressure at lower concentrations but a higher osmotic pressure at higher concentrations. This behavior arises because the avoidance model gives a lower free energy, except at extremely low concentrations (the ideal case) and very high concentrations (with no avoidance) where the two models yield identical free energies. Since the osmotic pressure is proportional to the derivative of the free energy per particle with respect to particle number concentration, the two osmotic pressure curves must cross.

CONCLUSIONS

In summary, we have proposed an avoidance model to calculate a variational free energy for hard particles with soft repulsions. The method is an improvement over the simple mean field approximation. We have applied the method to isotropic and parallel spherocylinders with screened electrostatic interactions and found extensive short-range order for parallel rods with large axial ratios.

ACKNOWLEDGMENTS

This work was supported by NIH grant HL-36546.

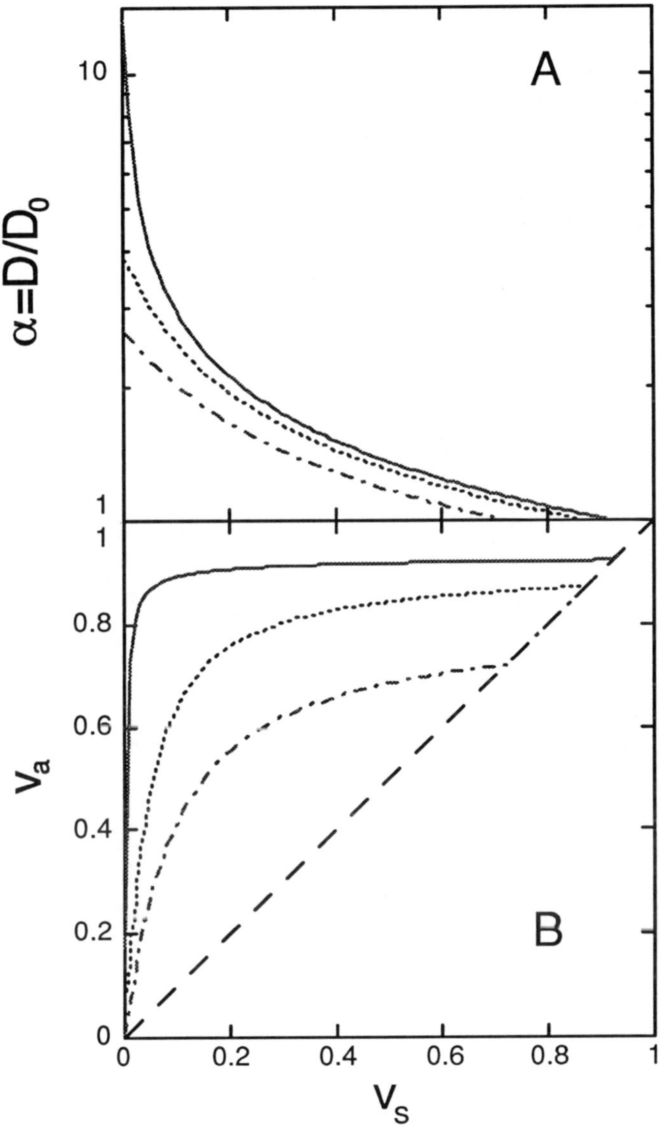

FIGURE 1. The ratio of avoidance diameter to hard core diameter (A) and the apparent volume fraction (B) vs. hard core volume fraction for parallel spherocylinders with $J = 2$ (using energy units of $k_B T$ and volume units of $\left(\pi D_0^3/6\right)$), and $n = 1000$ and $\lambda/D_0 = 2$ (——), or $n = 1000$ and $\lambda/D_0 = 0.5$ (-------), or $n = 100$ and $\lambda/D_0 = 0.5$ (–·–·–·). In the absence of short-range order $\alpha = 1$ and $v_a = v_s$ (– – – –).

REFERENCES

[1] L. Onsager, Ann. N. Y. Acad. Sci. **51**, 627 (1949).

[2] A. Stroobants, H. N. W. Lekkerkerker, and T. Odijk, Macromolecules **19**, 2232 (1986).

[3] A. P. Minton and H. Edelhoch, Biopolymers **21**, 451 (1982).

[4] A. P. Minton, Biophys. Chem. **57**, 65 (1995).

[5] J. J. van Laar, *Sech vortrage uber das thermodynamische potential* (Vieweg und Sohn, Braunschweig, 1906).

[6] P. J. Flory, Proc. Roy. Soc. London ser. A **234**, 73 (1956).

[7] M. A. Cotter and D. C. Wacker, Phys. Rev. A **18**, 2669 (1978).

COMPUTATIONAL MODELS FOR MULTICOMPONENT DIFFUSION IN POLYMERIC MATERIALS

G. ROSSI, M.A. SAMUS
Ford Research Laboratory, Ford Motor Company, P.O. Box 2053, Mail Drop 3083, Dearborn, MI 48121-2053.

ABSTRACT

Situations where a polymeric material is exposed to a solvent mixture so that the different components within the mixture can diffuse into the polymer are common both in industrial applications and in biological processes. Often one of the components is taken up preferentially and its presence affects the diffusion properties of the remaining components. The problem of accounting for processes of this type has not been dealt with in a systematic way. This may in part be due to the difficulty of characterizing experimentally the separate diffusion behavior of the various components: data of this kind are now becoming available for simple binary mixtures. In order to model this class of problems, a lattice model involving a polymer matrix (M) and two diffusing components (A and B) has been introduced. The Monte Carlo evolution of the system has been examined for different values of the local A–M, B–M and A–B interactions. These results shed light on the microscopic origin of selective uptake.

INTRODUCTION

Diffusion of solvents in polymeric materials continues to be an active field of investigation. Considerable progress has been made both in the experimental techniques used to study processes of this kind and in the theoretical interpretation of these experiments [1]. However, thus far the overwhelming majority of the research effort has been devoted to situations where a single diffusant penetrates a polymeric material. By contrast, in many practical industrial applications as well as in biological processes, the polymer is often exposed to a mixture of different chemical components, each of which can diffuse into the polymer.

Several experimental techniques are becoming available to study the separate diffusion of each component. For example, under appropriate conditions, FTIR spectroscopy can be used to obtain separate sorption curves tracking the weight uptake of each species within a polymeric sample as a function of time [2]. Techniques such as Rutherford backscattering spectroscopy [3] and NMR imaging [4], that were devised to obtain the concentration profiles of a single diffusant at a given time, may also be used in the case of mixed diffusion to obtain the concentration profile of a specific component that has been labelled for this purpose. Even in industrial applications, such as permeability testing of plastic films [5], use of gas chromatography or mass spectrometry is becoming increasingly common, since often there is a need to ascertain the composition of diffusant permeating across a film.

At least in principle, the information obtained from the aforesaid experimental techniques can be used to determine the phenomenological parameters, namely diffusion coefficients and solubilities (or final equilibrium concentrations), that enter a macroscopic theoretical description of the diffusion process. In the simplest situations, such a description consists of a set of differential equations for the local concentration $\phi_i(\mathbf{x}, t)$ of each diffusing component at time t and at location $\mathbf{x} \equiv (x, y, z)$ within the polymeric sample. For example, for two diffusing components A and B, the equations are:

$$\frac{\partial \phi_A(\mathbf{x}, t)}{\partial t} = \nabla \cdot \Big(D_A(\phi_A, \phi_B) \, \nabla \phi_A(\mathbf{x}, t) \Big) \qquad (1a)$$

$$\frac{\partial \phi_B(\mathbf{x}, t)}{\partial t} = \nabla \cdot \Big(D_B(\phi_A, \phi_B) \, \nabla \phi_B(\mathbf{x}, t) \Big) \qquad (1b)$$

Mat. Res. Soc. Symp. Proc. Vol. 463 © 1997 Materials Research Society

Here D_A and D_B are respectively the diffusion coefficients for species A and for species B. In general, these diffusion coefficients are functions of the local concentrations of both diffusants, and this concentration dependence couples the equations for ϕ_A and ϕ_B.

The physical assumptions underlying eqs. (1) are straightforward. In accordance with Fick's phenomenological law of diffusion, the local flux of each species is proportional to the gradient $\nabla\phi_i(\mathbf{x}, t)$ of the corresponding local concentration, with the proportionality constant given by the diffusion coefficient D_i. This, together with the equation of continuity for each species, yields eqs. (1). In the simplest case where equilibrium is established at the polymer surface as soon as the polymer is put in contact with the mixture, the boundary conditions will require that the concentration of each species at the surface be equal to the final equilibrium concentration Φ_i for that species. However, it should be noted that for two diffusing components, the assumption that equilibrium is reached at the sample surface as soon as the polymer is exposed to the mixture, is stronger than in the single component case, since it implies equilibration not only for the total amount of solvent within the polymer but also for the separate concentration of each species. In general, other processes, such as swelling [6], development and release of macroscopic elastic stresses [1,7], and changes of state such as plasticization [7–9], may accompany diffusion in polymeric materials. The simplified picture described above must then be modified to account for these phenomena as it is done in the case of a single diffusant [1].

It should be clear from this brief overview that in practice the task of obtaining from experiments the phenomenological input quantities that enter a macroscopic description may be quite complex. Yet there is a need for predictive tools capable of addressing practical questions of the kind encountered in polymeric barrier technology, *e.g.*, to predict both the amount and the composition of permeant that escapes across a given (single-layer or multi-layer) polymeric barrier in a given time. Molecular level computer simulations of the diffusion process [10] are at present of little help in this respect, due to the relatively small size of the systems that can be treated in this way. In the rest of this paper, we describe a class of lattice models that aim at bridging the gap between molecular level simulations and the macroscopic description outlined above. The goal of this work is to provide a simple tool capable of helping both in visualizing the diffusion process and in modelling how diffusion is affected by the interactions between the different species involved.

COMPUTATIONAL MODELS

Our starting point is the kind of lattice gas models that have been used extensively in the literature [11-13] to study cluster growth during phase segregation in a binary mixture. In the prototype such model [11], each of the N sites of a two dimensional square lattice is occupied either by a type A or type B particle. The microscopic state of the system can be specified by associating to each site a variable η_i ($i = 1, .., N$), such that $\eta_i = 1$ for sites occupied by A particles and $\eta_i = -1$ for sites occupied by B particles. The particles interact via an Ising hamiltonian:

$$H = -J \sum_{<i,j>} \eta_i \eta_j \qquad J > 0 \tag{2}$$

where $< i, j >$ means that the sum runs over nearest neighbors. At each time step, the particle occupying two randomly chosen nearest neighbor sites are exchanged with a probability W given by the usual Metropolis prescription:

$$W = Min\Big[1, e^{-\Delta H}\Big] \tag{3}$$

where ΔH is the energy difference between the updated and the starting configuration: this kind of Monte Carlo evolution preserves the number of particles of each species. For J larger than a critical (Onsager) value J_c, this system is known to phase separate. Much work has gone towards studying the time evolution of the structure function and of the cluster distribution following quenching above J_c [11-13]. If, by contrast, J is taken to be smaller than J_c but the system is started in a non equilibrium configuration (for example, one where the right half of the lattice is occupied by A particles and the left half is occupied by B particles), the model becomes a tool to study the kinetics of mixing, *i.e.*, interdiffusion, of A and B. In the non interacting case $J = 0$, each particle simply performs a random walk and the mutual diffusion coefficient coincides with the self diffusion coefficient of the single particle.

The models of diffusion of small molecules in a polymeric solid that we consider here are variants of the above interdiffusion model. In order to be able to deal with relatively large systems, we renounce a description of the motion of the separate chains that make up the polymeric material. Instead, the polymer is treated as a lattice background. Solvent particles move in this background according to the same kind of Monte Carlo kinetics outlined above. Specifically, consider a situation where two solvent species A and B can diffuse into the polymeric material M. An interaction energy χ_{RS} is associated with each lattice bond located between a site occupied by species R and a site occupied by species S: here R and S can take the values A, B or M. The total energy H is the sum of the interaction energies χ_{RS} associated with each bond. Again at each time step a bond is chosen at random and the particles occupying the sites joined by this bond are exchanged with a probability given by eq. (3).

Since we want to model practical experimental situations, such as sorption or permeation across a film, we must give appropriate prescriptions to deal with the lattice boundaries. Here we restrict ourselves to two dimensional lattices of $N = L_x L_y$ sites that are periodic in the y direction: *e.g.*, site $(m, 1)$ coincides with site $(m, L_y + 1)$ $(m = 1, .., L_x)$. To handle sorption we take the edge at $m = L_x$ to be reflecting: *e.g.*, any exchange involving particles located at sites (L_x, n) $(n = 1, .., L_y)$ is forbidden. On the remaining edge, $m = 1$, we fix the concentration of A and B: in other words any particle exchange involving sites $(1, n)$ and $(2, n)$ is accepted, however, when such an exchange takes place, a site along the edge containing a particle of the type moved to $(1, n)$ is chosen at random and updated to contain a particle of the type moved to $(2, n)$. To handle permeation across a film, separating a region where solvent is present from one devoid of solvent, we use the prescription given above for the edge at $m = 1$ (solvent side), and we take the edge at $m = L_x$ (empty side) to be absorbing: in other words, all exchanges involving an A or B particle moving to these sites are accepted, but the A or B particle is then destroyed, *e.g.*, updated to become an M particle. Both the sorption and the permeation boundary conditions described here imply that the numbers of particles of each species present in the lattice are not conserved.

RESULTS

In this section we present a few examples of the type of information that can be obtained using the procedures outlined above. Figure 1a and 1b display results from sorption simulations. Figure 1a shows how the final equilibrium uptake of A particles changes as the concentration of A particles at the $m = 1$ boundary increases, for two different values of the interaction parameter χ_{AM}: $\chi_{AM} = 0.6$ for the lower set of data and $\chi_{AM} = 0.8$ for the upper set of data. In both cases, no B particles were present and $\chi_{AA} = \chi_{MM} = 0$. Under these conditions, the concentration of A particles at the $m = 1$ boundary plays the role of the vapor pressure in an actual sorption experiment, so that the results shown in figure 1a can be viewed as the analog of sorption isotherms. As the (repulsive) interaction energy χ_{AM} between the diffusant and the polymer increases, marked departure from Henry's law behavior (shown by the broken line) is observed at lower and lower values of the boundary concentration. Henry's law holds independent of the boundary concentration for the ideal (non interacting) case $\chi_{AM} = \chi_{AA} = \chi_{MM} = 0$: in this case each A particle simply performs a random walk and the procedure described above amounts to solving Laplace's equation with a fixed concentration at the $m = 1$ boundary. These results can be qualitatively explained in terms of an effective particle–particle attraction resulting from the repulsive interaction between the diffusing particles and the polymer background.

Figure 1b shows how the final equilibrium uptake of B particles changes when, at the $m = 1$ boundary, the concentration of B particles is held fixed and the concentration of A particles is

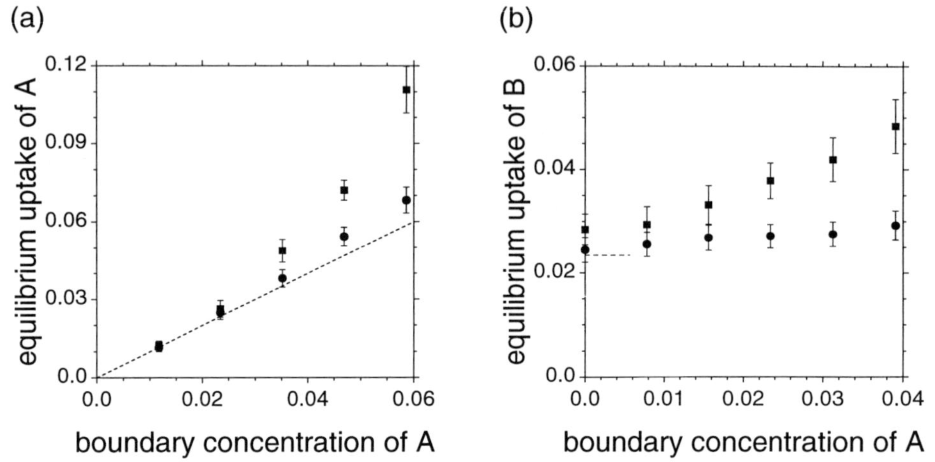

FIGURE 1. Results from sorption simulations: (a) gives the final equilibrium uptake of A particles as a function of the boundary concentration of A particles for $\chi_{AM} = 0.6$ (•) and $\chi_{AM} = 0.8$ (▪); no B particles were present and $\chi_{AA} = \chi_{MM} = 0$; the broken line corresponds to the behavior expected for the non interacting case $\chi_{AM} = 0$. (b) gives the final equilibrium uptake of B particles as a function of the boundary concentration of A particles for $\chi_{AM} = \chi_{BM} = 0.6$ (•) and $\chi_{AM} = \chi_{BM} = 0.8$ (▪); here the boundary concentration of B particles is $3/128$ (this value is indicated by the broken line) and $\chi_{AA} = \chi_{BB} = \chi_{AB} = \chi_{MM} = 0$.

increased. To obtain the results shown here we took $\chi_{AA} = \chi_{BB} = \chi_{AB} = \chi_{MM} = 0$ while $\chi_{AM} = \chi_{BM} = 0.6$ for the lower set of data and $\chi_{AM} = \chi_{BM} = 0.8$ for the upper set of data. Again, an effective A–B attraction resulting from the particle-polymer repulsion provides a rationale for these results.

Figure 2 shows results obtained from a permeation simulation. In this case A (black) particles and B (dark gray) particles permeate across a wall made of three different polymer layers. The χ parameters are chosen in such a way that there is an attractive interaction between the B particles and the polymeric material in the central layer (white in figure 2a) and also between the A particles and the polymeric material that makes up the two outer layers (light gray in figure 2a). Furthermore, there is a repulsive interaction between the A particles and the polymer in the central layer. Use of multilayer structures of this kind to reduce permeation of specific components is common in polymeric barrier technology [14]. In the present simulation when an A or B particle crosses the boundary between two layers, the site left empty reverts to the layer to which it initially belonged. Note that to handle this situation in the macroscopic description of eqs. (1), continuity of both the flux and the ratio between concentration and solubilities for each particle species must be imposed at the boundary between the two layers.

A snapshot of the system taken after steady state has been reached is shown in figure 2a: the central layer acts as a barrier for the A (black) particles, but takes up a significant amount of B (dark gray) particles. These features are also evident in the concentration profiles of figures 2b and 2c. In 2b the continuous and broken curves are within statistical uncertainty: this indicates that steady state is reached for the A particles much sooner than for the B particles. Attainment of saturation for the B particles in the central region is very slow due to the limited rate at which B particles can diffuse through the left layer. The steady state concentration profile of A displays a sharp drop at the interface between the left and center layers, but increases again slightly at the interface between the center and right layer. Although the boundary concentration of A at the left boundary is three times greater than that of B, the steady state permeation rate of A is much smaller than that of B, as seen by the slopes of the concentration profiles in the right layer. By choosing appropriately the solubilities of the two species in the different layers it is possible to

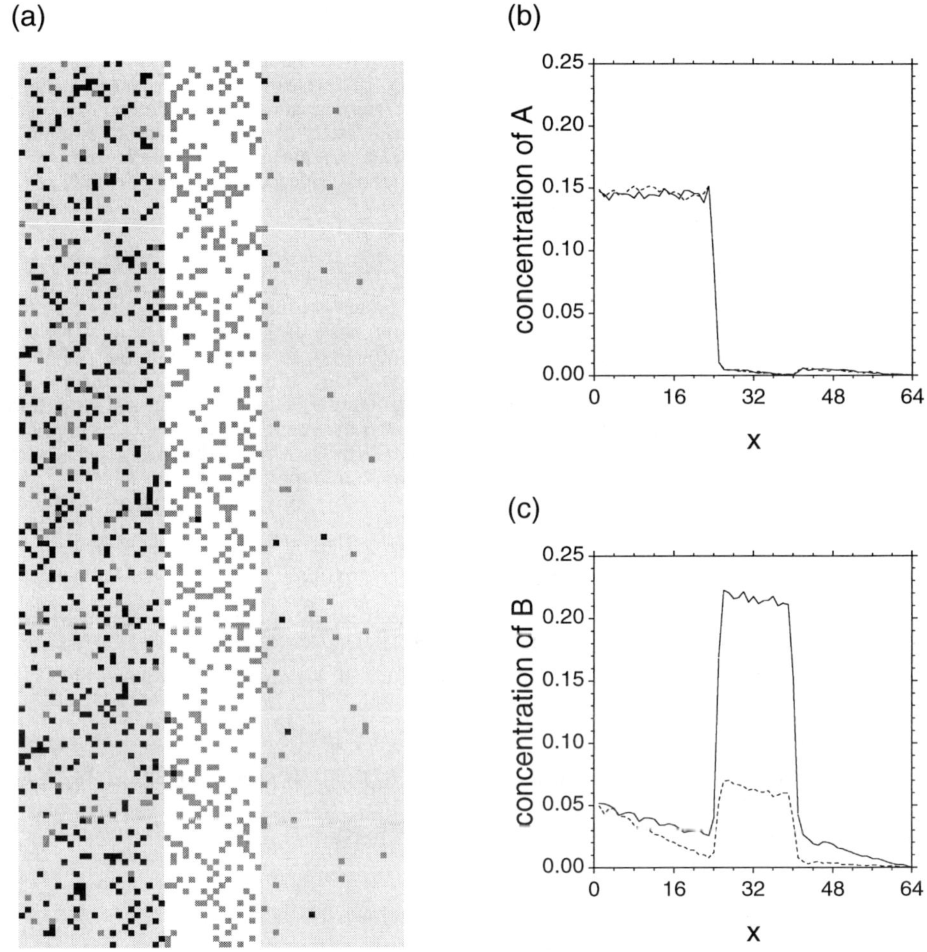

FIGURE 2. Results from a permeation simulation on a 64×256 lattice. A (black) and B (dark gray) particles diffuse across a three layer structure: the central layer is 16 lattice sites wide. At the left edge the concentration of A and B particles is fixed: it is 0.15 for the A particles and 0.05 for the B particles; particles can escape from the right edge. The interaction parameters between the diffusing particles are $\chi_{AA} = \chi_{BB} = 0.5$, $\chi_{AB} = -0.25$. Furthermore, $\chi_{AM} = -0.5$, $\chi_{BM} = \chi_{MM} = 0$ for the two outer (light gray) polymer layers, and $\chi_{AM} = 0.2$, $\chi_{BM} = -0.5$, $\chi_{MM} = 0$ for the central (white) layer. (a) is a snapshot of a portion of the lattice, taken after steady state had been reached. (b) and (c) give the concentration profiles of A and B particles respectively after 10^8 (broken curves) and $2 \cdot 10^9$ (continuous curves) time steps.

obtain similar qualitative concentration profiles from the macroscopic description of eqs. (1).

For the values of the interaction parameters used here there is no detectable dependence of the diffusion coefficients on particle concentration. Indeed the slope of the concentration profiles at steady state are constant through each layer, within statistical uncertainty. By contrast, in a practical experimental situation the diffusion coefficient can change substantially (for glassy polymers by orders of magnitude [7]) when a relatively small amount of solvent is taken up by the polymer. In principle this behavior can be included in the model described above by further biasing the exchange probability of each particle species. Likewise one can devise *ad hoc* rules to account for (passive) swelling. However, we stress that the motivation for introducing this model is not to mimic *a posteriori* specific features of experimentally observed behavior. Rather, as noted above, our aim is to gain understanding of the relation between local particle interactions and the macroscopic description surveyed in the introduction. Indeed, the present methods can be viewed as the lattice (microscopic) realization of eqs. (1).

CONCLUSIONS

In summary, we have described a class of simple Monte Carlo lattice models that we believe can be a useful arena to study the statistical (microscopic) basis of the diffusion process for both single component and multicomponent diffusion in polymers. In these models the polymeric material is treated as a passive lattice background, and no attempt is made to describe the motions of single polymer chains. As a result, properties that are determined by these motions, such as the actual values of the diffusion coefficients and their concentration dependence, cannot be predicted within this framework [15]. By contrast, equilibrium properties, such as sorption isotherms or the barrier effect due to repulsive polymer solvent interactions, are reproduced by our method.

REFERENCES

1. For a recent review see G. Rossi, Polym. Trends **4**, 337 (1996).
2. M.A. Samus and G. Rossi, in Multi-Dimensional Spectroscopy of Polymers, edited by M. W. Urban and T. Provder (ACS Symposium Series **598**, 1995), p. 535.
3. J.F. Romanelli, J.W. Mayer, E.J. Kramer and T.P. Russell, J. Polym. Sci.: Polym. Phys. Ed. **24**, 263 (1986).
4. M.A. Rana and J.L. Koenig, Macromolecules **27**, 3727 (1994).
5. R.L. Demorest, J. Plast. Film Sheeting **8**, 109 (1992); W.J. Kollen and P.L. Murthy, Annu. Tech. Conf. - Soc. Plast. Eng. **1**, 1108 (1992).
6. G. Rossi and K.A. Mazich, Phys. Rev. **E48**, 1182 (1993).
7. M.A. Samus and G. Rossi, Macromolecules **29**, 2275 (1996).
8. G. Rossi, P. Pincus and P.-G. de Gennes, Europhys. Lett. **32**, 391 (1995).
9. A. Friedman and G. Rossi, Macromolecules *in press.*
10. D.N. Theodorou, in Diffusion in Polymers edited by P. Neogi (Marcel Dekker, 1996), p. 67.
11. A.B. Bortz, M.H. Kalos, J.L. Lebowitz and M.A. Zendejas, Phys. Rev. **B10**, 535 (1974).
12. J. Marro, J.L. Lebowitz and M.H. Kalos, Phys. Rev. Lett. **43**, 282 (1979).
13. F.J. Alexander, C.A. Laberge, J.L.Lebowitz and R.K.P. Zia, J. Stat. Phys. **82**, 1133 (1996).
14. R. Colvin and S. Moore, Modern Plastics, October 1996, p. 67.
15. Note that these properties are, at least in principle, calculable using the type of molecular level computer simulations described in ref. [10].

COMPARISON OF EXPERIMENTAL AND THEORETICAL MELTING BEHAVIOR OF DNA

KENNETH A. MARX*, J. W. BIZZARO, IMAN ASSIL and R.D. BLAKE#

Department of Chemistry, University of Massachusetts, Lowell, MA. 01854 and
Department of Biochemistry, Microbiology and Molecular Biology#, University of Maine
Orono, ME 04469

ABSTRACT

High resolution melting curves of total nuclear *Dictyostelium discoideum* DNA (AX3 strain) are compared to theoretical melting calculated from GENBANK sequences(1.74 % of total) by the statistical thermodynamics program MELTSIM, parameterized for long DNA sequences(1,2). The lower and upper limits of simulated melting agree quantitatively with the experimental melting of total DNA. Calculated melting of coding, intron and flanking regions indicate that intron and flanking DNAs are extremely (A+T)-rich and account for the earliest melting DNA. There is no temperature overlap of these regions with coding DNA. A theoretical denaturation map of DNA containing the ribosomal DNA genes showed excellent agreement with subtransition positions of these genes in experimental curves. Agreement between these calculated and experimental melting data demonstrates our ability to accurately simulate DNA melting in complex eukaryotic genomes. This has important consequences for the understanding of sequence dependent energetic properties of nucleic acids and their potential use as biomaterials.

INTRODUCTION

Like other evolved biological macromolecules, DNA, by itself and in association with other molecules, possesses a number of properties that we associate with intelligent materials. We have previously described these properties(3). A requirement to rationally implement the use of DNA as a biomaterial is an understanding of the thermodynamic stability of the double helical form as a function of its sequence and environment. It is this consideration, along with the increasing availability of DNA sequences in the public databases, that has led us to undertake the following study comparing theoretical DNA melting behavior, computed from database sequence information, to experimental data.

The slime mold, *Dictyostelium discoideum*, is a lower eukaryote, long studied as a simple developmental model system, for which considerable genetic linkage information exists. There is developing a significant DNA sequence database for this organism in GENBANK. The genome of *Dictyostelium* is also of interest because of its extreme utilization of (A+T) base pairs. The genome has an average base composition of 24.1% (G+C) determined from the midpoint of the melting curve(4). This fact makes it of great interest from a physical chemical, information theory, as well biomaterials point of view.

Since *Dictyostelium* DNA begins melting as early as the poly (dA.dT) marker DNA added to the melt as an internal standard, one may ask whether the melting of this unusual base composition genome can be accurately described by theoretical methods. We utilized MELTSIM, a program based on the algorithm of Poland & Scheraga(5) and Poland(6), with the approximation of the loop function by a sum of exponentials, as proposed by Fixman and Friere(7). Parameters have been experimentally determined(1,2) for long DNA sequences. The program enables calculation of a denaturation map of positions along any DNA sequence.

In the present study, the experimental total genome melt was compared to the computed simulated melt of available GENBANK sequences. There was excellent agreement at the extremes

Mat. Res. Soc. Symp. Proc. Vol. 463 © 1997 Materials Research Society

of the melting spectrum. That is, both earliest and latest melting DNAs were found to coincide with the theoretical melt temperature extremes. The computational melt was also performed on the *Dictyostelium* files broken up into total coding, intron and flanking DNA files and interesting differences described. The use of MELTSIM and this methodology provides a convenient and visual thermodynamic representation of the sequence distribution found in the individual files. It should prove useful in future studies as more organisms become fully sequenced under the aegis of different Genome Projects.

MATERIALS AND METHODS

D. discoideum DNA Isolation. DNA was isolated from log phase *D. discoideum* AX3 strain cells following washing twice with 0.2 M NaCl, and resuspension in nuclei buffer(0.25 % ficoll, 1 M sorbitol, 0.0008% spermidine, 1 mM DTT, 0.05 M Tris buffer, pH 7.4). The cells were homogenized in a Dounce homogenizer with added NP-40 (2 %). Released nuclei were collected by centrifugation, resuspended and repelleted at 4 °C. Purified nuclei, in 10 mM Tris, 10 mM EDTA, pH 8.0, were treated sequentially with: NaCl, SDS and RNase added to 1 M, 1% and 50 ug/ml respectively and the mixture incubated for 2 hr at 65°C. DNA was extracted 3X with equal volumes of phenol:chloroform:isoamyl alcohol(:1:0.04). Following extraction with chloroform, DNA was ethanol precipitated overnight, washed 2X with 70 % ethanol, air dried and resuspended in 5 ml TE buffer. Only DNA with the characteristic EcoRI restriction pattern was used for further experiments(8). DNA was purified from RNA in CsCl gradients and dialyzed into E buffer: 0.0716 M NaCl, 0.0038 M cacodylic acid, 0.00015 M EDTA, pH 6.7 for melting. Immediately prior to melting, poly (dA.dT)marker was added and the DNA was filtered through a 0.45 um filter to remove any contaminating particulate matter.

Equilibrium Melting Curves. High resolution derivative melting curves were obtained by a difference-approximation method(9,10), using a modified double beam Cary Spectrophotometer as previously described(1,2).

Computational Melting Curves. The program MELTSIM was used to calculate the melting spectra for all files. This is a 32-bit Microsoft Windows operating system that calculates derivative melting curves with the algorithm of Poland and Scheraga(5) and Poland(6), with the Fixman and Friere(7) approximation of the loop entropy by a sum of exponentials. The model is the one-dimensional Ising lattice to which loop entropy has been appended(11-13). Neighboring nucleotide residues in this model exist in either the helical or coil state for each configuration of states. The probability the DNA chain will assume the kth configuration is given by Z_k/Z, where Z is the partition function over all configurations, and Z_k is the fraction of residues in the kth configuration in the helical state,

$$Z_k = \left(1 - \theta_{h_k}\right)/\theta_{h_k} \qquad [1]$$

The free energy for the kth configuration is therefore given by,

$$\Delta G_k = -RT \ln Z_k \qquad [2]$$

θ_{h_k} is proportional to the statistical weights of different configurations,

$$\theta_{h_k} = \sum (N_k Z_k / Z)/N \qquad [3]$$

where N_k is the number of paired residues in the kth configuration, and Z_k is the product of several weighting factors:

$$Z_k = \sigma \cdot f(N_k) \cdot s_{LM}{}^{N_k} \qquad [4]$$

and where σ is the cooperativity parameter. $f(N) = N_k^{-1.75}$ is the loop function, expressing the statistical weight for closed ring structures of N_k denatured base pairs, and s_{LM} is the unit equilibrium constant for unstacking-unpairing pair L on M at the end of a helical segment of length N_k,

$$s_{LM} = \exp\left[-\left(\Delta H_{m_{LM}} - T\Delta S_{m_{LM}}\right)/RT\right] \qquad [5]$$

$\Delta H_{m_{LM}}$ and $\Delta S_{m_{LM}}$ are values for each of the ten unique nearest stacked and paired neighbor enthalpies and entropies for pair L on M. The program MELTSIM is due to be released to the public by the summer of 1997 and will be available on various Internet servers.

RESULTS AND DISCUSSION

Experimental Melts. *Dictyostelium* DNA was isolated from a number of separate preparations with and without a CTAB(cetyltriethylammonium bromide) extraction step to remove contaminating carbohydrate. There was no consistent difference noted in the melting spectrum. The average of 14 spectra is shown in Figure 1. The midpoint($\pm$0.5%) of the melting curve occurs at a T_m of 74. 1 °C, indicating a genome composition of 24.1% (G+C), according to the MSD equation(2), in good agreement with the value, 23 % (G+C) measured by CsCl gradient buoyant density centrifugation(8). Sharp melting features are probably due to sequences in two satellite DNAs previously observed in this organism.

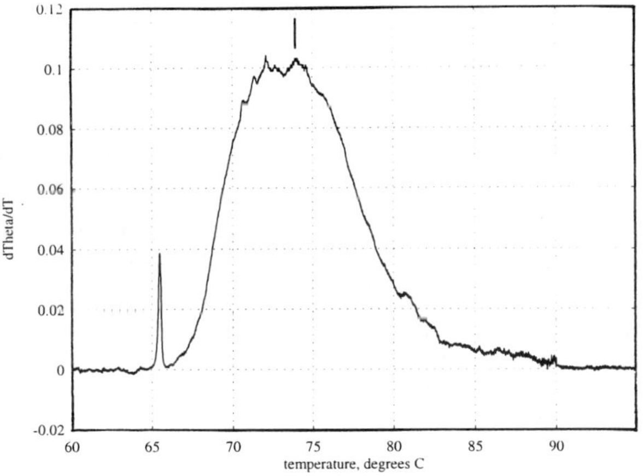

Fig. 1 Derivative melt of *D. disc.* total DNA. Line indicates T_m.

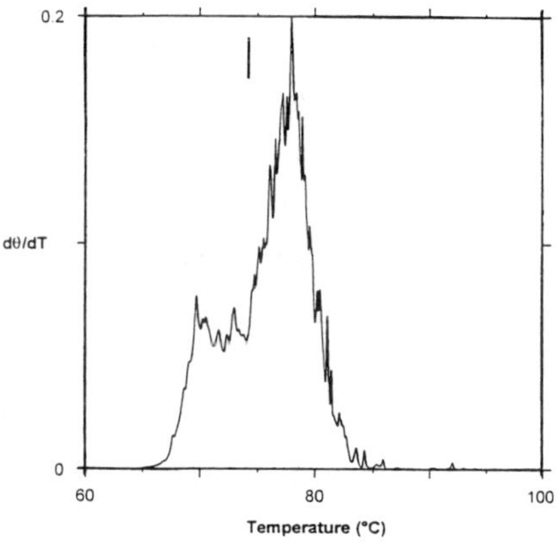

Fig. 2 Simulated derivative melt of *D. disc.* sequences using MELTSIM. Line indicates exp. T_m.

Melting begins at 66 °C, and is complete at 84 °C. Low temperature melting indicates a significant fraction of the pure homopolymer poly(dA.dT) exists in genomic DNA. This observation led us to examine the occurence of tract frequencies in *Dictyostelium* coding, intron and flanking DNA files as a function of their length N(14). These data described $(A.T)_N$ $(T.A)_N$ DNA homopolymers at higher than expected frequencies in flanking and intron sequences. We also observed that tracts of length 5 bp and longer were not spaced randomly in the genome but occured with an average periodicity of 185-190 bp(15), exactly the nucleosome size in Dictyostelium-187 bp. These data also suggest that the long tracts(>10 bp) may be lengthened by a slip strand type mechanism during replication(16).

Theoretical melts. Table I describes the *D. discoideum* GENBANK(11/96) files totalling 872,654 bp nuclear DNA, representing 1.74% of the genome. MELTSIM was applied and the resulting melting curve is presented in Figure 2. The asymmetric shape is due to the

TABLE I

file type	bp	T_m	%(G+C) #	% (G+C) range
Nuclear total(100%)	5.5×10^7	74.1	24.1	66-88
GENBANK				
total(1.74%)*	872,654	76.6	30.1(28.8)	66-86
coding(1.15%)	573,667	77.7	32.7(33.3)	70-85
intron(0.038%)	19,180	68.8	11.5(11.8)	65-75
flanking(0.31%)	153,735	70.3	15.1(15.5)	66-78
ribosomal	7,972	- -	----(43.2)	76-87
17 S DNA(0.37%)	1,870	- -	----(42.5)	80-83
26 S DNA(0.65%)	3,240	- -	----(43.3)	79-83

* This is found from the GENBANK(11/96) documents comprising that file type.
# Calculated from the T_m using the MSD equation. In parentheses are given the computed database values for comparison.

overrepresentation of coding sequences deposited in the database. Consistent with this interpretation is the 2.5 °C higher T_m for this file compared to the experimental T_m, which contains more (A+T) rich flanking DNA. Individual peaks in the curve represent authentic melting subtransitions in the total database examined.

Melting begins(66°C) and ends(86°C) exactly as observed experimentally. MELTSIM is therefore calculating melting properties accurately. Distinct peaks in the simulated melt represent dominant domains for sequences in the database of limited representation. Database sequences were fragmented into files according to their functional roles: coding, intron or flanking sequences(Table I). Simulated melts are presented in Figures 3 A & B. Coding DNA melts over a significantly higher range, corresponding to a mean (G+C) % of about 32 %. This was expected since triplet codons are fixed within proscribed limits of 30-70 % (G+C) across the biosphere in the coding sequence distributions of amino acid compositions found in all globular proteins.

In constrast, the melting of the flanking DNA occurs at significantly lower temperatures. Melting begins at 66°C and is nearly complete by about 78°C, where the coding sequences are just beginning their melting. The flanking DNA database shows a 'rougher' melting curve than coding DNA. This is due partially to the limited size of this database relative to the larger coding dataset. The lower melting temperatures, represent lower effective base compositions found in both these functional classes of sequences. A plausible explanation for the shift to lower % (G+C) in flanking and intron DNAs is due to biased DNA repair, and to the lack of biological constraints on the extension of (A.T) tracts.

Theoretical Denaturation Map. A valuable description of any DNA sequence lies in the positional melting behavior of the sequences in a given region. Of particular interest in the experimental melt were the minor peaks observed at the higher melting temperatures(80-87 °C). It has been known for a number of years that the ribosomal genes of eukaryotes are (G+C)

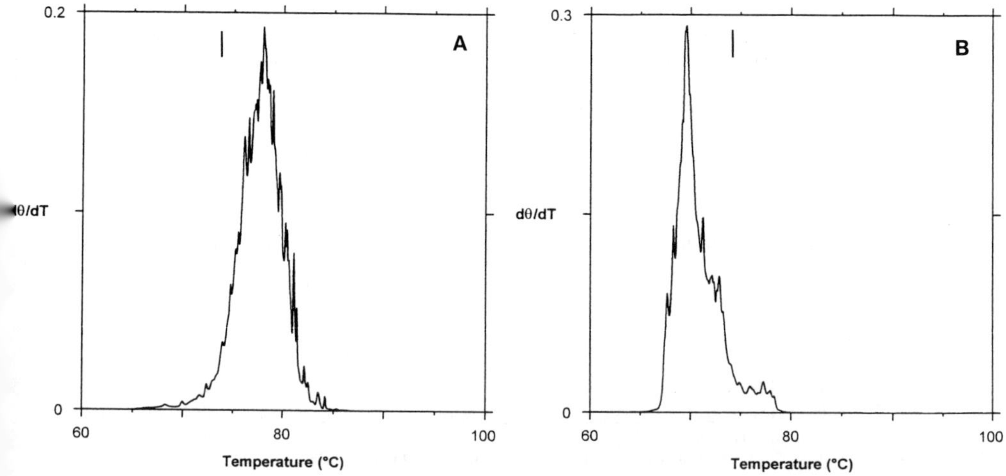

Fig.3 Simulated derivative melts of *D. discoideum*: A) coding sequences; B) flanking sequences, using MELTSIM. Lines indicate T_m of experimental melt.

rich sequences and that they occur in a unit interspersed with spacer DNA. This DNA segment is tandemly repeated hundreds of times/haploid genome in most eukaryotes. In the case of *Dictyostelium,* this repeating unit is known to be located on a large, linear plasmid repeated anywhere from 100-200 fold/cell(8). This level of repetition places it among the most highly repeated sequences found in *Dictyostelium* and thus capable of being responsible for experimental melt subtransitions.

A GENBANK file(AC# X00601 V00189) containing 7,972 bp of sequence which included the ribosomal DNA genes is a portion of the large linear *Dictyostelium* ribosomal DNA gene plasmid. The theoretical denaturation map of this sequence is presented in Figure. 4. The two highest theoretical melting subtransitions occur at the positions of the highest experimental subtransitions. Most likely these correspond to the ribosomal gene DNAs, whose

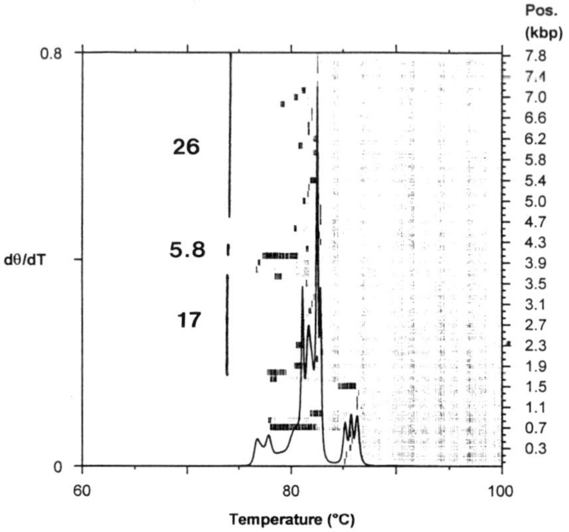

Fig. 4 Theoretical denaturation map and derivative melt of *D. disc.* DNA containing ribosomal genes calculated using MELTSIM. Positions of the ribosomal genes are shown relative to the total DNA file positions on the right axis. The melting of each DNA block in the sequence is indicated by the short closely spaced vertical lines at that position. The melting temperatures covered by each set of vertical lines correspond to the range of $\theta = 0.2$-0.8.

positions are shown. Regions known to be major ribosomal DNA gene sequence(17S and 26S) correspond to the (G+C) rich 79-83 °C melting subtransitions. The Non-Transcribed Spacer DNA (1-998) and External Transcribed Spacer(999-1793) accounts for the most (G+C) rich 85-87 °C subtransitions. These are evident even in the Figure 2 experimental melt. The remaining 76-78 °C subtransitions in Figure 4 result from the melting of more (A+T)-rich Internal Transcribed Spacer 1 DNA(3665-3994) interspersed between the major ribosomal DNA genes. These may correspond to low temperature subtransitions in the total Figure 2 experimental melt.

ACKNOWLEDGEMENTS

The authors acknowledge support from the Center for Intelligent Biomaterials at the University of Massachusetts.

REFERENCES

1. Delcourt, S.G. and Blake, R.D., J. Biol. Chem. **266**, 15160-15169 (1991).
2. Blake, R.D., in Encyclopedia of Molecular Biology and Molecular Medicine, Volume 2, (R.A. Myers, ed.) VCH Publ., Basel, 1-18 (1996).
3. Marx, K.A., Lim, J.O., Minehan, D, Pande, R., Kamath, M and Tripathy, S., J. Intel. Material Systems and Structures, 5, 447-454 (1994).
4. Patti, J.L., Marx, K.A. and Blake, R.D., J. Cell Biology **115**, 90 (1991).
5. Poland, D. and Scheraga, H.A., Theory of Helix-Coil Transitions in Biopolymers, Academic Press, New York, (1970).
6. Poland, D., Biopolymers, **13**, 1859-1871 (1974).
7. Fixman, M. and Friere, J., Biopolymers, **16**, 2693-2704 (1977).
8. Firtel, R., Cockburn, A., Frankel, G. and Hershfeld, V., J. Mol. Biol. **102**, 831-852 (1976).
9. Blake, R.D. and Hydorn, T.G., J. Biochem. Biophys. Methods, **11**, 307-316 (1985).
10. Yen, S-W.W. and Blake, R.D., Biopolymers **19**, 681-700 (1980).
11. Ising, E. Physik **31**, 253 (1925).
12. Hill, T.L., Statistical Mechanics, McGraw-Hill, New York (1956).
13. Wartell, R.M. and Benight, A.S., Physics Rep. **126**, 67-107 (1985).
14. Marx, K.A., Hess, S.T. and Blake, R.D., J. Biomol. Str. & Dyn. **11**, 57-66 (1993).
15. Marx, K.A., Hess, S.T. and Blake, R.D., J. Biomol. Str. & Dyn. **12**, 235-246 (1994).
16. Kunkel, T.A. and Bebanek, K., Biochim. Biophys. Acta. **951**, 1-15 (1988).

CONCENTRATED DNA RHEOLOGY AND MICRORHEOLOGY

T.G. MASON*, A. DHOPLE, and D. WIRTZ
Department of Chemical Engineering, Johns Hopkins University, 24 NEB, 3400 N. Charles St., Baltimore, MD 21218. *Email: tgmason@jhu.edu

ABSTRACT

We present mechanical measurements of the frequency-dependent linear viscoelastic storage and loss moduli, $G'(\omega)$ and $G''(\omega)$, and the yield stress, τ_y, and yield strain, γ_y, for calf thymus DNA (13 kbp) over a range of mitotically relevant concentrations from C_{DNA} = 1 to 10 mg/ml. For large C_{DNA}, we find a dominant plateau elasticity, G'_p, at high ω. As ω decreases, G' falls until it is equal to G'' at the crossover frequency, ω_c, below which G'' dominates. We measure $G'_p \sim C_{DNA}^{2.25}$ and $\omega_c \sim C_{DNA}^{-2.4}$, consistent with scaling exponents for classical polymer solutions. The mechanical $|G^*(\omega)|$ agree well with those measured using a new microrheological technique based on video tracking microscopy of thermally-driven fluorescent colloidal spheres and a frequency-dependent Stokes-Einstein equation. We have developed this technique to probe how enzymes, typically available in small quantities, can affect the rheology of the DNA. Using it, we report preliminary measurements of a higher ω_c for a DNA network in which the ATP-powered enzyme Topoisomerase II transiently cuts and rebinds the DNA, thereby relaxing entanglements.

INTRODUCTION

During the prometaphase of mitosis, DNA is locally concentrated within the cell nucleus before replication and separation occur. In a typical mammalian cell, 10^9 nucleotides occupy a typical volume of 100 μm^3 [1], yielding a concentration, $C_{DNA} \approx$ 10 mg/ml. At such high concentrations, DNA molecules having micron-sized contour lengths can become entangled, leading to a viscoelastic rheological response typical of polymers in their semidilute regime [2]. In cell division, the entanglement elasticity would normally inhibit separation of the DNA after replication through the contraction of microtubules connected to the mitotic spindles. However, the action of an ATP-powered enzyme, known as Topoisomerase II (Topo II), which transiently cuts and rebonds one of two crossed DNA at an entanglement site [3], allows the entaglements to relax and the DNA to be separated. A theory based on the release of topological constraints predicts that active Topo II in sufficient concentrations at entanglement points for megabase-sized DNA should cause a faster viscoelastic relaxation [4,5]. Experimentally, Topo II has been shown to relax supercoiled, double-stranded, circular plasmid DNA [6], but its effect on entanglement networks of linear, double-stranded DNA have not been measured.

We present mechanical measurements of the rheology of concentrated linear DNA in an aqueous buffer which can support the activity of Topo II. In the first part of this paper, we consider DNA in the absence of Topo II, and we report the frequency-dependent linear storage and loss moduli, $G'(\omega)$ and $G''(\omega)$, and the yield stress and yield strain, τ_y and γ_y, for a series of C_{DNA}. Since repeating these measurements with Topo II would require prohibitively large quantities of the enzyme, we have developed a new technique which can locally probe the microrheology around fluorescent colloidal spheres which are tracked using video microscopy. This is the subject of the second part of this paper. By calculating the time-dependent mean square displacement, $\langle\Delta r^2(t)\rangle$, from the particle trajectory, we extract the magnitude of the linear complex modulus, $|G^*(\omega)|$, using a frequency-dependent version of the Stokes-Einstein equation. This new technique permits us to reduce the sample volume to the microliter range where a reasonably high concentration of Topo II can be achieved. For one C_{DNA}, we present preliminary measurements of $|G^*(\omega)|$ without and with Topo II, which indicates there is a decrease in the relaxation time of the network.

I. RHEOLOGY OF DNA IN THE ABSENCE OF TOPOISOMERASE II

EXPERIMENT

We disperse linear, double-stranded DNA from calf thymus having an average number $N = 1.3 \cdot 10^4$ base pairs in an aqueous buffer containing: [Tris-HCl] (pH 7.9) = 10 mM, [NaCl] = 50

Mat. Res. Soc. Symp. Proc. Vol. 463 © 1997 Materials Research Society

mM, [KCl] = 50 mM, [$MgCl_2$] = 5 mM, [BSA] = 15 μg/ml, and [EDTA] = 0.1 mM. An optically homogeneous dispersion is attained through gentle mechanical agitation from three days to one week for C_{DNA} = 10 mg/ml. We employ a controlled strain mechanical rheometer equipped with a 25 mm diameter cone and plate geometry requiring a sample volume of about 200 μl. We enclose the sample with a vapor trap. The temperature is fixed at T = 25 C.

RESULTS

We measure the linear G' and G'' of the DNA solutions by setting the amplitude of the perturbative oscillatory strain at $\gamma = 0.02$ and sweeping from low to high ω; the results are shown in Fig. 1. At the highest ω, the plateau G' dominates G'', while at the lowest ω, G'' dominates G'. As C_{DNA} decreases, the plateau elasticity does also, while the crossover frequency becomes larger. To characterize the transition from linear to nonlinear flow, we measure G' and G'' at $\omega = 1$ rad/s from low to high γ, as shown in Fig. 2. For strains less than the yield strain of $\gamma_y \approx 0.7$, we find that G' and G'' are independent of γ, γ_y is nearly concentration-independent for $C_{DNA} > 3$ mg/ml,

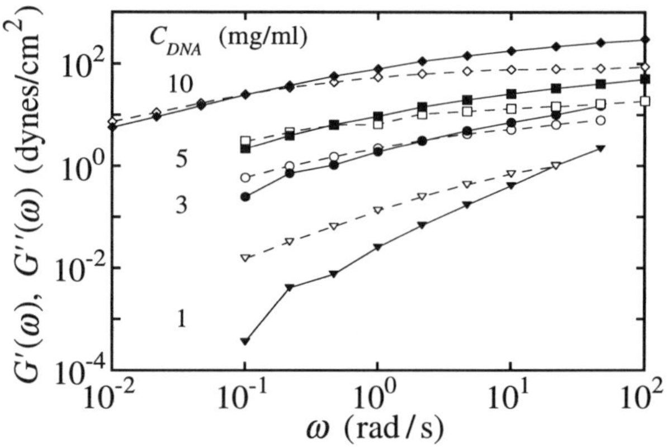

Figure 1. The frequency dependence of the storage modulus, G', (solid symbols) and the loss modulus, G'' (open symbols) for DNA concentrations of 1 mg/ml (triangles), 3 mg/ml (circles), 5 mg/ml (squares), and 10 (diamonds) mg/ml.

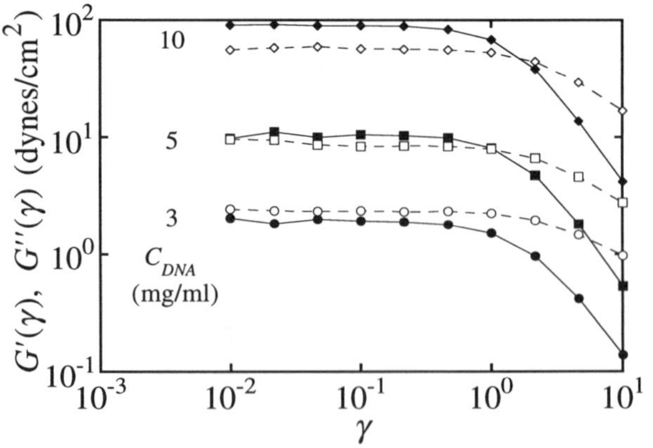

Figure 2. The strain dependence of the storage modulus, G', (solid symbols) and the loss modulus, G'' (open symbols) for DNA concentrations of 3 mg/ml (circles), 5 mg/ml (squares), and 10 (diamonds) mg/ml.

confirming that $G'(\omega)$ and $G''(\omega)$ have been obtained in the linear regime. By contrast, the yield stress, τ_y, rises about two orders of magnitude to C_{DNA} = 10 mg/ml.

DISCUSSION

If DNA behave as classical polymers, we expect the DNA to develop a dominant plateau elasticity for concentrations larger than a concentration corresponding to significant entanglements. We estimate this critical concentration by considering disordered random close packing (RCP) of DNA blobs (spheres) at a volume fraction of $\phi_{RCP} = 0.64$; we take the sphere's radius to be average radius of gyration, $R_g \approx 3\ N^{\nu}$ Å, where $\nu = 0.59$, and estimate $R_g = 0.1\ \mu$m for our DNA. The critical concentration is $C^* = \phi_{RCP}(M_{DNA}/V)$, where M_{DNA} is the average mass of a single DNA and $V = 4\pi R_g^3/3$ is its effective volume, yielding C^* = 3 mg/ml. The large rise in G' we observe between C_{DNA} = 1 mg/ml and C_{DNA} = 3 mg/ml is in excellent agreement with this estimate.

For $C_{DNA} > C^*$, a classical law for semi-dilute polymer solutions predicts that the plateau elasticity should scale with the concentration as $G'_p \sim C_{DNA}{}^{3\nu/(3\nu-1)} \sim C_{DNA}{}^{2.3}$ [2]. We extract the plateau elasticity at ω = 100 rad/s and plot it as a function of concentration, as shown by the squares in Fig. 3. Our measured exponent is 2.25, shown by the dashed line on the plot, in good agreement with this prediction. Given the empirical relationship that the bare low frequency viscosity, η_0, varies as $\eta_0 \sim C_{DNA}{}^{4.7}$ (without Rouse renormalization) [2], we estimate that the crossover frequency is where the viscous and elastic components are equal: $\omega_c \approx G'_p/\eta_0 \sim C_{DNA}{}^{-2.4}$. We plot the measured ω_c (circles) as a function C_{DNA} in Fig. 3. Our measured exponent is -2.4, again in excellent agreement with this semi-empirical prediction. Our interpretation of results for calf thymus DNA agree well with those for much longer T2 phage DNA [2]. We conclude that the linear viscoelastic behavior of the DNA can be treated as that of a classical semi-dilute polymer solution. However, we know of know predictions which explain the large magnitude and weak concentration dependence of γ_y, nor for the strong concentration dependence of τ_y as the strongly nonlinear regime is approached.

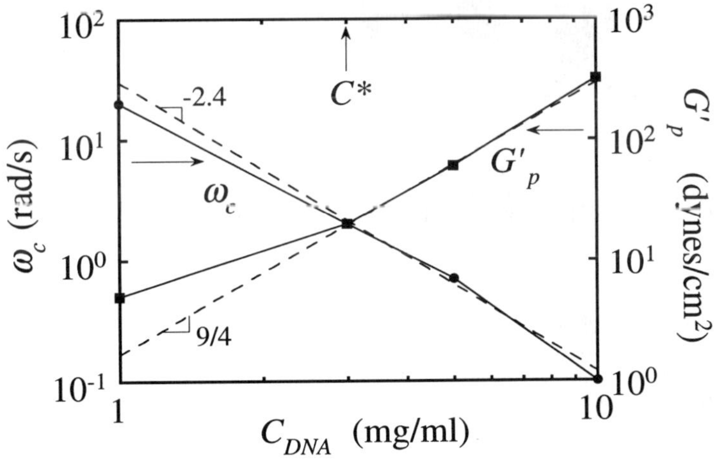

Figure 3. The DNA concentration dependence of the crossover frequency, ω_c, (circles) and plateau storage modulus, G'_p (squares).

II. VIDEO MICRORHEOLOGY OF DNA AND DNA+TOPOISOMERASE II

EXPERIMENT

Our particle-tracking apparatus consists of an inverted fluorescence microscope at 2500 magnification equipped with a SIT camera which is connected to a VCR and a computer-controlled frame grabber. Fluorescent spheres having radii $a = 0.52\ \mu$m are mixed into 30 μl of DNA solution at a volume fraction of 10^{-4} and loaded into a cell having thickness 100 μm. The illumination intensity and the voltage, gain, and black level of the SIT are adjusted to obtain intensity profiles of the spheres which exhibit a clear maximum at their centers. We set the focus in the middle of the sample cell to eliminate wall effects. Since uniform convection over the entire viewing area (caused by the gross mechanical contact of the oil-immersion objective with the cell) can disturb the diffusive motion of the spheres, we capture movies for pairs of spheres which are well-separated, yet in the same field of view. This allows us to eliminate this correlated convective motion in our later analysis. We obtain the two-dimensional trajectories for each of the two particles over 250 seconds at 20 frames/s using a centroid algorithm on the measured intensity profile of each frame. We subtract the two trajectories to eliminate convection, and calculate the mean square displacement between the particles by fixing the time interval and scanning all possible initial times. The actual three-dimensional mean square displacement for one particle, $\langle\Delta r^2(t)\rangle$, is three-forths of this result (6/4 from 2D to 3D times 1/2 from two to one particle). Here, the mean represents a time average; it is not a true ensemble average, since we have only considered two particles. However, a true ensemble average can be obtained by repeating the procedure for many pairs of spheres and averaging the results together.

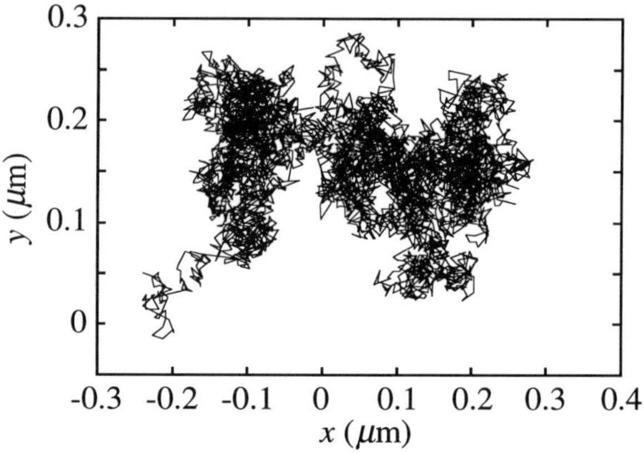

Figure 4. The separation trajectory between two spheres of radii $a = 0.52\ \mu$m measured with video fluorescence microscopy over about 250 seconds. Similar size in x and y indicate convection has been eliminated.

Assuming that the local viscoelasticity around a sphere is the same as the bulk viscoelasticity, the complex viscoelastic modulus can be calculated from the frequency-transformed $\langle\Delta r^2(t)\rangle$ using a generalized Stokes-Einstein equation [7]. This approach has been demonstrated to hold for light scattering measurements of entangled polymeric solutions in which the radii of the probe spheres are much larger than the polymer radius of gyration or blob interspacing [7]. However, taking the numerical transform introduces errors due to the finite size of the dataset at the highest and lowest frequencies. To eliminate this problem, we approximate the generalized Stokes-Einstein equation using a simple algebraic computation:

$$|G^*(\omega)| \approx \left.\frac{k_B T}{\pi a \langle \Delta r^2(t) \rangle}\right|_{t=1/\omega}, \tag{1}$$

where k_B is Boltzmann's constant. This equation is exact for a sphere in a viscous fluid with no-slip boundary conditions: $\langle \Delta r^2(t) \rangle = 6Dt$, where D is the diffusion coefficient, so $|G^*(\omega)| = [k_BT/(6\pi aD)]\omega = \eta\omega$, verifying the limiting result of the simple Stokes-Einstein equation. A comparison of Eq. 1 with the transformation method reveals that it is a surprisingly good estimate despite its simplicity; it underestimates $|G^*|$ by at most 40% for all ω. Most importantly, it yields accurate $|G^*|$ at the lowest frequencies where the transformation method fails. For data taken using video microscopy over a limited range of ω, Eq. (1) is more accurate, so we use it in our analysis.

For the composition with Topo II, we add [ATP] = 3 mM to C_{DNA} = 10 mg/ml, and mix this in a 2:1 volume ratio with 64 μg/ml Topo II from *Drosophila melanogaster* [6] supplied in a buffer of 15 mM sodium phosphate, 50 mM NaCl, 0.1mM EDTA, 0.2 mM DTT, 0.5 mg/ml BSA, and 10% glycerol. This yields 30 μl at C_{DNA} = 6.7 mg/ml and [Topo II] = 21 μg/ml. Although this concentration may seem low since it reflects effectively one Topo II molecule (molecular weight of $3.3 \cdot 10^5$ Daltons) for every four DNA molecules, it is actually very large compared to that we would have to use if we relied upon a mechanical technique. The microscopy was performed at T = 25 C after a ten minute incubation period.

RESULTS

The measured separation trajectory for a single pair of spheres at C_{DNA} = 6.7 mg/ml in the absence of Topo II is shown in Fig. 4. To our knowledge, this is the first extensive visualization measurement of the trajectory of thermally-driven spheres in a visoelastic complex fluid. The scale on the figure is less than a micron, so we are effectively resolving the displacement of a micron-sized sphere to an effective length scale of about 10 nm, much smaller than the wavelength of light. This resolution is achieved through the good signal-to-noise of the particle intensity profile provided by the fluorescence detection system and centroid averaging. No continuous or pronounced extension in either direction x or y is found, indicating that convection has been effectively eliminated.

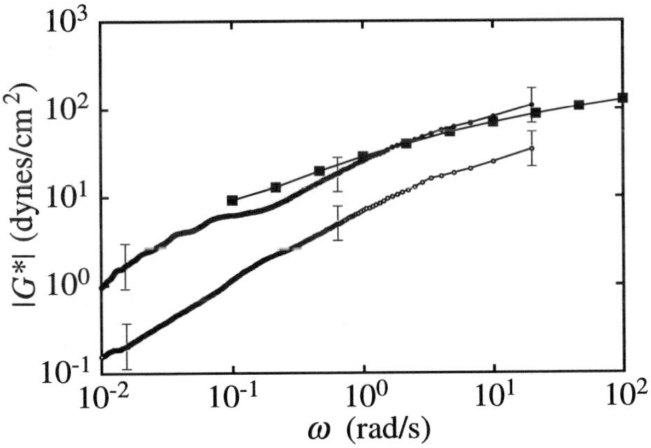

Figure 5. The ω-dependence of the magnitude of the complex shear modulus, $|G^*|$, measured mechanically (squares), and with video microscopy (circles) at C_{DNA} = 6.7 mg/ml. The solid circles reflect the absence of Topo II (upper data points), while the open circles are with Topo II at 21 μg/ml (lower data points).

DISCUSSION

From the trajectory of Fig. 4, we calculate the mean square displacement in three dimensions, and then determine $|G^*(\omega)|$ using Eq. (1). The results for DNA alone are shown by the small solid circles in Fig. 5, and mechanical measurements at identical C_{DNA} are co-plotted as the large squares. There is excellent agreement between the two measurements, supporting the mean field assumption which should be valid for $a > R_g$. The irregularities in the shape of $|G^*(\omega)|$ obtained by particle-tracking are a signature of the one specific trajectory we have measured, and they become more pronounced toward the lowest frequencies where fewer time averages are possible due to the limited duration of our experiment. The error bars we have drawn at high and low ω crudely indicate the polydispersity in a which directly propagates into the error for $|G^*(\omega)|$ through Eq. (1). Despite these sources of error, the good comparison in the magnitude and shape of the two sets of data gives us confidence that we are indeed measuring the rheology of the DNA solution.

For DNA in the presence of Topo II, the data for $|G^*(\omega)|$ are plotted in Fig. 5 as the small circular open symbols. They consistently fall below the data for DNA in the absence of Topo II, and there is a wider separation between the two datasets at low ω than at high ω. This reflects the larger displacements of the spheres in the DNA+Topo II solution; the visualized random walk is similar in form to Fig. 4, but larger in scale. Given the large error bars and uncertainties we have listed above, our data are insufficient to quantify the effect of the Topo II with much certainty. Given the small ratio of Topo II to DNA molecules, we would have expected a much weaker effect than the factor of three difference we observe in the magnitude. Part of this discrepancy may be attributed to differently sized particles. However, comparison of the shapes of the data rather than their absolute magnitudes reveals that the Topo II does appear to raise the crossover (indicated by the knee) to higher ω, consistent with its hypothesized transient cutting and rebinding activity. Although we cannot draw strong conclusions from this single preliminary result, we believe that video microrheology will enable us to precisely quantify changes in the rheology of concentrated DNA with sufficiently large Topo II concentrations that the action of the enzyme can be easily resolved.

CONCLUSION

Our mechanical measurements of DNA rheology reveal that it behaves as a semi-dilute polymer solution, exhibiting linear viscoelasticity which obeys known scaling laws. By contrast, the behavior of the yield strain and yield stress remains unexplained. The random structure of our artificial DNA network may not reflect the organized and time-dependent structures of real chromosomes, so we refrain from implying that our results reflect the true rheology within the nucleus. However, we have demonstrated that $|G^*(\omega)|$ can be determined from real-space measurements of $\langle\Delta r^2(t)\rangle$ for single particles in a viscoelastic medium, and we have demonstrated that fluorescence video microscopy can be used to visualize the the thermally-driven rms displacements of colloidal particles to an impressively small resolution of 10 nm. Although our preliminary results for the modification of DNA rheology by Topo II are not conclusive, we believe that this microrheological technique will be extremely beneficial to the biological community, since we ultimately envision its application *in vivo*, within a cell or its nucleus. For these applications, the assumption of homogeneity implied by Eq. (1) must be revisited.

ACKNOWLEDGEMENTS

We have benefitted from discussions with S. Kuo. We are grateful for support from the National Science Foundation and the Whittaker Foundation.

REFERENCES

1. B. Alberts, D. Bray, J. Lewis, M. Raff, K. Roberts and J.D. Watson, *Molecular Biology of the Cell*, (Garland, New York, 1990).
2. R. Musti, J.-L. Sikorav, D. Lairez, G. Jannink and M. Adam, *C. R. Acad. Sci. II* **320**, 599 (1995).
3. C. Holm, *Cell* **77**, 955 (1994).
4. J.-L. Sikorav and G. Jannink, *C. R. Acad. Sci. II* **316**, 751 (1993).
5. J.-L. Sikorav and G. Jannink, *Biophys. J.* **66**, 827 (1994).
6. N. Osheroff, E.R. Shelton and D.L. Brutlag, *J. Bio. Chem.* **258**, 9536 (1983).
7. T.G. Mason and D.A. Weitz, *Phys. Rev. Lett.* **74**, 1250 (1995).

Part V

Micromanipulation in Physics and Biology

DYNAMICALLY-STABILIZED PORES IN BILAYER MEMBRANES

J.D. MOROZ, P. NELSON
Department of Physics and Astronomy, University of Pennsylvania, Philadelphia, PA 19104

ABSTRACT

Zhelev and Needham have recently created large, quasi-stable pores in artificial lipid bilayer vesicles. Initially created by electroporation, the pores remain open for up to several seconds before quickly snapping shut. This result is surprising in light of the large line tension for holes in bilayer membranes and the rapid time scale for closure of large pores. We show how pores can be dynamically stabilized via a new feedback mechanism. We also explain quantitatively the observed sudden pore closure as a tangent bifurcation. Finally we show how Zhelev and Needham's experiment can be used to measure accurately the pore line tension, an important material parameter. For their SOPC/CHOL mixture we obtain a line tension of 3.0×10^{-6} dyn.

INTRODUCTION

Lipid bilayer membranes have remarkable physical properties. One of the most important among these properties is a membrane's resistance to rupture. In the body, this resistance is critical to the maintenance of well defined and properly-functioning cells. Indeed, when a cell needs to undergo a topological change (as it does during cell division, cell fusion, endocytosis, and exocytosis) it usually has to make use of specialized machinery which carries out the change at the cost of chemical energy. This cost is largely determined by the material properties of the lipid membranes in question.

We can quantify a membrane's resistance to rupture in terms of a line tension (γ), the free energy cost per unit length of exposed edge. Edges are disfavored due to the high cost of either exposing the hydrophobic lipid chains to water, or creating a highly-curved rolled edge to hide them. Many authors have devised ingenious indirect measurements of γ in various lipid systems [1, 2, 3], but direct measurement has proven difficult. Among the biologically-relevant questions which require such measurements is the variation of γ with lipid shape [4].

Recently Zhelev and Needham have found a new technique allowing direct mechanical measurement of the line energy [5, 6]. In this paper we will present a new analysis of their experimental data. The experiment revealed some surprising qualitative phenomena involving pores, which we will explain. Briefly (see below), they created long-lived quasistable pores about a micron in radius. After persisting for up to several seconds, the pores snapped shut in just one video frame. We will quantitatively explain the longevity of the pores and their sudden demise, fitting several quite different events with a common value of γ and two auxiliary parameters.

To see why long-lived pores are surprising, consider the usual energy of a circular hole in a flat bilayer membrane [1]. This energy can be written as a line tension term which is linear in the pore radius minus a surface tension term which is quadratic:

$$E(r) = 2\pi r\gamma - \Sigma\pi r^2. \tag{1}$$

This energy has only one stable minimum (at $r = 0$). There is a critical radius ($r = \gamma/\Sigma$) above which the pore is unstable to rupture. To cross this critical point, the system must surmount a significant energy barrier ($\Delta E = \pi\gamma^2/\Sigma$). For typical estimates of the line tension (10^{-6} dyn) thermally-driven rupture thus requires a surface tension on the order of one dyn/cm, as observed

Mat. Res. Soc. Symp. Proc. Vol. 463 © 1997 Materials Research Society

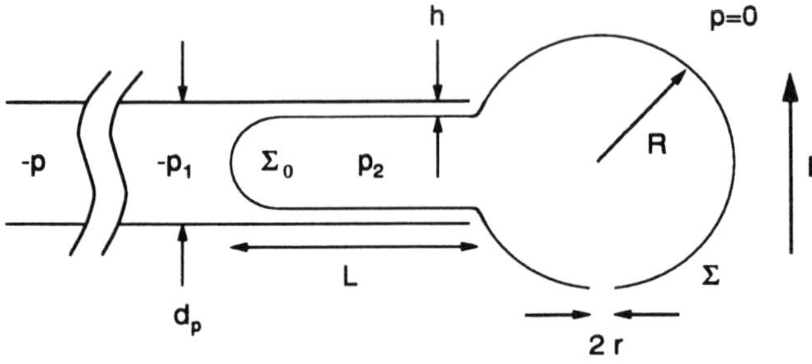

Figure 1: Geometry of the stabilized pore experiment of Needham and Zhelev.

[7]. For lower tensions, any transient pore will reclose rapidly, while for larger tensions it will grow rapidly and lyse the vesicle; in either case, one does not expect large stable pores to exist.

Zhelev and Needham surmised that the pores somehow sit at the unstable point of Eq. 1 before suddenly falling off. Inferring Σ and r from the data then let them find the line tension from $\gamma = \Sigma r$. It seems unlikely, however, that pores would remain in unstable equilibrium for so long. In this paper we will find a feedback mechanism that dynamically stabilizes pores. Our explanation will involve an effective pore energy with a minimum that can exist for several seconds before disappearing suddenly. From the critical point at which stability is lost, we will be able to produce estimates of various parameters in the theory; in particular, we will accurately determine the line tension of the bilayer membrane.

EXPERIMENT

Figure 1 defines our notation. A micropipette several microns in diameter is used to immobilize a chosen vesicle using a suction $(-p)$. The suction pressure is held constant throughout the experiment at a distant manometer. Initially a small amount of membrane is pulled into the micropipette, leaving a tense spherical outer bulb of radius $R_{\rm init}$. A square-wave electric field pulse is then applied across the vesicle: the effect of this field is to open a hole in the membrane by electroporation [8]. Occasionally the pore so created stabilizes and the vesicle moves slowly down the pipette in a controlled fashion.

What is measured is then the constant applied suction $(-p)$ at the manometer, the initial bulb size $(R_{\rm init})$, the micropipette diameter $(d_{\rm p})$, and the location $(L(t))$ of the leading edge of the membrane as it advances down the micropipette. Figure 2 shows a typical time course. Other quantities in figure 1, such as the bulb radius (R), the pore radius (r), the lubrication layer thickness (h), the surface tensions $(\Sigma,\ \Sigma_0)$, and the pressures $(-p_1,\ p_2)$ are all time dependent and must be inferred from the directly measured data.

STABILIZATION MECHANISM

Here we outline the stabilization mechanism that is described more fully in [9]. The main ingredients of this mechanism are two frictional terms which oppose the aspiration of the vesicle into the pipette. The first such term arises from a decrease in pressure along the length of the micropipette due to Poiseuille loss. The corrected pressure is given by $-p_1$:

$$-p_1 = -p + \frac{32\eta}{d_{\rm p}^2}(L_{\rm eff} - L)v \ . \tag{2}$$

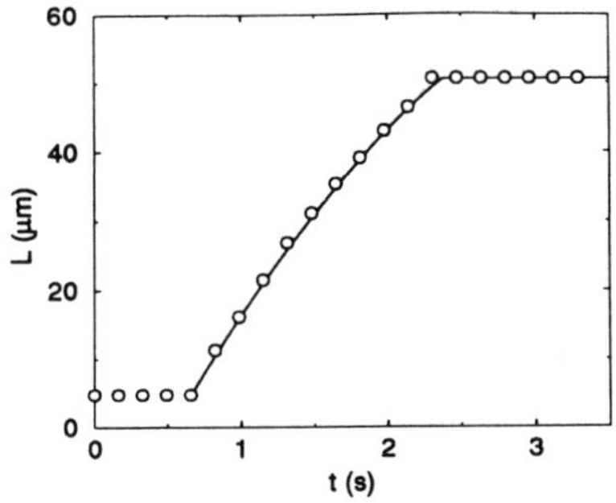

Figure 2: Progress of the leading edge of a SOPC/CHOL mixed lipid vesicle down the pipette (reproduced from figure 3 of [5]). For this event $R_{\text{init}} = 17.6\,\mu\text{m}$ and $p = 353\,\text{dyn/cm}^2$. The initial velocity (v_{init}) is approximately $35\,\mu\text{m/s}$. The solid line depicts the theoretical $L(t)$ curve for our stabilized pore model.

Here η is the viscosity of the sugar solution, L is the projected length of membrane in the micropipette (figure 1), and $v \equiv \dot{L}$ is the velocity of the vesicle's leading edge. Eq. 2 should really be regarded as a definition of the effective length (L_{eff}) which accounts for several unmeasureable effects in a phenomenological way [9]. As we shall see later, L_{eff} can be determined from the data of figure 2 using our theoretical approach.

There is a second velocity-dependent friction term which arises due to shear in the lubricating layer sandwiched between the membrane and the micropipette wall. This frictional force creates a difference between the surface tension (Σ_0) at the leading edge and (Σ) on the exterior bulb:

$$\Sigma = \Sigma_0 - \frac{\eta L}{h} v \,. \tag{3}$$

Here h defines the thickness of the lubrication layer. This parameter, like L_{eff} will be fixed from the data in figure 2 using our model. These two sources of friction along with the manometer suction control the full dynamics of the inhalation process.

We can use Eqs. 2 and 3 along with the two equations generated by the Laplace equation applied to the leading edge and the bulb of the vesicle to eliminate p_1, p_2, and Σ_0 to give

$$\Sigma = \frac{2R}{2R - d_{\text{p}}} \left(\frac{p d_{\text{p}}}{4} - \eta v \left(\frac{8(L_{\text{eff}} - L)}{d_{\text{p}}} + \frac{L}{h} \right) \right) \,. \tag{4}$$

This equation for the surface tension now has explicit dependence on the projected membrane length (both through L itself and through the function $R(L)$ which can be obtained from the constraint of fixed membrane area). We still need to eliminate the dependence of Σ on the velocity of the leading edge.

This elimination is achieved by equating two expressions for the outflow of solution through the open pore. The first of these expressions is obtained by considering the time derivative of the total volume contained inside the vesicle. The second describes the outflow as a function of the pressure inside the vesicle which drives the flow. From these two calculations of the outflow the velocity is obtained:

$$v = \frac{8}{3\pi\eta d_{\text{p}}} \frac{\Sigma r^3}{R(2R - d_{\text{p}})} \,. \tag{5}$$

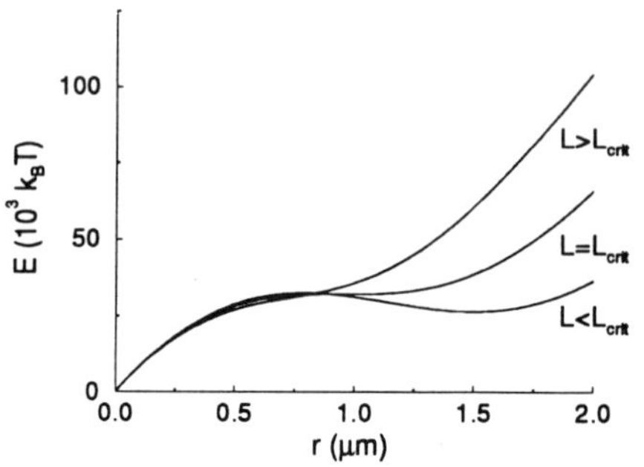

Figure 3: Energy as a function of pore size for supercritical, critical, and subcritical lengths of aspirated membrane: $L = 0, 45.8, 90\,\mu\text{m}$. ($L_{crit} = 45.8\,\mu\text{m}$.)

This equation can now be substituted back into Eq. 4 to yield a new expression for the surface tension. This form has the surface tension as a function of the projected length (L) and the pore size (r):

$$\Sigma(r, L) = \frac{\alpha_1}{1 + \alpha_2 r^3}\,, \tag{6}$$

where

$$\alpha_1 \equiv \frac{p d_\mathrm{p}}{4} \frac{2R}{2R - d_\mathrm{p}}\,, \quad \text{and} \quad \alpha_2 \equiv \frac{16}{3\pi(2R - d_\mathrm{p})^2 d_\mathrm{p}} \left(\frac{8(L_\mathrm{eff} - L)}{d_\mathrm{p}} + \frac{L}{h} \right)\,. \tag{7}$$

These equations define the feedback mechanism through which the effective surface tension is modified by the friction terms outlined above. The next step is to see how this feedback mechanism accounts for the experimentally observed behavior — that is, to see how it stabilizes pores.

Our main physical hypothesis is now that at each moment the pore adjusts quickly to minimize an *effective energy* similar to Eq. 1, but with the tension replaced by the varying quantity just found (Eq. 6). As long as this effective energy has a nontrivial minimum, the pore size will track it. This gives the pore size, and hence $v \equiv \dot{L}$ via Eq. 5 in terms of L. We can then solve this ordinary differential equation to obtain the time course $L(t)$ and compare to the data in Figure 2. This program relies on the presence of two different time scales: a slow scale for changes of L and R, and a much faster time scale on which the pore size r adjusts and the membrane tension equilibrates. We are adiabatically eliminating the fast variable to obtain a simple dynamics for the slow one. Thus our effective pore energy depends on L:

$$E_\mathrm{eff}(r, L) = 2\pi r \gamma - \frac{\pi \alpha_1 r^2}{1 + \alpha_2 r^3}\,. \tag{8}$$

This new form of the energy can indeed have two minima: the trivial one ($r = 0$) that appeared before and a new one at a finite pore size, depending on L. As long as L does not exceed some critical value (L_crit), there will be a local minimum in the effective energy at some finite pore size ($r_\mathrm{stable}(L)$). When the projected length reaches this critical value, the second stable minimum disappears and the pore collapses (figure 3).

As promised, we have succeeded in reducing the full dynamics of the problem to only one variable, the projected length (L). This can be seen by substituting the surface tension formulae (Eqs. 6 and 7) along with $R(L)$ and the newly determined pore size ($r_\mathrm{stable}(L)$) into the velocity

Table 1: Comparison of experimental [5] and theoretical (this paper) closure times for SOPC/CHOL vesicles. The fourth event was used to determine L_{eff}, h, and γ. A bar (-) indicates that the pore did not reseal.

$R_{\text{init}}(\mu\text{m})$	p (dyn/cm^2)	t_{exp}(s)	t_{th}(s)	*Bilayers*
18.3	399	–	2.7	1
28.1	310	2.6	3.3	1
14.3	331	–	–	1
17.6	353	1.7	1.7	2
17.0	324	0.33	0.48	2
16.3	337	0.33	0.62	2
16.0	355	2.0	2.9	1
25.6	168	11.0	11.2	1
22.5	240	5.5	5.6	1
16.2	320	2.5	3.7	1

(Eq. 5):

$$\dot{L} \equiv v(L) = \frac{8}{3\pi\eta}\left(\frac{\alpha_1(L) r_{\text{stable}}(L)^3}{1+\alpha_2(L) r_{\text{stable}}(L)^3}\right)\frac{1}{R(L)d_{\text{p}}(2R(L)-d_{\text{p}})} \,. \tag{9}$$

This equation for $\dot{L}$ as a function of L can now be integrated to obtain $L(t)$.

EXTRACTION OF PARAMETERS

The model that we propose has seven parameters in all. Of these, two are measured directly from microscope images, namely the initial bulb radius (R_{init}) and the pipette diameter (d_{p}). The pressure (p) at the manometer and the viscosity (η) of the solvent are also determined experimentally.

Of the remaining three parameters, Zhelev and Needham estimated L_{eff} and h via auxiliary experiments, then deduced γ. Because these values were found through rather different experiments experiments however, we will consider them undetermined and fit them to experimental data using our model. Finally we will deduce the last remaining parameter in the model: the line tension (γ).

Zhelev and Needham [5] give the full time course for one event, which we will use to find the three undetermined parameters listed above. This event is reproduced in Figure 2 and appears as the fourth entry in table 1 below. From the experimental $L(t)$ curve we first extract the projected length $L_{\text{crit}} = 45.8\,\mu\text{m}$ at pore closure, the initial velocity $v_{\text{init}} = 35.5\,\mu\text{m/s}$, and the final velocity $v_{\text{crit}} = 15.7\,\mu\text{m/s}$. These initial and final conditions can then be used to fix the three parameters: we find $L_{\text{eff}} = 317\mu m$, $h = 0.47\mu m$, and $\gamma = 3.0 \times 10^{-6} erg/cm$ [9].

With the parameters so determined, it is now possible to determine the time evolution of the other events in [5] from the initial conditions. For each of the other nine SOPC/CHOL events reproduced in table 1, we integrated Eq. 9 for the given initial bulb radius (R_{init}) and pressure (p) to obtain the critical time at which the pore closed. In some cases a double-bilayer with twice the nominal line tension was needed to fit the data [5].

Although the parameters were fixed by data from a single event, they produce reasonable critical times for the entire data set. The first event in table 1 is clearly an exception which we have no explanation for. Perhaps there was a large fluctuation in the manometer pressure or perhaps the electric field generated a pore so large that relaxation to the stable pore size was impossible. Despite this anomaly we are confident that we have faithfully determined the line tension of the

SOPC/CHOL membrane making up the vesicles in question.

CONCLUSION

In this paper we have explained the existence of the large dynamically-stabilized pores observed by [5]. By separating the fast timescale on which the membrane relaxation occurs from the slow one associated with the motion of the aspirated vesicle down the pipette, we have been able to establish a modified pore energy function with a stable minimum in the one micron range. In addition, we have described a new mechanism by which this minimum disappears destabilizing the pore.

The theory that we have developed permits an accurate determination of an important membrane parameter: the line tension. From a single event in Zhelev and Needham's work, we were able to determine this parameter and two auxiliary parameters. The values so determined were then used to reconstruct all ten of the published SOPC/CHOL mixed lipid events: our theoretical post-prediction for the critical time at which pore stability is lost agreed well with the experimental result for all but one event. The agreement supports the value of our theory as a method for experimentally determining the line tension of bilayer membranes.

ACKNOWLEDGEMENTS

We would like to thank R. Kamien, U. Seifert, and D. Zhelev for helpful discussions. This work was supported in part by the US/Israeli Binational Foundation grant 94–00190 and NSF grant DMR95–07366. JDM was supported in part by an FCAR Graduate Fellowship from the government of Quebec.

REFERENCES

1. C. Taupin, M. Dvolaitzky, and C. Sauterey, *Biochem.* **14**, 4771 (1975).
2. W. Harbich and W. Helfrich, . *Z. Naturforsch.* **34a**, 1063 (1979).
3. L. Chernomordik, M. Kozlov, G. Melikyan, I. Abidor, V. Markin, and Y.Chizmadzhev, *Biochim. et Biophys. Acta* **812**, 643 (1985).
4. S. Leikin, M. Kozlov, L. Chernomordik, V. Markin, and Y. Chizmadzhev, *J. Theor. Biol.* **129**, 411 (1987).
5. D. Zhelev and D. Needham, *Biochim. et Biophys. Acta* **1147**, 89 (1993).
6. D. Zhelev and D. Needham, in Biological effects of electric and magnetic fields, edited by D. Carpenter and S. Ayrapetyan (Academic Press, 1994), pp. 105-142.
7. E. Evans and W. Rawicz, *Phys. Rev. Lett.* **64**, 2094 (1990).
8. D. Chang, B. Chassy, J. Saunders, and A. Sowers, *Guide to Electroporation and Electrofusion* (Academic Press, 1992).
9. J. Happel and H. Brenner, *Low Reynolds Number Hydrodynamics* (Prentice-Hall, 1965).
10. J.D. Moroz and P. Nelson, "Dynamically Stabilized Pores in Bilayer Membranes", submitted to *Biophys. J.* (cond-mat/9610072).

LATE STAGES OF THE PEARLING INSTABILITY IN LIPID BILAYERS

J.L. GOVEAS*, S.T. MILNER**, W.B. RUSSEL*
*Department of Chemical Engineering, Princeton, NJ 08544
**Exxon Research and Engineering, Route 22 East, Annandale, NJ 08801

ABSTRACT

We have studied the late stages of the "pearling instability" in lipid bilayers, which is brought on by applying laser tweezers to a cylindrical vesicle. This produces a front that propagates down the vesicle, leaving behind it bilayer-covered droplets separated by thin tubes, which appear under the microscope as pearls on a string. At later times, the "pearls" are observed to drift slowly towards the "trap" (the spot where the tweezers are applied, into which the surfactant is drawn). We model the hydrodynamics of the drifting pearls as a combination of translation of the string of pearls, and "slipping" of the bilayer skin over the pearls, to relate the speed of the pearls to the underlying flux of the surfactant into the trap.

INTRODUCTION

Laser "tweezers" are becoming a useful tool for probing the dynamics of bio-membranes in a controlled way. An interesting example of this is the experiments of Bar-Ziv and Moses [1] where the application of laser tweezers to lipid tubules produces a peristaltic shape instability by inducing an effective surface tension in the membrane, ending in a beaded tube.

The metastable cylindrical vesicles are made of unilamellar lipid bilayers (DMPC or DGDG) with water forming the inner and outer fluids. They are anchored at both ends by massive globules which seal off the inner volume and act as lipid reservoirs.

Tweezing the vesicle causes a front to propagate out from the laser spot, leaving behind it a modulated state with a well-defined wavelength; this state coarsens to form nearly spherical shapes separated by thin tethers, giving the appearance of pearls on a string. The "pearls" migrate towards the laser spot where they jam up but do not coalesce. This process of pearl migration continues after the laser is shut off, the pearls forming a long-lived metastable state that relaxes slowly back to the original cylinder. Here we focus on the late stages of the instability where the pearled state is observed and describe the mechanism by which the pearls move towards the laser trap.

The effect of the laser is to induce a sudden tension in the membrane. At optical frequencies the lipid has a higher dielectric constant than that of water making it energetically favorable for surfactant to fall into the trap, displacing water. The laser-induced tension Σ is the energy gain per unit area bilayer sucked into the trap and is estimated to be 2×10^{-3}ergs/cm^2 for an applied laser power of 50mW, which is far in excess of the critical tension for shape transformations [2].

In the early linear stages, this large tension applied by the laser results in a Rayleigh-like instability [2, 3, 4, 5]. The system minimizes surface area at fixed volume by creating the modulated state, but incurs viscous losses in the water by doing so. When the modulation of the cylindrical vesicle saturates to the pearled state, the spherical shapes of the pearls that minimize surface tension are stabilized against pinching off by the bending rigidity κ of the membrane. This provides a lower cut-off for the radius a of the thin tethers connecting the pearls, which is estimated to be $\sqrt{\kappa/\Sigma} \approx 0.1\mu$m (where $\kappa \approx 0.6 \times 10^{-12}$ ergs for DMPC)[2].

Mat. Res. Soc. Symp. Proc. Vol. 463 © 1997 Materials Research Society

The tension in the surfactant skin is greatest at the laser spot and decreases along the string with distance from the spot [5], due to the viscous drag from the water surrounding the skin. At the high temperature of the experiment ($\approx 45°C$) the membrane is in a pure fluid state, and we regard the surfactant skin as incompressible (the two-dimensional bulk compressibility $K = 10^{-1}\text{ergs/cm}^2 >> \Sigma$), which in flowing towards the trap entrains the interior water causing the observed migration of the pearls. If the pearls were rigid they would translate along in solid body motion with the surfactant skin: instead, their water cores are able to exert viscous stresses on the interior pearl surfaces, slowing them down relative to the moving skin. As water is not drawn into the trap with the surfactant, the pearls come to a halt at the spot, thus forming the jam. Our goal shall be to find the speed of the surfactant skin and the related speed of the pearl drift towards the trap.

THEORY

We divide the string of pearls into the stationary pearls that form the jam and the moving pearls. The pearls are assumed to be perfectly spherical with radius R and are connected by thin cylinders of radius a and length l, where (R/a) and (l/a) are large compared to unity. We neglect the surfactant viscosity throughout our analysis. Figure 1 shows the spherical co-ordinate system we use.

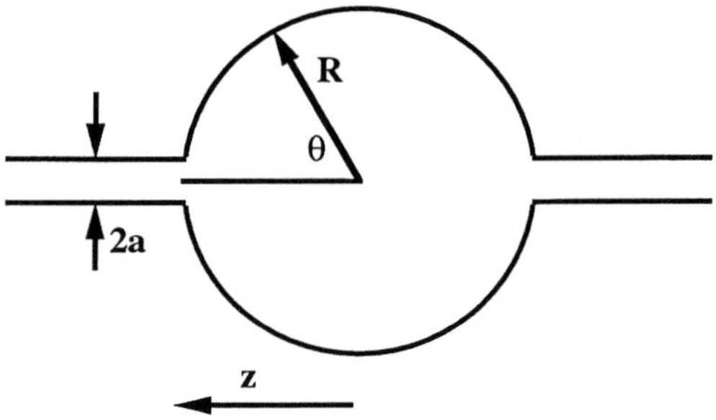

Figure 1: Spherical co-ordinate system of the pearl.

We represent the pearl motion as a superposition of two flows:
i) Rigid translation of the pearl towards the trap, at a speed denoted by U. This is simply the well-known Stokes flow for a sphere.
ii) The slipping of the surfactant skin over the surface of a stationary pearl. We shall assume that the water in the thin tethers is entrained in plug flow at a speed v_T ($U + v_T$ in the laboratory frame), which will be the characteristic scale for the motion of the surfactant skin. Poiseuille flow of the interior fluid relative to the tether skin would create a large pressure drop across them and a correspondingly large dissipation, which seems physically unlikely.

We calculate the flow field for this skin-slip motion using the Lamb solution [6], which is a prescription for finding the pressure and velocity fields inside and outside a sphere given the surface velocity field. Solving Stokes' equations for an incompressible fluid, the velocity field may be constructed as (see [7] for details)

$$\mathbf{v} = \sum_{n=-\infty}^{\infty} \nabla \times (\mathbf{r}\chi_n) + \nabla\phi_n + \frac{(n+3)}{2(n+1)(2n+3)}\frac{r^2}{\eta}\nabla p_n - \frac{n}{(n+1)(2n+3)}\frac{\mathbf{r}}{\eta}p_n \quad (1)$$

Here χ_n,ϕ_n and p_n are solid spherical harmonics of degree n, which may be found directly by considering three independent combinations of the harmonics on the sphere surface that are related to natural boundary conditions on the velocity:

$$\frac{\mathbf{r}}{r}\cdot\tilde{\mathbf{v}}=\sum_{n=1}^{\infty}X_n,\ -r\nabla\cdot\tilde{\mathbf{v}}=\sum_{n=1}^{\infty}Y_n,\ \mathbf{r}\cdot\nabla\times\tilde{\mathbf{v}}=\sum_{n=1}^{\infty}Z_n \tag{2}$$

where $\tilde{\mathbf{v}}(\theta,\varphi)$ is the surface velocity field, and X_n, Y_n and Z_n are surface harmonics. Summing the p_ns produces the pressure field. Expanding $\mathbf{r}\cdot\tilde{\mathbf{v}}, \nabla\cdot\tilde{\mathbf{v}}$ and $\nabla\times\tilde{\mathbf{v}}$ in a series of spherical harmonics and equating term by term gives us p_n, χ_n and ϕ_n in terms of X_n,Y_n and Z_n. In order that the velocity be bounded at the center of the sphere and decay to zero at infinity, we keep only positive harmonics for the interior solutions, and negative harmonics for the exterior.

Boundary conditions for the skin-slip problem

Since the surfactant skin is incompressible, there is a constant flux of lipid fluid through any cross-section of the pearl. This equals the lipid flux $2\pi a v_T$ in the thin tethers, where we impose a no-slip boundary condition at the tether surface so the water velocity tracks the lipid velocity. For a plane cutting through the pearl at a given θ, the surface velocity is

$$\tilde{\mathbf{v}}=-\frac{a}{R}\frac{v_T}{\sin\theta}\hat{\theta} \tag{3}$$

where $\hat{\theta}$ is the unit tangent vector at the surface. Applying equation 2 we see that
i) The radial component of the curl of this flow-field is zero by azimuthal symmetry.
ii) Incompressibility of the surface fluid implies that the divergence of the surface velocity is everywhere zero, except where surfactant is injected into and withdrawn from the sphere. We model the inlet and outlet as a source and sink of known strength of surfactant so that

$$-r\nabla\cdot\tilde{\mathbf{v}}=\frac{a}{R}v_T[\delta(\cos\theta-1)-\delta(\cos\theta+1)] \tag{4}$$

iii) The radial velocity of the fluid is also zero on the surface except where water is pumped into and out of the pearl. We treat the inlet and outlet as a source and sink of water, consistent with the surfactant flux of equation 4

$$\frac{\mathbf{r}}{r}\cdot\tilde{\mathbf{v}}=\frac{1}{2}\left(\frac{a}{R}\right)^2 v_T[\delta(\cos\theta-1)-\delta(\cos\theta+1)] \tag{5}$$

The delta-functions above may be expanded in the spherical harmonics; however the finite radius of the thin tether implies that there is an upper cut-off R/a to the infinite series, which smears the delta-functions over the inlet and outlet of the sphere. Specifying these boundary conditions thus determines the interior and exterior skin-slip flow-fields completely.

Determining the pearl migration speed

We determine the pearl migration speed, U, by balancing forces on the moving pearl. These forces are due to i) viscous stresses exerted on the membrane by the surrounding water, ii) tangential stresses in the membrane due to the tension gradient across the pearl and iii) normal stresses in the membrane which arise in response to i) and ii).

We shall call these membrane normal stresses "constraint forces": a device we use to keep the pearl spherical, which otherwise must deform slightly with the flow. We assume that the surface tension is high enough that this deformation is small. These constraint forces increase across the pearl in proportion to the tension, and must sum to zero for the membrane to be in mechanical equilibrium, since they arise physically from small deformations of the surfactant skin and hence from intramembrane forces.

The normal stress balance on the membrane is given by the Laplace relation [8]

$$(\sigma_{rr,o} - \sigma_{rr,i}) + f_c = \frac{2T(\theta)}{R} \tag{6}$$

where the σ_{ij} are components of the three dimensional stress tensor of the water evaluated at the sphere surface (calculated from the Stokes' and skin-slip pressure and velocity fields), and f_c denotes the constraint forces in the membrane.

To calculate $T(\theta)$ consider a spherical ring at some θ on the surface of the membrane. The tension force acting on the ring to the left must be balanced by viscous and constraint forces on the cap to the right:

$$0 = -2\pi R \sin\theta T(\theta)\hat{\theta} + 2\pi R^2 \int_\theta^\pi (\sigma_{r\theta',o} - \sigma_{r\theta',i})\hat{\theta}' \sin\theta' d\theta' + 2\pi R^2 \int_\theta^\pi [(\sigma_{rr,o} - \sigma_{rr,i}) + f_c]\hat{n} \sin\theta' d\theta' \tag{7}$$

where $\hat{n}$ is the outward pointing unit normal vector. Taking the z-component of the forces in this equation, differentiating with respect to θ and using the Laplace relation gives us

$$\frac{dT}{d\theta} = -R(\sigma_{r\theta,o} - \sigma_{r\theta,i}) \tag{8}$$

which may be integrated to give the tension variation across the pearl surface.

The constraint forces for each of the two flows have no net vertical component from azimuthal symmetry, but non-trivial z-components, since the tangential forces in both upper and lower hemispheres of the pearl increase from right to left. The sum of these z-components is negative for the skin-slip flow and positive for the Stokes flow. These two sums must add to zero for the membrane to be in equilibrium, which gives us U in terms of v_T

$$U = \frac{29}{18}\frac{a}{R}v_T \tag{9}$$

Since $a/R << 1$ this means that the pearl drifts towards the trap more slowly than the skin slips over it, as we would qualitatively expect.

Determining v_T

While U is determined from a local balance on a single moving pearl, the transport of surfactant toward the trap is communicated through the whole string of pearls. Hence v_T must be determined from global considerations. This is done by calculating the viscous dissipation of the entire string and equating it to the rate of work done by the laser in pulling surfactant into the trap. We shall need to consider the dissipation due to the moving pearls, the pearls in the jam and the thin tethers.

Using the Stokes' and skin-slip flow-fields the dissipation for a single moving pearl is

$$\dot{E}_m = \frac{77}{3}\pi\eta a v_T^2 + 6\pi\eta a U v_T + 6\pi\eta R U^2 \tag{10}$$

The skin-slip term scales as $\eta a {v_T}^2$ showing that the bulk of the dissipation due to this motion comes from the mouth of the pearl where $|\mathbf{v}| \sim v_T$ and $\nabla \sim \frac{1}{a}$ over a region of volume a^3.

The jammed pearls have the same skin-slip motion but are stationary. The water in the narrow neck between two jammed pearls is locally in a shear flow, so we use the lubrication approximation to calculate the important contribution to the exterior dissipation. With the same interior skin-slip solutions as for the moving pearls, the dissipation for a jammed pearl is $(47/3)\pi\eta a v_T^2$.

We use a result from slender body theory [9] to find the dissipation outside a thin tether moving at velocity $U + v_T$, which is given by $4\pi\eta l(U + v_T)^2/\ln(l/a)$.

RESULTS

To sum the dissipation for the string we consider as an example a state of the string of pearls in which 1/8 of the length consists of those which have reached the jam. For cylinders of initial radii 1μm, and string lengths in the range $100 - 200\mu$m with $R = 1.6\mu$m,$a = 0.1 - 0.2\mu$m and $l = 5\mu$m , this represents a total of $15 - 30$ pearls (and hence 4-8 jammed pearls). The rate of work done by the laser is the applied tension times the flux of lipid into the trap, equal to $2\pi a v_T \Sigma$. Equating the total dissipation to the rate of work done by the laser, (for laser power in the range $50 - 100$ mW) and substituting equation 9 for U allows us to solve for a skin-slip speed of $v_T \approx 2 - 13\mu$m/sec, and a pearl drift of $U \approx 0.2 - 3\mu$m/sec. The high-end estimates are based on higher laser power, shorter strings and thicker tethers.

Existing experimental data is for pearl speeds *after* the laser is switched off. This is because the same microscope is used to focus the laser beam and observe the experiment, and so the field of view need always include the spot when the laser is on. Our analysis depends on the operation of the trap as a sink of surfactant, and so only applies to pearl motion while the laser is on. An additional mechanism (as yet unknown) is required to explain why the pearls continue to move towards the spot after the laser is turned off. For pearl radii and laser power in the ranges we have used above, these speeds average 5μm/sec [1], which are the same order of magnitude as our high-end estimates for U.

We can use the calculated v_T to estimate other experimental observables. Since the modulation of the cylinder requires loss of area, the time to transform it into the final string of pearls is controlled by how fast lipid flows into the trap. Using the calculated v_T for the flux of surfactant into the trap, we find coarsening times in the range of 12 to 340 seconds, where the high-end estimate for the flux gives the shorter time. Experimental coarsening times are on the order of 30 seconds [1].

The driving force for the instability is the drawing of surfactant into the laser trap. However the trap has a finite size, and so if it were to become full, the instability would stop. We use the surfactant flux calculated above and assume that lipid packs densely into the trap to set an upper limit on the filling time. For a spherical trap of diameter $d = 0.8\mu$m, and a lipid molecule of length $D = 20\AA$ we calculate a time of $(\frac{4}{3}\pi(d/2)^3)/(2\pi a D v_T)$ which is 8-97 seconds. Yet the experiment generally goes on for much longer than this, so lipid must be leaking out of the trap.

Nelson [10] has pointed out that surfactant micelles may be thermally excited out of the trap. A spherical micelle with a radius on the order of the length of a lipid molecule, has an energy in the trap of $\Sigma\frac{4}{3}\pi D^2 \approx 3 \times 10^{-16}$ergs $= 7 \times 10^{-3} k_B T$. A more accurate calculation of the binding energy of a micelle requires knowledge of the bilayer structure inside the trap. It would be interesting to confirm experimentally that lipid is released into solution, perhaps by tagging lipid molecules.

CONCLUSIONS

We have identified two characteristic velocity scales in this problem: v_T, the surfactant velocity in the tether, and a significantly slower pearl migration speed U which scales as $(a/R)v_T$. The most direct test of these predictions would be a measurement of the surfactant flux itself in conjunction with a measurement of the pearl drift during laser operation.

We have presented a simplified view of the hydrodynamics to elucidate the underlying physics. While we expect that the scaling of our predictions is robust, more refined calculations for the hydrodynamics will likely produce $O(1)$ corrections to the prefactors. We do not explain why the pearls continue to drift after the laser is switched off, nor their subsequent slowing down and relaxation back to the initial state. This remains an important and intriguing phenomenon.

ACKNOWLEDGEMENTS

This work was supported primarily by the MRSEC program of the National Science Foundation under Award Number DMR-9400362. We thank P. Nelson, E. Moses and P. Olmsted for useful discussions.

REFERENCES

1. R. Bar-Ziv and E. Moses, Phys. Rev. Lett. **73**, p.1392 (1994).

2. P. Nelson, T. Powers and U. Seifert, Phys. Rev. Lett. **74**, p.3384 (1995).

3. R. Granek and Z. Olami, J. Phys. II (France) **5**, p.1349 (1995).

4. R. Goldstein, P. Nelson, T. Powers and U. Seifert, J. Phys. II (France) **6**, p.767 (1996).

5. P. Olmsted and F. Mackintosh, submitted to J. Phys. II (France)

6. H. Lamb, Hydrodynamics, Dover,1945 Ch.11, Art. 336.

7. J. Happel and H. Brenner, Low Reynolds Number Hydrodynamics, Noordhoff International Publishing, 1973 Ch.3, Section 2.

8. L. Landau and E. Lifshitz, E., Hydrodynamics Pergamon Press, 1987, p.241.

9. G. Leal, Laminar Flow and Convective Transport Processes, Butterworth-Heinemann, 1992, 251.

10. P. Nelson, personal communication

Front propagation in laser-tweezed lipid bilayer tubules

PETER D. OLMSTED* and FRED C. MACKINTOSH**

*Department of Physics, University of Leeds, Leeds, LS2 9JT, United Kingdom
**Department of Physics, University of Michigan, Ann Arbor, MI 48109

Abstract We study the mechanism of the 'pearling' instability seen recently in experiments on lipid tubules under a local applied laser intensity. We argue that the correct boundary conditions are fixed chemical potentials, or surface tensions Σ, at the laser spot and the reservoir in contact with the tubule. While most qualitative conclusions of previous studies remain the same, the 'ramped' control parameter (surface tension) implies several new features. We also explore some consequences of front propagation into a noisy unstable medium.

1 Introduction

Recent experiments [1] have demonstrated a dynamic instability induced on tubules of single lipid bilayers by application of laser 'tweezers', whereby the cylindrical tubule of radius R_0 modulates with a wavenumber given by $q^*R_0 \simeq 0.8$. This instability has been attributed to an excess surface tension due to the gain in electrostatic energy when lipid molecules, of higher dielectric constant than water, displace water in the region of the laser. This phenomenon resembles the Rayleigh instability [2, 3] of a thin cylindrical thread of liquid with positive surface tension, which reduces its surface area at fixed volume by modulating and evolving towards a string of beads. The present problem, however, requires a different dynamical analysis which relates the flow of lipid molecules in the interface to the bulk flow in the surrounding fluid.

At present there are (at least) two theoretical treatments of these experiments [1]. Bar-Ziv and Moses [1] and Nelson and co-workers [5] suggested that the surface tension rapidly equilibrates everywhere to an induced value Σ_0, and the instability proceeds from this state. Granek and Olami [4] postulated a constant rate-of-suction of lipid molecules into the trap. This loss of lipid is accommodated by stretching out small wavelength fluctuations, leading to a uniform surface tension Σ_0. Goldstein, *et al.* (GNPS) [6] demonstrated how the equilibration of the tension in the tube stays 'ahead' of a shape change, and argued that the primary loss of area is in the shape instability itself.

We propose a slightly different picture. The tubules are several hundred microns long and are attached at either end to lipid globules of order 10μm in diameter [1]. If we assume the system is equilibrated, it follows that the chemical potential for exchange between the tubule and reservoir (lipid globules) vanishes [7], and we may assume a reference chemical potential of zero or, equivalently, zero surface tension. This coincides with the observation of visible thermal fluctuations on the tubules [1]. Treating the anchoring globules as reservoirs presupposes that damping processes retarding the transfer of lipid to and from the globules is negligible relative to other damping processes. We expect this to arise from the same source as two-dimensional surface viscosities, which we argue below to be negligible.

In the electric field of a laser the chemical potential of a lipid molecule is lowered by $\delta\varepsilon\mathcal{E}v_0$, where v_0 is the molecular volume, $\delta\varepsilon$ is the dielectric constant relative to water, and $\mathcal{E}$ the energy density deposited in the trap. Nelson *et al.* [6] calculated an energy gain per area of bilayer of $\Sigma_0 \sim 2\cdot10^{-3}\,\mathrm{erg\,cm^{-2}}$, for a laser power of 50 mW.

This reduction in the local chemical potential induces lipid motion into the laser spot. Loss of lipid adjacent to the spot may be interpreted, mechanically, as an increased surface tension. Because the other end of the tubule is in contact with a reservoir at zero chemical potential, the final state (prohibiting, for the moment, surface undulations) must be a non-equilibrium steady state in which: (1) Lipid flows at constant velocity from the reservoir to the laser trap; and (2) the chemical potential drops linearly along the tubule, and its

Mat. Res. Soc. Symp. Proc. Vol. 463 © 1997 Materials Research Society

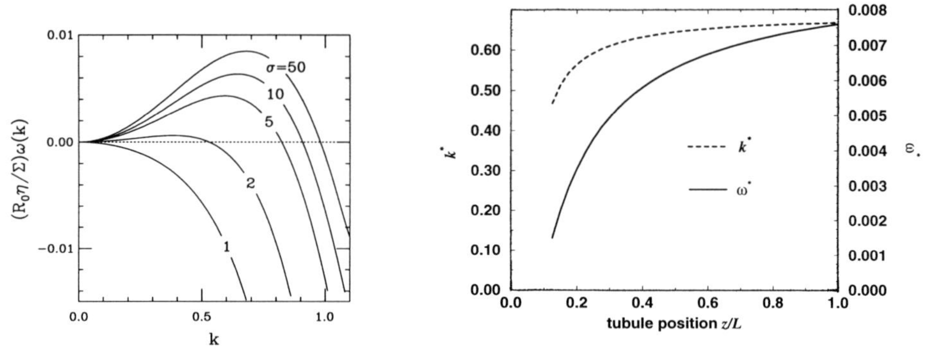

Figure 1: (Left) Dispersion relation $\omega(k)$, with parameters as calculated by GNPS, $\beta = 3.5, \epsilon = 0.5$. Reprinted from Ref. [7]. Alternatively, we may consider this as a plot of $\omega(k, z)$ for $\sigma = 50$ and $z = L, L/5, L/10, L/25$, and $L/50$. Here, k is in units of $2\pi/R_0$. (Right) Frequency ω^* and dimensionless wavenumber k^* of fastest growing mode as a function of position, obtained from (left) by taking $\sigma(z = L) = 20$.

gradient balances the frictional drag of the bulk fluid. This differs from the treatments of Nelson *et al.* and Granek and Olami in that lipid must flow out of the anchoring globules.

Several consequences follow from this observation. We predict a speed of propagation and characteristic wavenumber which are largest near the laser spot and decreases to zero somewhere near the anchoring reservoirs (see Fig. 1). We also argue that, to initiate the pearling instability, an instability must occur to allow a continual flux into the trap. We have estimated the criterion for buckling inside the trap, although other mechanisms such as ejection of micelles or formation of multilamellar phases cannot yet be discounted. We discuss the implications of a small spatial variation in the surface tension on the calculation of GNPS [5, 6], and on front propagation [9]. This issue is delicate, and our results suggest front propagation which is either (a) characterized by by the so-called Marginal Stability Criteria (MSC) [6, 9], or (b) dominated by amplification of existing 'noise', which can lead to behavior reminiscent of front propagation for a steep enough ramp.

2 Steady state

First we calculate the steady-state flow profile in the tubule, prohibiting shape undulations, assuming the laser enforces a chemical potential $-\Sigma_0$ at the laser spot. This implies a lipid reservoir in the trap.* An estimate of a buckling instability inside the supports this boundary condition [15]. Changes in chemical potential $\delta\mu$ are related to changes in surface tension $\delta\Sigma$ by $\delta\mu = -\phi^{-1}\delta\Sigma$, where ϕ is the lipid concentration ($\phi^{-1} = a$ is the area per lipid). Also, $\Sigma = -p$, the 2-dimensional pressure of the fluid of lipid molecules. The boundary conditions are $p(z = 0) = p_0$ (reservoir) and $p(z = L) = p_0 - \Sigma_0$ (laser spot), where Σ_0 is the surface tension induced by the laser; and the cylinder axis is parallel to the z-axis.

The Navier-Stokes and continuity equations for the 2D fluid of lipid molecules are

$$\partial_t \phi = -\nabla\cdot(\phi \mathbf{v}) \qquad \text{and} \qquad \rho_s(\partial_t + \mathbf{v}\cdot\nabla)\mathbf{v} = -\nabla p + \Delta\mathbf{T}^b\cdot\hat{\mathbf{r}}. \tag{1}$$

We have ignored 2D surface viscosities (which can be shown to be negligible); ρ_s is 2D lipid mass density, and $\Delta\mathbf{T}^b \cdot \hat{\mathbf{r}}$ is the viscous drag acting on the surface from the dissipative

*After a long time the trap will fill up with lipid and the chemical potential return to zero everywhere. However, for strong enough laser power the shape instability will have occurred by this time.

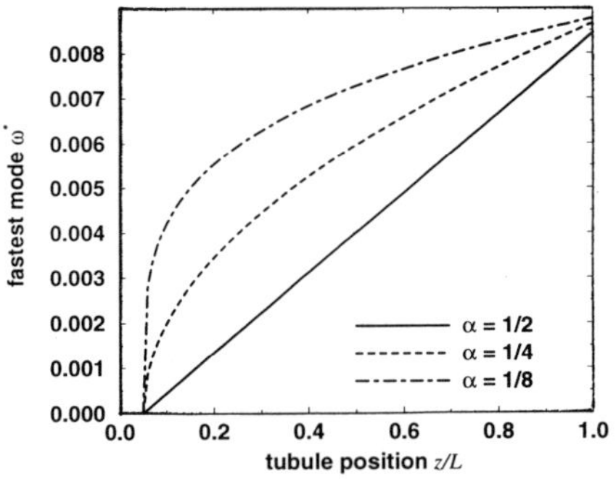

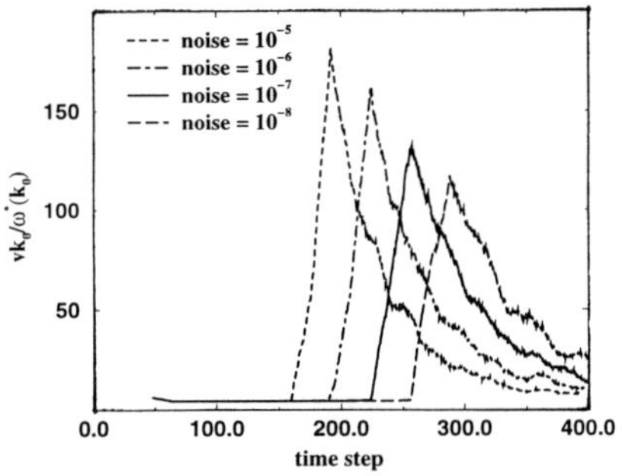

Figure 2: (Left) Local dispersion relation ω^* for Eq. (4), for $\alpha = 1/2, 1/4, 1/8$. The trap is at $z/L = 1$. (Right) Front velocity vs. time, averaged over 100 realizations of initial noise. The velocity was measured by tracking the leading edge of the envelope of wavelets, and is plotted in units of the characteristic velocity $\omega(k_0)/k_0$, where $\omega(k) = 2\,k_0^2\,k^2 - k^4$ is the dispersion relation for Eq. (4) for $a(x) = 1$. Ramp parameter $\alpha = 1/8$. The noise is given as a fraction of the final amplitude.

stress tensor $T^b_{\alpha\beta} = \frac{1}{2}\eta(\nabla_\alpha u_\beta + \nabla_\beta u_\alpha)$ in the surrounding fluid. Flow is established in a vorticity diffusion time $\tau_v \simeq \rho R_0^2/\eta \sim 10^{-7}$s, which is much smaller than other times in the problem. The boundary stress is given by the shear stress in the tube. For uniform flow $\mathbf{v} = v\hat{\mathbf{z}}$, the interior flow is Poiseuille [10] and the boundary stress is $\Delta\mathbf{T}^b\cdot\hat{\mathbf{r}} = -2\eta v/R_0$. The pressure obeys the constitutive relation $p = p_0 - B\phi_0\delta a$, where B is the bulk modulus. The time to attain steady state after turning on the laser may be calculated from Eqs. (1) [15]; $\tau_{ss} \sim 2L^2\eta[(2\pi)^2 R_0 B]^{-1} \sim 10^{-5}$ s, for $B \sim 150\,\mathrm{erg\ cm^{-2}}$ [11] and $L \sim 100\mu$m. This estimate ignores thermal fluctuations which soften B [12]. As shown in Ref. [6], fluctuations can reduce B by up to three orders of magnitude, increasing τ_{ss} to of order 10^{-2} s.

Solving Eqs. (1) yields the steady-state profile

$$\delta\Sigma = \Sigma_0\frac{z}{L}, \tag{2}$$

with an average lipid velocity $\bar{v} = R_0\Sigma_0/(2\eta L)$. We estimate $\bar{v} \sim 1\mu\mathrm{m\ s^{-1}}$, using $\Sigma_0 \sim 10^{-3}\mathrm{erg\ cm^{-2}}$ [1].

3 Microscopic Picture

Lipid flow, which accompanies an effective non-zero surface tension far from the trap, follows from a boundary condition of fixed chemical potential at the trap. Here we argue that this boundary condition requires an instability in the trap, and we suggest buckling as a possible scenario. Other mechanisms include ejection of micelles, or formation of multilamellar structures or 'cancerous' membranes. This is surely not an exhaustive list.

The laser spot has a finite side $\Delta \simeq 0.15\mu$m in Ref. [1]. Upon applying the laser lipid can enter the trap until the electrostatic energy gain balances the cost of compressing the molecules in the bilayer. At this point flow stops and the chemical potential (and surface tension) of the tube reverts back to zero. However, for some critical tension Σ^* the trap may become unstable with respect to buckling. For higher intensities the trap continues to fold to accommodate more lipid, allowing flow along the tubule. This flow is accompanied by a chemical potential (or surface tension) gradient, which drives the shape instability. A detailed calculation [15] or a simple comparison of bending and electrostatic energies yields a

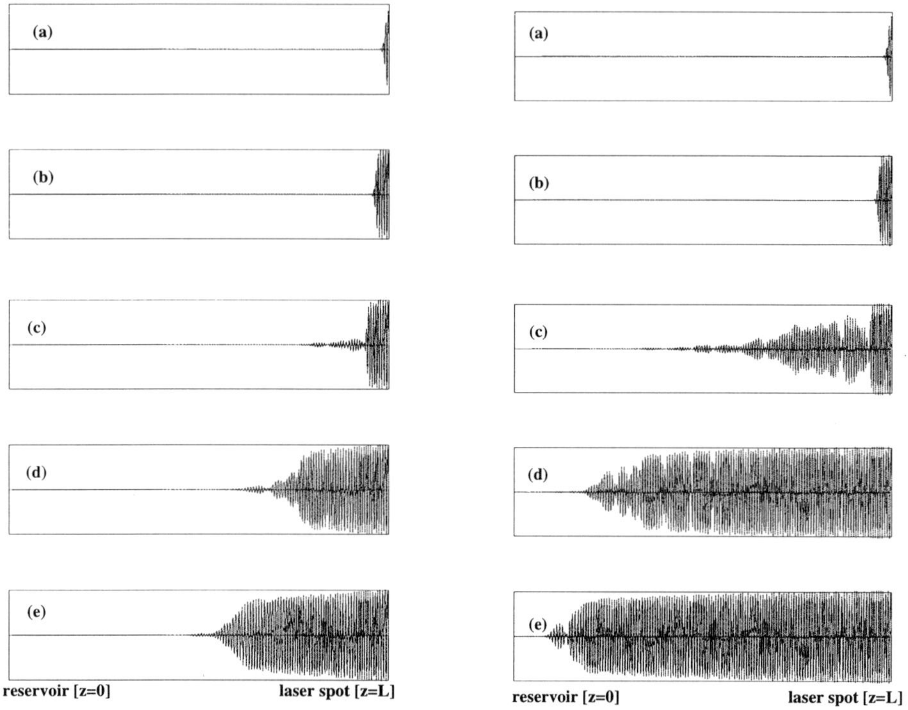

Figure 3: (Left) Evolution of $u(x,t)$ from Eq. (4), with: convection $\bar{v} = 0$; initial perturbation $u(L, t = 0) = 0.1u(L, \infty)$; $\alpha = 1/2$; initial mean noise amplitude a fraction 10^{-6} of the final amplitude. Intervals are every 80 time steps. The vertical and horizontal scales are not the same. (Right) Same parameters, except ramp parameter $\alpha = 1/8$.

critical intensity $\sigma^* \equiv \Sigma^* R_0^2/\kappa \simeq 1$. A typical experimental value is $\sigma \simeq 20$ [1]. Upon turning off the laser, the system could revert to the original tubule by unfolding or, if topological changes have occurred (*e.g.* budding in the laser spot or the creation of metastable 'pearls'), perhaps attain some other long-lived metastable state.

4 Dispersion Relation and Front Propagation

We turn to the effects of a surface tension gradient on growth. Rather than repeating the analysis of GNPS [6] with a non-uniform surface tension, we note that the characteristic wavenumber at which the instability occurs is typically $q^* R_0 \simeq 0.8$. Since $R_0 \ll L$, we suspect that the assumption of a locally constant surface tension is a good first step.

The growth rate $\omega(q)$ of a Fourier mode q of an undulation $u(q,t)$ is defined by

$$(\frac{\partial}{\partial t} + iq\bar{v})u(q,t) = \omega(q)u(q,t), \tag{3}$$

where the convective term arises because the lipids have an average velocity.

GNPS [6] calculated $\omega(q)$ for a uniform surface tension $\bar{\Sigma}$ (Fig. 1), including the effects of bending and intra-bilayer friction. Changing the boundary conditions to allow for flux from the reservoir adds the convective term above and, aside from the local approximation, changes nothing else. Since σ is z-dependent (Eq. 2), the growth rate ω^* and wavenumber q^* of the fastest growing mode are z-dependent, and are greatest near the laser spot (Fig. 1).

A position-dependent dispersion relation has several immediate consequences: The local wavenumber and apparent growth rate should decrease, with the instability vanishing at a point close to the reservoir where the induced surface tension is not strong enough to overcome the barrier due to bending, $\Sigma^* \simeq \kappa/R_0^2$. Hence, in experiments with $\sigma \simeq 20$ [5], this occurs at $1/20th$ of the distance from the anchoring globule to the laser trap.

GNPS argued that the MSC hypothesis provides a reasonable estimate for the propagation speed v_f and the selected wavenumber k^* [6]. Extending this calculation to a local dispersion relation implies a spatially varying front speed and selected wavenumber. However, front propagation implies: (1) the existence of a propagating front, and (2) the absence of noise. Rather than discuss the effect of a spatially varying control parameter on v_f and k^* [9], we focus on the issue of the *existence* of a propagating front.

In the presence of 'noise' (*i.e.* thermal fluctuations) a propagating front can be expected to exist for times less than the characteristic growth times of fluctuations 'close' to the most unstable mode. Hence, after a quench into an unstable 'ramped' state with an initial perturbation (a 'pinch' induced by the laser) propagation occurs initially, followed by rapid growth everywhere as the initial conditions are amplified to visible length scales. The effect of a 'ramped' control parameter should be most dramatic after the noise overwhelms the front propagation: for no ramp the noise grows randomly everywhere, and the 'front' should break down when the noise has grown to visible amplitudes. For a steep enough ramp the non-uniform amplification of the noise could resemble front propagation.

To check these conjectures we have employed a simple dynamics, specified by

$$(\frac{\partial}{\partial t} + \bar{v}\partial_x)u(x,t) = \left[a(x)2k_0^2\partial_x^2 + \partial_x^4\right]u(x,t) - gu(x,t)^3, \tag{4}$$

where $a(x)$ mimics the dispersion relation and z-dependence of Fig. 1. A choice which gives reasonable agreement is $a(x) = |x - x_0|^{\alpha+1}/(x - x_0)$, where x_0 is the position at which the system is absolutely unstable. This is a toy model whose details do not correspond to the Bar-Ziv *et al.* experiments, but which we believe contains the essential physics of front propagation into an unstable inhomogeneous medium. For Fig. 1, $x_0 \sim 0.05L$ and $\alpha = 1/8$ are reasonable. Fig. 2 shows the local dispersion relation $\omega^*(x)$ for various α.

Fig. 3 shows the evolution of Eq. (4) for an initial perturbation at the trap of 1% of the final amplitude, and noise comprising 300 harmonics with a Boltzmann weight corresponding to a non-zero surface tension (*i.e.* with an energy proportional to q^2). The general features are as described above. We find a dimensionless initial front velocity of $v\,\omega(k_0)/k_0 = 3.3$, while the MSC [8] predict 4.6. A similar agreement was found by GNPS [6]. The steeper ramp has well-defined growth in the noisy region which almost resembles a front, while growth into the shallow ramp looks little like a front. The shallower ramp has a very faster 'propagation' speed in the noisy regime, a consequence of the faster overall growth rate for a shallower ramp, because a larger fraction of the tubule is more unstable. For a noisier system the effective propagation speed in the noisy regime is faster, and the breakdown of the MSC propagation occurs earlier (Fig. 2). The initial propagation velocity is independent of the noise amplitude. The delay before noise-dominance increases logarithmically with increasing noise amplitude, consistent with a delay until the noise has grown (exponentially) to a given amplitude. This study raises several questions. If the noise is weak enough and an apparent front exists, can this be understood quantitatively in terms of the ramp, and how does this relate to previous investigations of 'ramped' control parameters [9]?

5 Conclusion

We have given the following picture of the action of lipid tubules upon the application of laser tweezers. In the absence of buckling the laser induces a local compression of lipid molecules in the laser spot. This takes place in a time of order $\tau_{ss} \sim 10^{-5}$s. A sufficiently large laser intensity induces an instability such as buckling of the membrane in the trap, which initiates flow down the tubule from the reservoir. We do not have an estimate of the delay time for the instability in the trap. In the absence of undulations *outside* the trap, this flow would build up to a steady-state value $\bar{v} \sim 1\,\mu$m s^{-1} in a time order τ_{ss}. Including the softening of B due to thermal fluctuations [12, 6] increases τ_{ss} to $\tau_{ss} \sim 10^{-2}$ s, so that an accurate quantitative calculation must include the dynamics of the increase in surface tension. This leads to, effectively, a smaller applied tension σ and hence a slower propagation speed. This spirit was followed in the approach of Granek and Olami [4].

Given the steady-state tension profile within the membrane and the reasonable assumption that the gradient occurs over a length (L) much larger than the critical wavelength ($\sim R_0$), the analysis of GNPS [6] leads to a Rayleigh-like instability to undulations. This instability initiates near the laser spot where the chemical potential (or surface tension) is lowest, and propagates away from the spot to a point along the tubule at which the local surface tension falls below the critical tension $\Sigma_{cr} = \kappa/R_0^2$ which characterizes the instability. Typical growth frequencies are $\omega \sim 25\,\mathrm{s}^{-1}$ [6], which corresponds to times $\tau_\omega \sim 10^{-1}$s. Experiments find a significant time delay of order seconds before the instability, [14], a still-unexplained observation.

Front propagation and the effects of propagating into a spatially-varying medium have only been touched upon in our numerical treatment. We have not studied non-linear effects. This has been addressed by: Olami and Granek [4], who considered the effect of removing lipid from the membrane; GNPS [6], who added the correct non-linear terms in the bending energy; and Goveas *et al.* [16], who studied the motion of fully-developed pearls.

Our theory differs from previous theories in several respects. Obviously, we expect flow out of the reservoir. The inhomogeneous surface tension implies that the velocity of front propagation v_f and characteristic wavenumber should decrease farther away from the laser spot. It would also be interesting to test whether fluctuations are strong enough to destroy the front-like character, and whether two characteristic regimes exist in the experiments, as suggested in Fig. 2.

Acknowledgements

We thank E. Moses, R. Granek, P. Nelson, T. Powers, C.-M. Chen, S. Milner, and W. van Saarloos for helpful conversations and correspondence. This work was supported in part by NSF Grant No. DMR 92-57544; The Donors of the Petroleum Research Fund, administered by the American Chemical Society; and NATO Grant CRG-960678.

References

[1] R. Bar-Ziv and E. Moses, Phys. Rev. Lett. **73** (1994) 1392.
[2] Lord Rayleigh, Proc. Lond. Math. Soc. **10** (1879) 4; Phil. Mag. **34** (1892) 145.
[3] S. Tomotika, Proc. Roy. Soc. Lond. **A150** (1932) 322.
[4] R. Granek and Z. Olami, J. Phys. II (France) **5** (1995) 1349.
[5] P. Nelson, T. Powers, and U. Seifert, Phys. Rev. Lett. **74** (1995) 3384.
[6] R. E. Goldstein, P. Nelson, T. Powers, and U. Seifert, J. Phys. II (France) **6** (1996) 767.
[7] J. H. Schulman and J. B. Montagne, Ann. N.Y. Acad. Sci. **92** (1961) 366.
[8] W. van Sarloos, Phys. Rev. **A37** (1988) 211.
[9] L. Kramer, E. Ben-Jacob, H. Brand, and M. C. Cross, Phys. Rev. Lett. **49** (1982) 1891; L. Kramer and H. Riecke, Z. Phys. B - Cond. Matt. **59** (1985) 245.
[10] L. D. Landau and E. M. Lifschitz, *Fluid Mechanics* (Pergamon, Oxford, 1959).
[11] E. Evans and D. Needham, J. Phys. Chem. **91** (1987) 4219.
[12] W. Helfrich and R.-M. Servuss, Nuovo Cim. **3D** (1984) 137.
[13] R. Bar-Ziv, T. Frisch, and E. Moses, Phys. Rev. Lett. **75** (1995) 3481.
[14] E. Moses, private communucation.
[15] P. D. Olmsted and F. C. MacKintosh, *Journal de Physique* to be published (January, 1996).
[16] J. Goveas, S. T. Milner, and W. B. Russel, submitted for publication (1996).

COLLISION OF A FIELD-DRIVEN POLYMER WITH A POST: IMPLICATIONS FOR ELECTROPHORESIS IN MICROLITHOGRAPHIC ARRAYS

E.M.SEVICK* AND D.R.M. WILLIAMS**
*Research School of Chemistry and **Research School of Physical Sciences & Engineering, The Australian National University, Canberra ACT 0200 AUSTRALIA, sevick@rsc.anu.edu.au and drw110@rsphy3.anu.edu.au

ABSTRACT

We study the simplest model of a polyelectrolyte impinging upon a point, frictionless obstacle in the presence of a field. Using numerical simulation, we show that the wide range of impacts, ranging from direct impact forming a long-lived hairpin conformation, to glancing impacts where the chain slides off of the obstacle in short time, can be described universally. In strong field, the average collision time, $\langle t_c \rangle$, and average distance traveled during collision, $\langle z_c \rangle$, depend upon the impact and follow universal curves over a large range of molecular weights and field strengths. This result provides analytic formulas for the chain's mobility in an array of posts and yields insight into the effect of post spacing.

INTRODUCTION

Gel electrophoresis is one of the most widely used techniques for size separating charged molecular chains such as DNA or synthetic polyelectrolytes. The separation is achieved by driving the chains through a gel, usually agarose or polyacrylamide with an electric field. As a result of the field-driven mobility and the obstacles that the gel provides, small chains pass through the gel quickly while longer molecules move more slowly. In constant field, very long chains, e.g., DNA in excess of 30 kilobase pairs, exhibit weak length-dependent mobility and consequently, separation of these chains is not possible. In a pulsed field, this saturation zone is postponed to longer chain lengths, but the mobility can be non-monotonic and can prevent a simple separation. Most recently, Austin has introduced a new electrophoretic medium fabricated from lithography, introducing the possibility of tailoring the medium to enhance size-separation. Endless possibilities exist including arrays of posts, perforated barriers, or complicated mazes. However, at present, there exist no predictions of what array geometry is best for chain separation.

In this paper, we investigate the mobility of a charged chain through an array of posts in the single post approximation, i.e. where a chain interacts with one post at a time. We model the chain as a sequence of freely-jointed rods of fixed length ℓ and charge per rod Q, residing in an electric field, E. We define a dimensionless field strength $\varepsilon = QE\ell/kT$ and assume that the force required to maintain an average rod projection ℓ_z in the field direction, is

$$f(\langle \ell_z \rangle) = (kT/\ell)\mathcal{L}_d^*(\langle \ell_z \rangle/\ell) \qquad (1)$$

where $\mathcal{L}_d$ is the inverse Langevin function in d-dimensions. In the first section we review the conformation of an end-tethered chain in an electric field, providing some of the expressions which are used to construct the numerical simulation of the unhooking process of a free chain straddling an obstacle, which is described in the second section. In the third section, we construct analytic formulas for the characteristic time scale of unhooking in the strong field and weak field limits. In the final section, we describe a set of numerical simulations where chains more realistically interact with the post in a wide variety of ways, where partial hooking, glancing impact and incomplete unwinding predominate. Despite the wide range of possible interactions, we find a set of universal relations which describe the dynamics of the chain, and which can be used to predict length dependent mobility of the chains as a function of the spacing between post obstacles.

EQUILIBRIUM: A POLYELECTROLYTE END-TETHERED TO A STATIONARY OBSTACLE

Mat. Res. Soc. Symp. Proc. Vol. 463 © 1997 Materials Research Society

Consider a freely jointed chain of N charged rods whose ends are labeled consecutively from $n=0$ at the tether site to $n=N$ at the free end. The downfield position of bead n is $z(n)$, measured from the stationary point in units of link length. The projection of each link onto the field direction, or the extension of the spring, $z(n+1)-z(n)$, is found from minimization of the free energy of the chain, F. F is comprised of the field energy of the charged beads, which promote downfield advance of each bead, and the stretching energy, which prohibits dramatically different advances of neighboring beads: $F = F_{field} + F_{stretch}$. The field energy is

$$F = -QE\ell \int_0^N dnz(n) + kT \int_0^N dn \int_0^N d(\langle \ell_z \rangle) f(\langle \ell_z \rangle) \tag{2}$$

where the first and second terms are the field and stretching energies of the chains. The integral over the spring extensions in Equation (2) can be simplified by noting that in the limit of large N, the average extension $\langle \ell_z \rangle$ is $lim_{n \to \infty} \langle \ell_z \rangle = (\ell z(n+1) - \ell z(n)) / ((n+1) - n) = \ell\, dz(n)/dn$, where the stretching variable is $y \equiv dz(n)/dn$, and the force law is used in the definition for S,

$$F/kT = \int_0^N dn\{\varepsilon(n-N)y(n) + S[y]\} \text{ where } S[u] = \int_0^u dx \mathcal{L}_d^*(x). \tag{3}$$

The equilibrium extension of an end-tethered chain is found by minimization of the chain's free energy, Equation (3), with respect to the stretching profile $y(n)$, providing the equilibrium stretching profile

$$y(n) = \mathcal{L}_d[\varepsilon(N-n)] \tag{4}$$

According to Equation (4) the tethered chain is maximally stretched at the tether point, $n=0$, and unstretched at the free end of the chain, $y(N)=0$. This agrees with fluorescence microscopy images of individual stained DNA molecules where the free ends of the chain appear brighter, indicative that the free ends are less stretched in comparison with the rest of the chain.

DYNAMICS: A NUMERICAL SIMULATION OF THE UNHOOKING OF A CHAIN

The dynamics of a polyelectrolyte draped over a frictionless obstacle in an electric field is modeled as the response of a linear sequence of springs straddling a pivot obstacle, Figure 1. The unhooking process is assumed to begin with the P_o-th bead or monomer hooked by the obstacle, with $P_o > N/2$. Thus, the longer arm contains P_o monomers and the shorter one, $N- P_o$ monomers, respectively. Unhooking corresponds to the pivot monomer (P) advancing through each monomer from P_o (initial pivot point) through N until the chain is released, i.e. the pivot monomer is the last or the Nth labeled monomer. During the whole unhooking process, the hooking obstacle is taken as a point that transmits forces between neighboring pivot segments, but does not introduce additional frictional forces. The lengths of the arms depends not only upon the magnitude of the field force, but also upon how fast each part of the chain responds to the force. Consequently, the stretching profiles of the arms of the hooked chain are generally not at equilibrium. However, we assume that the force law (Equation 1) is still strictly obeyed by each bead or monomer during the dynamics. The governing equation of motion for each monomer, n, is found by equating the hydrodynamic drag on the nth monomer moving at speed $d\ell z(n)/dt$, to the conservative force:

$$\varsigma \frac{\partial \ell z(n)}{\partial t} = -\frac{\delta F(z(n))}{\delta \ell z(n)} \tag{5}$$

Here $\varsigma \approx A\eta\ell$ is the friction coefficient of the polymer where A is a dimensionless drag constant and η is the viscosity of the medium, and δ denotes a functional derivative. Equation (5) is made dimensionless through $\tau \equiv t/\tau_o$ where the timescale τ_o is $\tau_o \equiv \varsigma\ell^2$:

$$\frac{\partial z(n)}{\partial \tau} = \varepsilon + \frac{\partial}{\partial n} \mathcal{L}_d^*(y(n)) \tag{6}$$

Equation (6) written for each nth monomer, $1<n<N$, provides a complete set of N equations of motion for solution. This set of first order differential equations is solved numerically using Euler's method. We adopt an initial configuration of a hooked chain with pivot monomer $P_o=N/2+2$. with arms containing $N/2+2$ and $N/2-2$ springs and stretched according to the equilibrium profiles of the preceding section. The extension and retraction of arms as a function of time characterizes the unhooking process. Figure 2 shows the dynamics of arm extension/retraction versus dimensionless time, τ, for a chain of $N=100$ links in fields of dimensionless strength e=0.02 and 0.002 (weak field). At early stages of the unhooking process, the difference in arm lengths $\Delta=z(1)-z(N)$ is exponential in time and expressed as

$$d\Delta/d\tau = \Delta/\tau_c \tag{7}$$

where τ_c is the characteristic unhooking time. The total length of the chain, or the sum of the arm lengths, $z(1)+z(N)$, is almost invariant throughout the early unhooking process, but decreases at the end of the unraveling period sharply. In the late stage of unhooking (approximately the last 25% of the whole unhooking process), the retraction of the shorter arm is exponential while the advance is slower and linear in time. The total length of the chain decreases approximately 20% from its initial hooked extension upon release from the obstacle. All of these observations coincide with the experimental observations presented by Song & Maestre, who interpreted these features as being related to chain elasticity. However, in interpreting the faster retraction, Song & Maestre assumed friction at the pivot point and assumed that the stretching of the arms followed that of the end-tethered arms.

DYNAMICS: AN ANALYTIC APPROACH FOR UNHOOKING IN STRONG & WEAK STRETCHING REGIMES

The equation of motion can be cast into any general variable q (such as Δ or P) rather than the function $z(n)$, using the Rayleighian dissipation function,

$\mathcal{R} = \frac{1}{2}\varsigma\ell^2\int_0^N dn\dot{z}(n)^2$: the equation of motion appears as $\frac{\partial\mathcal{R}}{\partial q} = -\frac{\partial F(q)}{\partial q}$.

In the limits of strong field, $N\varepsilon >>1$, and weak field, $N\varepsilon<<1$, the equation of motion can be solved exactly with q set to Δ and P, respectively.

A. Field-driven dynamics

When the field is fairly strong, the extension of the chain is great and roughly a constant over the unhooking process, particularly the initial unhooking. Thus, to first order all monomers translate at the same rate and the Rayleighian term reduces to reduces to $\delta R(z)/\delta\ell\dot{z} = \varsigma\ell N\dot{z}$. Thus, equation of motion reduces to

$$\frac{1}{4}\varsigma\ell^2 N\frac{\partial\Delta}{\partial t} = -\frac{\partial F(\Delta)}{\partial\Delta} \tag{8}$$

which can be recast into an equation of the form $d\Delta/d\tau = \Delta/\tau_c$ by expressing the free energy $F(\Delta)$ in terms of an invariant stretching profile. An obvious choice for the invariant stretching profile is the equilibrium stretching profiles (Equation (4)) of each of the initial arms, $y(n, P_o)$ and $y(n, N-P_o)$ where P_o is the initial pivot monomer. Δ varies with transient pivot monomer, P, according to

$$\Delta = \int_0^P dny(n,P_o) - \int_0^{N-P} dny(n,N-P_o) \tag{9}$$

and the free energy of the chain is expressed in P as

$$\frac{F(\Delta)}{kT} = \int_0^P dn\{\varepsilon(n-P)y(n,P_o) + S[y(n,P_o)]\} \tag{10}$$

$$+ \int_0^{N-P} dn\{\varepsilon(n-N+P)y(n,N-P_o) + S[y(n,N-P_o)]\}.$$

In the initial stages of unhooking of an centrally straddled chains, the springs which straddle the obstacle can be approximated as $y(P_o - 1, P_o) \approx y(N - P_o + 1, P_o) \approx \mathcal{L}_d(\varepsilon N/2)$. This and eqs. (8-10) yield the characteristic time scale in the strong field limit, $N\varepsilon \to \infty$, $\tau_c = N/(2\varepsilon)$.

B. Weak-field or Quasi-equilibrium dynamics

When the field is very weak, i.e. $N\varepsilon << 1$, the arms are not very stretched and the elastic energy plays an important part in the chain unraveling. In this case, we can assume the stretching profile varies with time in such a way that at any instant of time, each arm posses its equilibrium stretching profile. In weak fields, the stretching profiles of the retracting arm ($0<n<P$) and advancing arm ($P<n<N$) are $y(n,P) = \varepsilon n/d$ and $y(n,N-P) = \varepsilon(N-n)/d$, respectively. The Rayleighian can be expressed as

$$R = \frac{1}{24}\varsigma(\varepsilon\ell\dot{P})^2(P^3 + (N-P)^3) \tag{11}$$

while the free energy is

$$\frac{F(P)}{kT} = \int_0^P dn \;\; \varepsilon(n-P)y(n,P) + \frac{d}{2}y(n,P)^2 + \int_P^{N-P} dn \;\; \varepsilon(n-N+P)y(n,N-P) + \frac{d}{2}y(n,N-P)^2$$

By noting the relation between Δ and P,

$$\Delta = \int_0^P dny(n,P) - \int_0^{N-P} dny(n,N-P) \tag{12}$$

we can recast the equation of motion into the form $d\Delta/d\tau = \Delta/\tau_c$ where the characteristic time scale, in the limit of large N, is $\tau_c = N^2/12$.

UNIVERSAL SCALING: A NUMERICAL SIMULATION ANALYSIS OF A WIDE RANGE OF CHAIN -OBSTACLE INTERACTIONS, FROM GLANCING TO DIRECT IMPACT

The dynamics of chains trespassing through an array of posts cannot be described solely as a simple unhooking process as described above. There exist a wide range of interactions between post and chain, varying from a glancing impact with minimal change in conformation of the chain, to direct impacts with incomplete hairpin formation. It is unreasonable to assume that the simple unhooking process above describes the full range of chain dynamics. We describe the range of impacts in the following way. Consider that a chain impacts the posts over a range of impact parameters, b, defined as the lateral displacement of the chain's center of mass from the post. At short times, portions of the chain extend downfield on either side of the post, forming a suspended conformation over a short time: as such, the hooking process itself does not significantly slow the chain. However, subsequent motion may or may not be seriously frustrated as different arms compete for length and exchange of monomers across the pivot obstacle.

In this section we present the results of a simulation study of the wide range of interactions experienced by a chain and post, allowing for glancing as well as head-on impacts. The release dynamics, i.e., dynamics by which a chain loses its contact with the obstacle, is solved by integrating the set of equations, Equation (6), starting from a random chain conformation with impact parameter b. The duration time of the interaction, t_c, and the downfield advance of the chain's center of mass during the collision, z_c, are found for each chain-post interaction. We have found that the release dynamics can be described for the full range of impacts using a set of universal functions with time and length scaling parameters, τ_c and ω. The time scaling parameter is simply the characteristic time scale of simple unhooking in strong field, $\tau_c = N/(2\varepsilon)$, and the length scaling parameter is the characteristic size or extent of the chain in the direction perpendicular to the field, $\omega = R_x \approx N^{1/2}$.

Figures 3 & 4 demonstrates the scaling behavior of collision time and downfield advance during release over a range of strong field strengths and chain molecular weights. The average time of collision varies with impact parameter according to $\langle t_c \rangle / \tau_c \approx f(bN^{-1/2})$ where f is a simple exponential for b less than a few chain radii. The function f is a complex convolution between the probability of collision and the various dynamical process occurring during the collision - f is not a simple multiplication of the collision probability and a constant time, as the probability of collision for $0<b<1$ is nearly a constant (unity) and falls significantly for $bN^{-1/2} > 1$. For large values of b, the time of collision decays much more rapidly than this simple exponential; indeed, for $b > N/2$, there is no collision time as there is no collision. The average advance of the chain's center of mass during collision also follows a universal curve over a range of field strengths and molecular weights, Figure 4. The average downfield advance varies with impact according to $\langle z_c \rangle / N \approx g(bN^{-1/2})$ where g is linear for about two radii of gyration.

Apart from the average collision time and distance, non-averaged values are also of interest. Figure 5 plots the scaled center of mass distanced traveled after impact, z_c / N, versus the scaled collision time, t_c / τ_c. Although there is some scatter to the plot, the numerical data all fall close to a simple universal curve: $z_c / N \approx h(t_c / \tau_c)$. The function h is linear at small times, corresponding to short interaction times for glancing impacts; but for long times, the function saturates at $h=1/2$, indicative of a single hairpin where release occurs when one full arm length advances downfield, or $z_c = N/2$. Another saturation zone is discernible for $h=1/4$, corresponding to another frustrated conformation, a double hooked chain, or double hairpin, which is less frequent that a single hairpin.

Although the universality of the numerical results is interesting in itself, these scaling relations are useful in deriving a general formula for the mobility of a chain in an array of posts. Let the posts be arranged in rows separated distance Z apart in the field direction and X apart in the perpendicular direction. The time taken for a chain to traverse a row of posts is the time for collision plus the time to drift to the next row: $t = t_c + (Z - z_c)/v_o$. The ratio of the mobilities of the chain with and without obstacles is then $\mu/\mu_o = Z/(Z + t_c v_o - z_c)$. The process of chain impact is stochastic so that there is a distribution of times t_c and distances z_c which we characterize with the probability distribution P_b defined such that $P_b(t_c, z_c)dbdt_cdz_c$ is the probability that a collision occurs with impact parameter between b and $b+db$, with collision time between t_c and $t_c + dt_c$, and center of mass advancement between z_c and $z_c + dz_c$. The average mobility is then given by $\langle \mu \rangle = \int dbdz_c dt_c P_b(t_c, z_c)\mu$. This expression can be simplified by invoking the universal relation between t_c and z_c and noting that our data for $b < R_x$ show that the standard deviation in t_c is less than $\langle t_c \rangle$ and can be approximated with a delta function centered at $\langle t_c \rangle$. In the limit that $N << Z$, we obtain

$$\langle \mu \rangle / \mu_o = 1 - BN^{3/2} / XZ \qquad (13)$$

where B is a numerical constant. The decrease in mobility $N^{3/2} / XZ$ has a simple physical explanation. It arises from a factor $N^{1/2} / X$ representing the probability of collision, multiplied by a factor N, representing the decrease in speed of the chain, and a factor $1/Z$ which is the number of obstacles per unit length in the Z direction. The mobility is a strongly decreasing function of N, and thus an array of posts should provide a reasonable means of chain separation.

The mobility formula given above is based upon the single-post approximation, where a chain interacts solely with one post at a time and the interaction is complete before impinging upon an additional post. In practice, however, the post-chain interaction will interfere with subsequent interactions. This interference can take two forms. First, if the radius of gyration is larger than the X spacing, then a chain can interact with more than one obstacle at one time. The second effect deals with conformational relaxation of the newly released chain. After unhooking from a hairpin, the chain conformation is strongly distorted. It first relaxes rapidly due to nonlinear elasticity, but then much more slowly due to linear elastic effects. In this latter stage, the relaxation is controlled by the Rouse time, $t^R \propto N^2/kT$, which can be compared with the time taken to travel between rows of posts, $\propto Z/(QE\ell)$. Thus for the chain to fully relax between rows requires $Z > \varepsilon N^2$. Row spacing less than this would violate the single post-approximation and render the process non-Markovian, requiring a more sophisticated treatment.

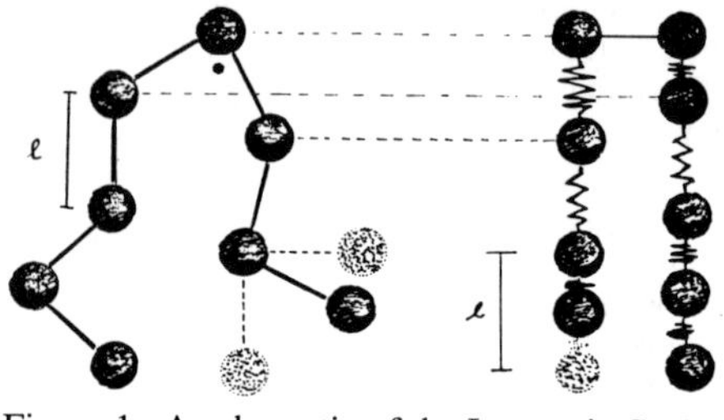

Figure 1: A schematic of the Langevin Spring model used for a polyelectrolyte draped over a frictionless obstacle.

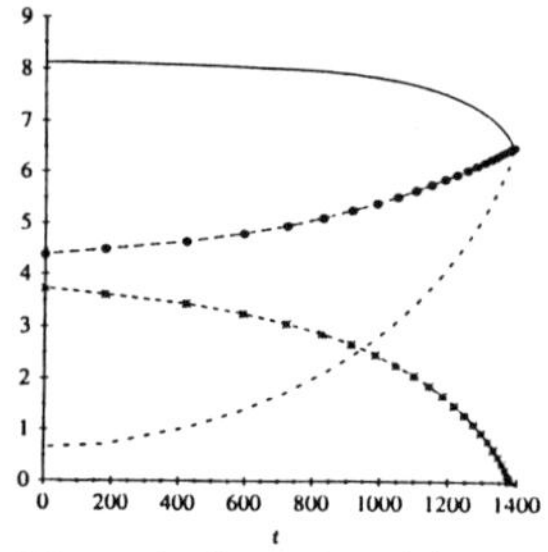

Figure 2: Numerical results of the dimensions of a single hairpin during the unhooking process in a strong field. The length of the advancing/retracting arms are indicated by filled circles/stars. The dashed line represents the difference in arm lengths, which appears exponential in time. The solid line is the total length or sum of arm lengths.

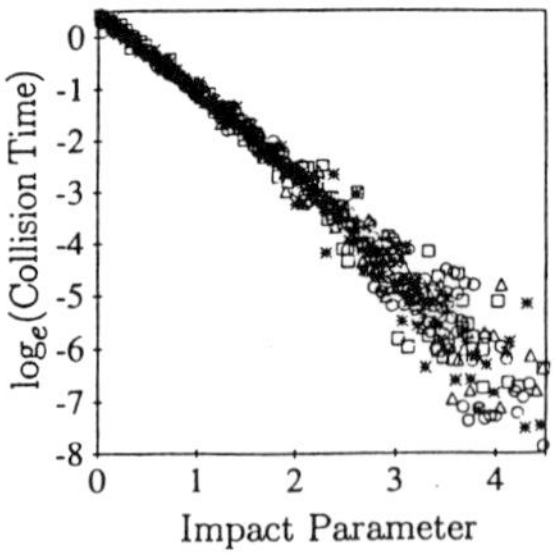

Figure 3: A log-linear plot of the scaled average impact time, $\langle t_c \rangle / \tau_c$, versus the scaled impact

figure 3 continued

parameter, $b/\sqrt{N}$. The average is taken over a large number of chains with a prescribed length, $20<N<80$, and dimensionless field, $10<\varepsilon<100$. Chains which do not collide with the post are assigned $t_c = 0$ and $z_c = 0$. The data collapse onto a single universal curve which appears exponential for $b/\sqrt{N} < 2$. At large $b/\sqrt{N}$ there is much scatter in the data as collisions are rare.

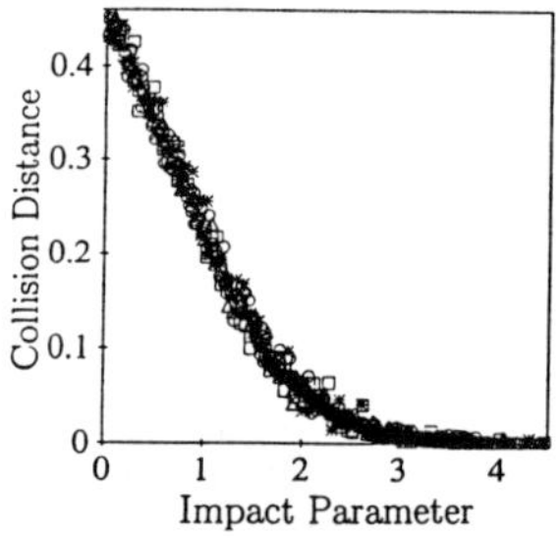

Figure 4: A plot of the scaled average distance moved by the chain's center of mass during collision, $\langle z_c \rangle / N$, versus the scaled impact parameter, $b/\sqrt{N}$. The data from various (N,ε) collapse onto a single universal curve which is linear for nearly direct impacts.

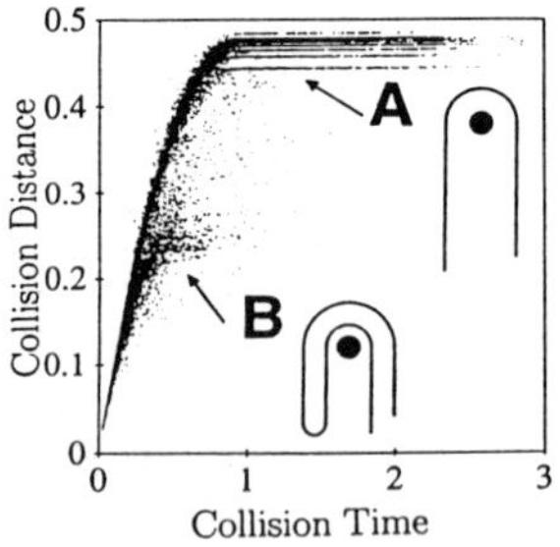

Figure 5: A plot of the scaled center of mass advance, z_c/N, versus the collision time, t_c, for each of the *300,000* simulated chains. Chains which interact with the post for significant time have frustrated, hairpin conformations: (A) single hairpins with z_c/N near *0.5* (the discrete levels of z_c/N less than *0.5* are due to short chains and lower fields, i.e. $N\varepsilon < \infty$.) and (B) double hairpins with $z_c/N = 0.25$.

Part VI

Friction and Fracture

CRACKING A TOUGH NUT* WITH A BIG COMPUTER

FARID F. ABRAHAM* AND D. SCHNEIDER, B. LAND, D. LIFKA, J. SKOVIRA, J. GERNER, M. ROSENKRANTZ†

*IBM Research Division, Almaden Research Center, 650 Harry Road, San Jose, CA 95120-6099, farid@almaden.ibm.com

†Cornell Theory Center, Cornell University, Ithaca NY, 14853-3801

ABSTRACT

We have simulated the dynamical failure of three-dimensional notched solids under tension using molecular dynamics and up to 100 million atoms. We discovered a dynamical brittle-to-ductile transition in the rapid cleavage of rare-gas solids when the crack velocity approaches one-third of the Rayleigh sound speed. At this transition, the crack tip has already begun to roughen on the atomic scale. This suggests that the brittle crack undergoes a dynamic instability which immediately leads to the initiation of plastic failure by the spontaneous emission and proliferation of dislocations and crack arrest.

* NUT: nanocrystal under tension. We make note of P-G. de Gennes comment "...refrain from immediately resorting to a jackhammer to break open a hazelnut without first checking for an incipient crack on the surface of the shell." in Fragile Objects (Copernicus, NY, 1996).

INTRODUCTION

"At liquid hydrogen temperatures, ... imperfect crystals of krypton have about the hardness of butter on a cold day," so state Kelly and Macmillan [1]in their now classic text "Strong Solids". However, even at liquid hydrogen temperature, krypton is at two-tenths of its melting temperature and therefore not exceptionally cold. This, along with the imperfection in crystal structure, can explain the softness of the krypton solid at twenty degrees Kelvin. A perfect krypton crystal at zero temperature may fracture in a brittle fashion, but theory suggests that it will not. Theoretical support for the inherent ductility of the rare-gas solids and, indeed, of most face-centered cubic solids at low temperature has been given by Rice and Thomson [2]this being in sharp contrast to our earlier simulation findings in two dimensions where the cold rare-gas films failed brittlely [3,4]. Our two-dimensional simulations revealed a very rich dynamics in the failure process (see Figure 1).

Pulling apart the simulated rare-gas film, the crack started out straight, then became rough, but finally started zigging and zagging. Also, the fracture advanced at considerably less than the speed of sound. Because the simulated system is perfect, we proved that the instability and the speed limitation have nothing to do with impurities or imperfections in the material or difficulties in the experimental setup. Rather, they are inherent properties of the fracture dynamics itself.

We have now simulated the dynamical failure of three-dimensional notched solids under tension using molecular dynamics and up to 100 million atoms and find that the rare-gas solid begins to fail by rapid brittle fracture [5]. When the crack velocity approaches one-third of the Rayleigh sound speed. the crack tip begins to roughen on the atomic scale. This suggests that the brittle crack undergoes a dynamic instability. This immediately leads

Mat. Res. Soc. Symp. Proc. Vol. 463 © 1997 Materials Research Society

Fig. 1. The time evolution of the propagating crack in a 2D solid is seen using a gray-scale rendering of the instantaneous local velocity, going from dark gray for the most negative velocity to light gray for the most positive velocity. Initially, the brittle crack propagates in a straight line. At the onset of the instability, the crack first begins to roughen, and then to oscillate back and forth. The zigs or zags in the crack direction are accompanied by propagating dislocations that are thirty degrees to the vertical and travelling at approximately the speed of sound for the bulk solid. These dislocations appear as slanted, inverted Vs being emitted from the moving crack tip, first to the right, then to the left and are simply an acoustical wake created by the moving dislocation.

to a dynamic brittle-to-ductile transition, the initiation of plastic failure, the spontaneous proliferation of dislocations and crack arrest. The details follow.

METHOD AND RESULTS

It has been suggested that atomic systems on the order of 10^8 to 10^{11} atoms could begin to bridge the gap between numerical experiments on the computer and "real" experiments in the laboratory [6]. In this study, we describe a simulation of 100 million atoms. The van der Waals bonding giving the cohesion of the rare-gas solid can be modelled accurately by two-body Lennard-Jones potential and has served as a paradigm for studying classical many-body phenomena of atomistic systems in computational physics [7]. At the time when the Lennard-Jones potential was being adopted for the simulation of the liquid state, and hence rare-gas liquids, Feynman [8]suggested that the single most important statement describing our real world is "that all things are made of atoms, little particles that move around in perpetual motion, attracting each other when they are a little distance apart, but repelling upon being squeezed into one another." This continues to be a rich prescription and certainly proved to be a good choice for studying the generic features of rapid fracture dynamics in two-dimensions. The Lennard-Jones potential has the form

$$\phi_{LJ}(r) = 4\epsilon\left[(\sigma/r)^{12} - (\sigma/r)^{6}\right], \tag{1}$$

where ϵ is the LJ well depth, σ is where the potential goes through zero. We express quantities in terms of reduced units; lengths are scaled by σ, energies by ϵ. The system is a three-dimensional slab with $\mathbf{L}_x = 336$ atoms by $\mathbf{L}_y = 336$ atoms by $\mathbf{L}_z = 896$ atoms for the three orthogonal sides. The notch is a slit beginning midway along $\mathbf{L}_x fory = 0$, having a y extension of length $\mathbf{l}_t = 120$ which extends through the entire thickness $\mathbf{L}_z$. The exposed notch faces are in the y-z planes with (110) faces, and the notch is pointed in the $< 1\bar{1}0 >$ direction. We note that the (110) face does not have the lowest surface energy. This choice is contrary to conventional wisdom which would identify the lowest energy surface, i.e., the (111) surface, as the cleavage plane for brittle fracture. The choice of the (110) surface is based our recent finding that the nonlinear elastic anisotropy of the crystal, and not the anisotropy of surface energy, dictates the cleavage behavior [9]. This continues to be true in three dimensions.

Periodic boundary conditions are imposed between the x-y faces at $z = 0$ and $z = \mathbf{L}_z$. This notched slab geometry has a total of 100,509,696 atoms, and the total simulation time for this study is 30,000 time-steps or 135 in reduced units. It takes 12 CPU seconds per time-step for a 306-node spatial decomposition simulation on the IBM SP2. This timing compares very favorably with other timing studies reported for parallel molecular dynamics programs for comparable system sizes (for example, see reference 10). However, the physical size multiplied with the total simulation time span is the proper measure for the computational burden (remember space-time is relevant for dynamics) and equals $\sim 10^{12}$ for this study. A crack would traverse a laboratory nanocrystal in approximately a nanosecond; our crack in the computer nanocrystal takes one hundred clock-hours. The slab is initialized at zero reduced temperature, and an outward strain rate $\dot{\varepsilon}_x$ is imposed on the outer most columns of atoms defining the opposing vertical faces of the slab. A linear velocity gradient is established across the slab, and an increasing lateral strain with time occurs in the solid slab with an applied strain rate of $\dot{\varepsilon}_x = 0.00033$. With this choice, the solid fails at the notch tip after 10,000 time-steps when the solid has been stretched by $\sim$ 1.5 percent, in agreement with the Griffith energy-balance criterion. The imposed strain rate is set to zero after another 3,500 time-steps, and the simulation is continued a further 20,000 time-steps. We adopt the initiation of crack tip motion as zero time.

In Figure 2, snapshot pictures of the notched solid dynamics are presented for early times. In order to see into the interior of the solid, we show only those atoms that have a potential energy greater than -6.1, where the ideal bulk value is -6.3; this trick was used earlier in our 2D work. This reduces the number of atoms seen by two orders of magnitude; the visible atoms are associated with faces of the slab and initial notch, surfaces created by crack motion, local interplanar separation associated with the material's dynamic failure at the tip, and topological defects created in the otherwise perfect crystal. In this figure, the crack's depth, projected on to the x-y plane with an off-diagonal perspective view, is only one third of the total actual slab thickness. This cut-away of depth was done in order to emphasize the important features of the early-time dynamics which are associated with

Fig. 2. Early-time sequence of the propagating crack is shown as overlapping partial landscapes of the growing surface due to brittle fracture in the first three images (times 1, 16, 32) and the subsequent appearance of dislocations in the fourth image (time 48) after the transition to ductility. Only atoms with a potential energy less than 97 percent of the bulk value are displayed, resulting in the selected visualization of atoms neighboring surfaces and dislocations. Approximately a third of the crack's depth is shown, and the slab's rear exterior surface is included. The other exterior surfaces are omitted.

the crack's brittle motion in the forward direction. The sequence is for times 1, 16, 32 and 48 (reduced time unit). In the time interval 0 to 42, the crack motion is representative of brittle fracture, with planar cleavage of the bonds between the two (110) neighboring atomic sheets defined by the initial notch. The small bud at the crack tip represents local expansion, and we will refer to this region as the "process zone." During brittle fracture, this process zone represents no plasticity, but, at a time of 43, the bud begins to blossom into a "flower of loop dislocations." This is apparent from the snapshot picture at time 48. Also, the crack slows to a stop and continues to dissipate elastic energy through the continued creation and motion of dislocations. Hence, a "dynamic brittle-to-ductile" (B-D) transition occurs in the fracturing of this rare-gas nanocrystal.

Figure 3 shows a late-time sequence of the propagating crack for the entire depth of the 100 million atom slab at times 43, 52, 68 and 90. We see the time evolution of the loop dislocations emitted from the arrested crack surface neighboring the vertical crack edge and propagating in the (111) family of slip planes through the solid slab. The dislocations move at approximately a quarter of the longitudinal sound speed ($c_s = 8.5$) and eventually terminate as steps on the front surface of the slab (not shown). For a comparison with laboratory experiment, we refer the reader to the transmission electron micrograph of an arrested crack in silicon above the ductile transition temperature [11], where very similar dislocation activity is visible in the presence of the silicon crack tip. However, unlike our computer experiment, it was not possible to determine the dynamic origin of these dislocations.

Figure 4 shows magnified off-diagonal perspective views of the process zone during the time period of the brittle-to-ductile transition (at times 36, 43, 49 and 54). We note that just prior to the B-D transition (time 34) the process zone is symmetric. The zone

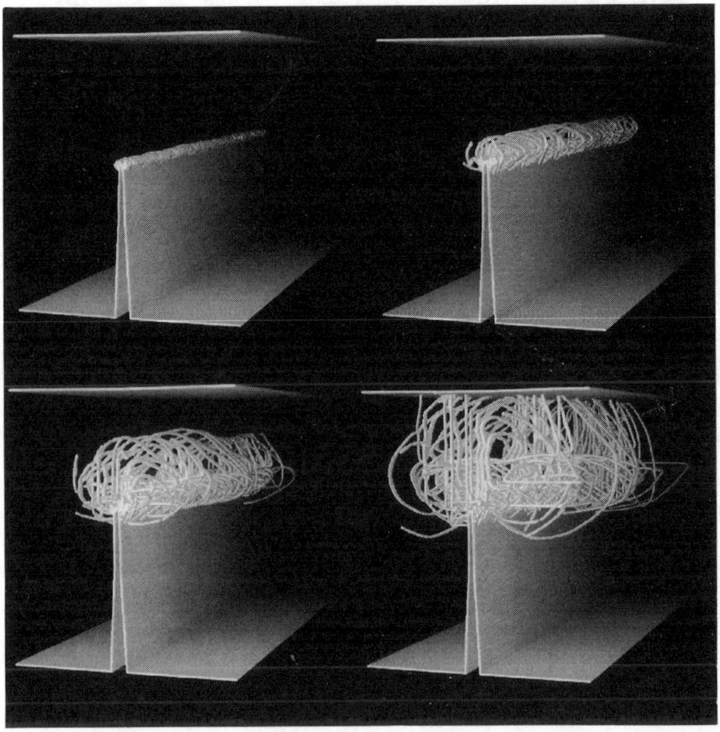

Fig. 3. Late-time sequence of the dislocations propagating through the 100 million atom slab is shown for times 43, 52, 68 and 90. With the exceptions that the entire depth and the top surface are shown, the rendering procedure described in the previous figure applies.

then becomes asymmetric with the onset of a roughening as a quasi one-dimensional strip (time 43) and rapidly broadens into a wall of embryonic dislocations (time 54). The crack surface begins to roughen prior to dislocation emission. In our two-dimensional fracture simulations [3], we saw a similar roughening which was identified as the onset of an intrinsic dynamical instability of the brittle fracture process. For the three-dimensional simulation, this roughening extends in depth as a one-dimensional step which is a few atom diameters wide. However, the three-dimensional B-D fracture transition immediately follows, in sharp contrast to the fracture of the two-dimensional rare-gas films where brittle fracture prevails. Furthermore, the surface roughening (i.e., the brittle instability dynamics) occurs when the crack tip velocity equals ~ 0.33 of the Rayleigh sound speed for the rare-gas solid, the same as we found in the two-dimensional systems.

We construct the following scenario for the dynamic failure process in the 3D rare-gas solids. From zero velocity, the crack accelerates smoothly by brittle fracture along the (110) plane until it approaches one third of the Rayleigh sound speed. At this point, a dynamic

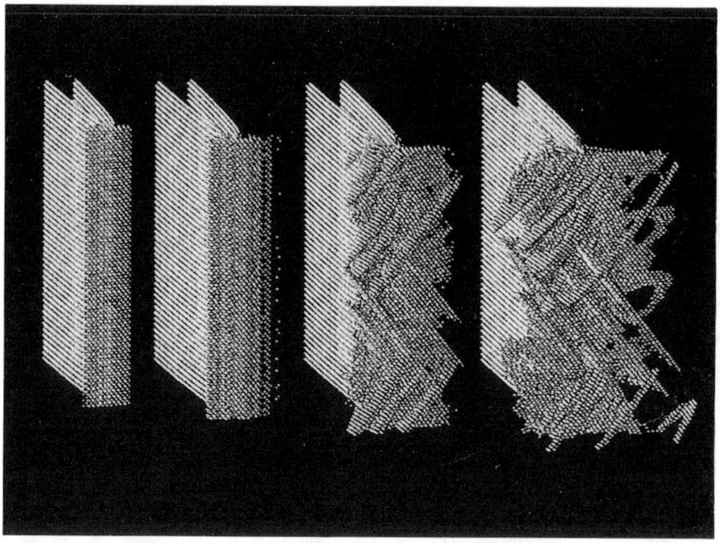

Fig. 4. Magnified off-diagonal perspective views of the process zone during the time period of the brittle-to-ductile transition (at times 36, 43, 49 and 54).

instability in the brittle fracture process occurs and manifests itself as a atomic surface roughening of the created surfaces. The brittle nature of this failure process marginally exists for the perfect (110) planar cleavage, and the deviation from planar cleavage gives rise to plasticity, and a spontaneous proliferation of dislocations. We are now studying whether the nonlinear elastic anisotropy for different fracture planes gives markedly different fracture behavior, similar to our findings in two-dimensions [9].

Finally, we considered a 3D thin film, only twelve atomic layers thick with approximately five million atoms (see Figure 5). By 'film', we mean that periodic boundary conditions are not imposed across the x-y faces, hence they are free surfaces. As time progresses, grooves appear at the free surfaces in the region of the crack tip and grow outward from the notch as well as into the slab's depth. The notch "grows" in length as the depth of the grooves reaches half of the slab's thickness, which occurs farther away from the original tip position as time increases. This mechanical grooving is a consequence of loop dislocations being emitted from the crack edge, meeting the free surfaces, and forming steps that move away from the stationary tip as the dislocations move down stream. The original notch is not of sufficient length (for the imposed strain) to fail by breaking bonds; however, the strained slab still fails by dislocation grooving.

A multimedia version of this atomistic simulation of fracture is available via World Wide Web:

http://www.tc.cornell.edu/ ~farid/fracture/100million/

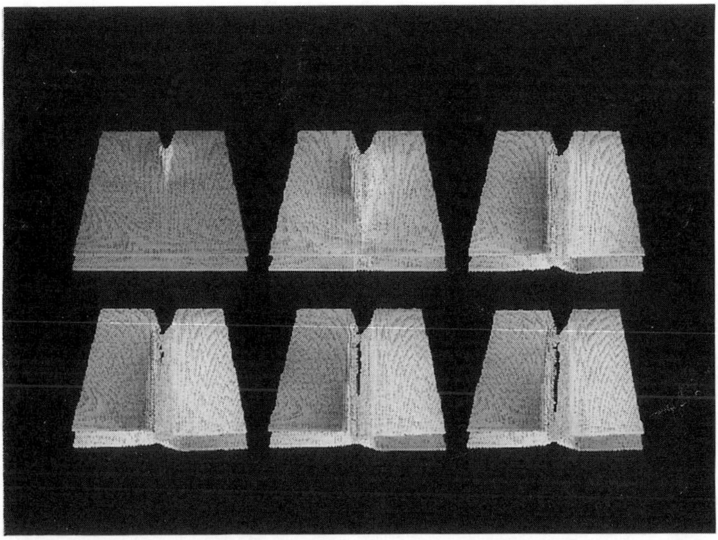

Fig. 5. Figure 5. Mechanical grooving of a very thin 3D film, twelve atomic layers thick. The shading is according to depth; the interior bulk atoms are not shown.

ACKNOWLEDGEMENTS

This research was conducted in part, using the resources of the Cornell Theory Center, which receives major funding from the National Science Foundation and New York State with additional support from the Advanced Research Projects Agency, the National Center for Research Resources at the National Institutes of Health, IBM Corporation and members of the Corporate Research Institute. I am grateful for collaborative efforts with Dominique Brodbeck and William Rudge at the IBM Almaden Research Center and with D. Schneider, B. Land, D. Lifka, J. Skovira, J. Gerner, M. Rosenkrantz at the Cornell Theory Center, Cornell University.

REFERENCES

1. A. Kelly and N. H. Macmillan, Strong Solids, Clarendon Press, Oxford, 1986, p.10.

2. J. R. Rice and R. Thomson, Phil. Mag. 29, **73** (1974).

3. F. F. Abraham, D. Brodbeck, R. Rafey and W. E. Rudge, Phys.Rev.Lett., **73**, 272 (1994).

4. F. F. Abraham, D. Brodbeck, W. E. Rudge and X-P. Xu, J.Mech.Phys.Solids, to appear.

5. F. F. Abraham, D. Schneider, B. Land, D. Lifka, J. Skovira, J. Gerner, M. Rosenkrantz, to be published.

6. R. M. Lynden-Bell, Science, **263**, 1704 (1994).

7. M. L. Klein and J. A. Venables, Rare Gas Solids, **Vol 1**, Academic Press, London, 1976.

8. R. P. Feynman, R. Leighton and M. Sands, The Feynman Lectures on Physics, Addison-Wesley Publishing Co., Reading, 1963, p.1-2.

9. F. F. Abraham, Phys.Rev.Lett., **77**, 869 (1996).

10. S. Plimpton, J.Comp.Phys. **117**, 1 (1995).

11. B. Lawn, B. Hockey and S. Wiederhorn, J.Mater.Sci., **15**, 1207 (1980); Fig.15a.

DEPINNING OF FRACTURE FRONTS

P. DAGUIER†, B. NGHIEM‡, E. BOUCHAUD†, F. CREUZET‡
†O.N.E.R.A. (OM), 29 Avenue de la Division Leclerc,
B.P. 72, 92322 Châtillon Cedex, FRANCE
‡Laboratoire CNRS/Saint-Gobain *"Surface du Verre et Interfaces"*,
39, Quai Lucien Lefranc, B.P. 135, 93303 Aubervilliers Cedex, FRANCE

ABSTRACT

By studying the morphology of the fracture surfaces of both a metallic alloy and of a silicate glass for various very low crack propagation velocities, it is shown that the pinning/depinning scenario of a line moving through randomly distributed obstacles seems well adapted to describe fracture of heterogeneous materials. The crossover length separating the large length scales regime characterised by a roughness index $\zeta \simeq 0.78$ and the small length scales one, for which $\zeta_c \simeq 0.5$, is shown to decrease with the measured crack velocity as a power law, $\xi_c \propto v^{-\phi}$, with $\phi \simeq 1$. ξ_c and v also vary with the pulling force as power laws, and the measured values of exponents ν and β are close to: $\nu \simeq 2$ and $\beta \simeq 2$.

It has been proposed in [1] that the recent progresses made in statistical physics in the understanding of the problem of lines moving through randomly distributed obstacles [2, 3, 4] could be useful to describe fracture in heterogeneous materials. The line being there the fracture front and the obstacles elements of microstructure, the fracture surface is just the *trace* left behind it during its propagation.

Experimental results concerning the morphology of the fracture surfaces of two very different materials – a metallic alloy and a silicate glass – are analysed here within the framework of these models [5, 6].

After the pioneering work of Mandelbrot *et al* [7], the fracture surfaces of many heterogeneous materials have been investigated thanks to various experimental techniques. These surfaces are usually self-affine [8], with a roughness index ζ close to the value 0.8, which was conjectured to be a universal value [9, 10], i.e. independent of the material and of the fracture mode. This exponent has recently been measured over five decades of length scales (0.5 nm-0.5 mm) [11] on the Superα_2 Ti_3Al-based alloy. However, Milman *et al* [12] have reported significantly smaller values (close to 0.5) of the roughness exponent from Scanning Tunneling Microscopy experiments, where the fracture surfaces of metallic materials were investigated at the nanometer scale. This effect has been interpreted [11, 13] as a kinetic effect similar to the one expected for a moving line near its depinning transition.

Various models have been proposed in that context. Ertaş and Kardar studied a local nonlinear three-dimensional Langevin equation to describe the morphology of polymers in shear flows or the motion of flux lines in superconductors [5]. The line is pulled away with a constant force F. Non linearities account for the variations of the local crack speed with the local orientation of the front. This equation leads to a large number of regimes, depending on the relative values of the prefactors of the non linear terms. For some values, this model predicts that for a finite velocity v, the roughness exponent is 0.75 at "large length scales" and 0.5 at "short length scales", the two regimes being separated by a crossover length ξ_c. The short length scales regime corresponds to the vicinity of the depinning transition [2, 3] where the crack front is just able to free itself from the pinning microstructural obstacles. In this case, i.e. when F is close although higher than a critical

Mat. Res. Soc. Symp. Proc. Vol. 463 © 1997 Materials Research Society

force F_c under which the line remains still, the velocity v tends to zero, $v \propto (F - F_c)^\beta$, and ξ_c diverges as $\xi_c \propto v^{-\phi}$. In this particular model [5], $\phi = 3$. It will be shown in this paper that this scenario is qualitatively correct [11], although the actual values of the various exponents are not the ones predicted.

It is worth noticing also that recent large-scale molecular dynamics simulations [14, 15] for amorphous materials are in remarkable agreement with the experimental results.

In this paper, ξ_c is shown to decrease with the crack velocity as a power law $\xi_c \propto v^{-\phi}$, with $\phi \simeq 1$, both for the fatigue fracture of the Superα_2 and for the stress corrosion fracture of soda-lime silica glass, which is heterogeneous at a small scale [16, 17].

Two notched compact tension specimens of Superα_2 are broken in fatigue, thanks to an electro-servohydraulic testing machine, operating under load control. The test is performed in air with a constant stress ratio $R = \sigma_{min}/\sigma_{max} = 0.1$ (σ_{max} and σ_{min} are respectively the minimum and the maximum stresses), at a frequency $f = 30Hz$. The evolution of the crack length a with time is measured with the potential drop method [18]. The fracture surfaces are observed for 4 different velocities spanning a wide range, using both an atomic force microscope (AFM) and a standard scanning electron microscope (SEM). The SEM observations consist in cutting and polishing the NiPd-plated fractured samples perpendicularly to the direction of crack propagation and registering images of the profiles at various magnifications for each of the 4 regions (see [19]). For AFM observations, five profiles of length 1 μm are registered in each region.

Fracture surfaces of soda-lime silica glass have been prepared by controlling the crack propagation with a four points bending system. The average crack velocity is measured by imaging the crack tip with an AFM at different times [20]. The humidity rate has been kept between 37 and 41%. The controlled crack propagation is maintained over a distance of about 30μm, so that fracture surfaces can be easily probed with AFM. Crack velocities range from 2 10^{-9} to 10^{-7} ms^{-1}. Ten AFM height profiles of length 1.5μm are registered perpendicularly to the direction of crack propagation ($\perp$), on three samples, and along this direction ($||$) for four other specimens. No significant anisotropy has been detected.

In order to determine the roughness exponents ζ and the crossover length ξ_c of the profiles recorded, the Hurst method is used [21]: $Z_{max}(r) = \langle max[z(r')]_{r_o<r'<r_o+r} - min[z(r')]_{r_o<r'<r_o+r}\rangle_{r_o} \propto r^\zeta$ is computed for each record.

As already shown previously [11], the fracture surfaces of the Superα_2 can be investigated over 5 or 6 decades of length scales thanks to the simultaneous use of AFM and SEM. Within the whole range of observations, $Z_{max}(r)$ is very well fitted by the sum of two power laws, $Z_{max}(r) = A((r/\xi_c)^{0.5} + (r/\xi_c)^{0.78})$ with exponents chosen to fit the results of previous experiments [11]. In the case of glass, the crossover is much sharper, and ξ_c is determined as the intersection of the two asymptotic power law regimes with exponents 0.5 and 0.78. Once the crossover lengths have been determined in each case, Z_{max} is plotted as a function of r/ξ_c. In Fig. 1, the curves $Z_{max}(r)/\sqrt{\xi_c}$ relative to each material are plotted as a function of r/ξ_c and shown to collapse on the same master curve. In both cases, the asymptotic regimes are well described by power laws with exponents 0.5 for $r/\xi_c \ll 1$, and 0.78 for $r/\xi_c \gg 1$. Thus:

$$Z_{max}(r) \propto r^{0.5}\, f(\frac{r}{\xi_c}) \tag{1}$$

with $f(x \to 0) \sim 1$ and $f(x \gg 1) \sim x^{0.28}$, showing that the amplitude of the small length

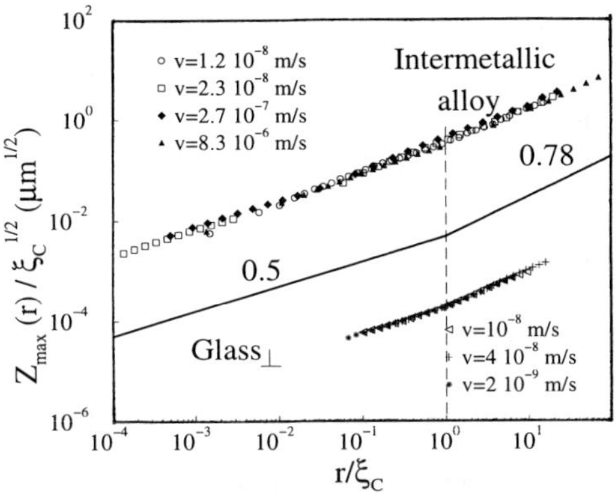

Figure 1: Plot of $Z_{max}(r)$ / $\xi_c^{1/2}$ against r / ξ_c. for the two materials separately. In these reduced units, the plots corresponding to the various velocities collapse on the same master curve (see Eq. (1)). Although the crossover regions are quite different for the two materials, the asymptotic regimes are well described by power laws with exponents 0.5 (r / $\xi_c \ll 1$) and 0.78 (r / $\xi_c \gg 1$). It is worth noticing that three orders of magnitude separate the results on glass from those on the Superα_2

scales contribution is independent of crack velocity.

The results obtained on materials as different as an intermetallic alloy and a glass thus confirm previous observations [11, 13], where the short and large length scales regimes were interpreted, respectively, as a "quasi-static" and a "dynamic" regime.

As can be seen in Fig. 2, ξ_c decreases with the crack velocity v in both cases, although the measured values of ξ_c are approximately 1000 times larger in the case of the Superα_2 than in the case of glass. However, both series of results are compatible with a power law decrease $v^{-\phi}$ with $\phi \simeq 1$.

As far as the fatigue of metallic materials is concerned [22], the low crack velocity regime: $\Delta K = (\sigma_{max} - \sigma_{min})\sqrt{a} \simeq \Delta K_{Th}$ is known as being intermittent. As a matter of fact, when ΔK is close to the threshold value ΔK_{Th} below which the external load is unable to move the crack, the crack tip opens and closes many times before it can extend over a small distance. This process is repeated several times and the crack propagates by successive "jumps". The number of cycles required to get the crack to advance decreases as ΔK increases, and crack motion is more and more continuous, microstructural obstacles being efficient at smaller length scales. At a given time, it can be considered that the force F exerted on the fracture front is proportional to ΔK, while the threshold force F_c is proportional to ΔK_{Th}. The frequency of oscillation of these forces (f=30Hz) being far more rapid than crack propagation, one can consider only the average force, which legitimates the analogy with the above-quoted models.

In the case of glass, F is proportional to the stress intensity factor K, while F_c is proportional to the threshold K_{Th}. Preliminary results indicate that, in the sub-critical regime, the crack velocity is not uniform, and intermittency is likely to occur. Thus, in both cases, the pinning/depinning scenario is qualitatively satisfactory.

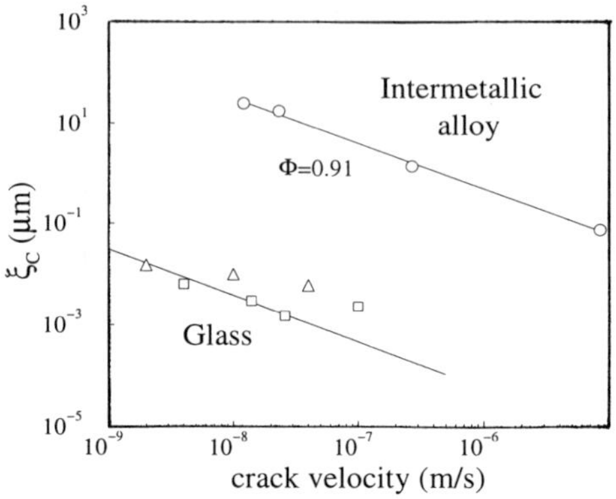

Figure 2: The cross over length ξ_c is plotted against the crack velocity v on a log-log diagram for the Superα_2 (∘) and for soda-lime silica glass (△/□: perpendicular [⊥]/parallel [||] to the direction of crack propagation). The data are well fitted by a power-law with exponent $\phi \simeq 0.91$. It can be noticed that ξ_c strongly depends on the considered material (it is approximately 10^3 times smaller for glass), although the exponents are very likely to be the same.

Fig.3 shows the evolution of the crack velocity v as a function of $\Delta K - \Delta K_{Th}$ for the Superα_2 and as a function of $K - K_{Th}$ for glass. When static fracture occurs, a clear deviation from the power law can be observed for high values of $\Delta K - \Delta K_{Th}$. A fit of these data gives $\beta \simeq 2$. This value is compatible with the measurements on glass (Fig. 3).

In Fig. 3, ξ_c is plotted for both materials as a function of $F - F_c$. A power law decrease can be observed, and the fit of the data relative to the metal gives $\nu \simeq 2$, compatible with the results on glass.

It is expected that the exponent n characterising the range of interactions ($n = 2$ for local models [2] and $n = 1$ when the non-locality of elasticity is taken into account [6]) is related to the exponents ν and $\zeta_{||}$ through the relation:

$$n = \zeta_{||} + \frac{1}{\nu} \tag{2}$$

where $\zeta_{||}$ is the in-plane roughness exponent [24], shown previously on the Superα_2 to be close to: $\zeta_{||} \simeq 0.54$. This leads to a value of $n \simeq 1.03$, very close to unity, as expected for elastic interactions [6, 23].

Knowing β and ν, one can in principle deduce the value of the dynamic exponents $z_{||}$ and $z_{\perp}$ describing the short time evolution of the front, repectively in the direction of crack propagation and perpendicularly to it. $z_{||} = \zeta_{||} + \frac{\beta}{\nu}$ leads to $z_{||} \simeq 1.5$, while $z_{\perp} = z_{||} + \frac{1}{\nu}$ should indicate that $z_{\perp} \simeq 2$.

For ductile materials as the Superα_2, the plastic zone size should be a relevant length scale as well. Although the same regimes are observed within the whole range of ΔKs, it can be noted that R_{plast} overpasses ξ_c for the two experiments corresponding to higher

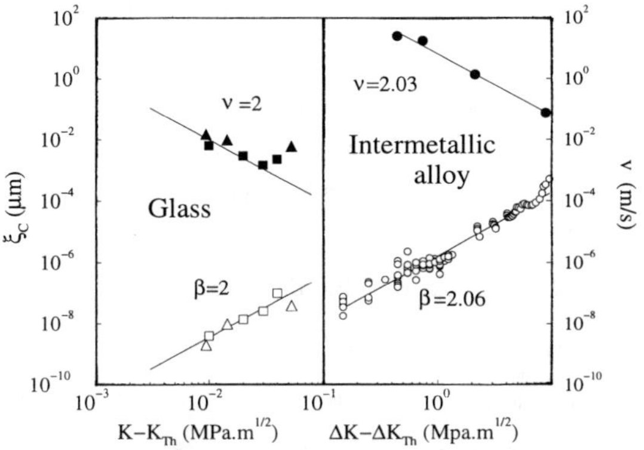

Figure 3: Superα_2 (○): the fatigue crack velocity (white symbols) is plotted against $\Delta K - \Delta K_{Th}$ on a log-log plot, as well as the crossover length ξ_c (black symbols). Glass (△ / □ perpendicular [$\perp$]/parallel [||] to the direction of crack propagation): the crack velocity is plotted as a function of $K - K_{Th}$ (white symbols), as well as ξ_c (black symbols).

velocities. When ξ_c becomes negligible in comparison to the plastic zone size, different mechanisms [25] might be relevant at short length scales.

The results presented here strongly support the interpretation of crack propagation at very low crack speed as a pinning/depinning phenomenon, close to those described by theoretical models [1 6,23]. Further experiments on different materials are needed in order to confirm the general character of this scenario, and to allow for a more precise determination of the critical exponents. In order to investigate the role of plasticity, experiments will be performed on an aluminium alloy, for which plastic zones are much larger than in the case of the Superα_2.

ACKNOWLEDGMENTS: The authors are particularly indebted to S. Navéos and G. Marcon for their technical help, to J.-P. Bouchaud, E. Orignac and S. Roux for enlightening discussions.

References

[1] J.-P. Bouchaud, E. Bouchaud, G. Lapasset, J. Planès, Phys. Rev. Lett., **71**, 2240, (1993).

[2] O. Narayan, D. Fisher, Phys. Rev. Lett. **68**, 3615 (1992); Phys. Rev. B**46**, 11520 (1992); O. Narayan, D. S. Fisher, Phys. Rev. B **48**, 7030 (1993).

[3] T. Nattermann, S. Stepanow, L-H. Tang, H. Leschhorn, J. Phys. II (France) **2**, 1483 (1992).

[4] T. Halpin-Healy, Yi-Cheng Zhang, Physics Report, **254**, 215 (1995), and references therein.

[5] D. Ertaş, M. Kardar Phys. Rev. Lett. **69**,929 (1992); D. Ertaş, M. Kardar Phys. Rev, E **48** 1228 (1993).

[6] D. Ertaş, M. Kardar Phys. Rev. Lett. **73**, 1703 (1994); D. Ertaş, M. Kardar, Phys. Rev. B **53**, 3520 (1996).

[7] B.B. Mandelbrot, D.E. Passoja, A.J. Paullay, Nature **308** 721 (1984).

[8] B.B. Mandelbrot, Phys. Scr. **32**, 257 (1985).

[9] E. Bouchaud, G. Lapasset, J. Planès, Europhys. Lett. **13**, 73 (1990).

[10] K.J. Maloy, A. Hansen, E. L. Hinrichsen, S. Roux, Phys. Rev. Lett. **68**, 213 (1992).

[11] P. Daguier, S. Hénaux, E. Bouchaud and F. Creuzet, Phys.Rev. E **53**, 5637 (1996).

[12] V.Y. Milman, N.A. Stelmashenko, R. Blumenfeld, Prog. Mater. Sci. **38**, 425 (1994).

[13] E. Bouchaud, S. Navéos, J.Phys. I France **5**, 547 (1995).

[14] A. Nakano, R. K. Kalia, P. Vashishta, Phys. Rev. Lett. **73**, 2336 (1994).

[15] A. Nakano, R. K. Kalia, P. Vashishta, Phys. Rev. Lett. **75**, 3138 (1995); R. K. Kalia, A. Nakano, A. Omeltchenko, K. Tsuruta, P. Vashishta, preprint (1996).

[16] S. Sen, J.F. Stebbins, Phys. Rev. B **50**, 822 (1994).

[17] P.H. Gaskell, M.C. Eckersley, A.C. Barnes, P. Chieux, Nature **350**, 675 (1991).

[18] R.O. Ritchie, K. J. Bathe, Int. J. of Fracture **15**, 47 (1979).

[19] E. Bouchaud in *Size-scale effects in the failure mechanisms of materials and structures*, IUTAM Symposium Turin October 1994, Edited by A. Carpinteri, E&F Spon (London), p. 121 (1996).

[20] E. Guilloteau, H. Charrue, F. Creuzet, Europhys. Lett. **34**, 549 (1996).

[21] J. Schmittbuhl, J.P. Vilotte, S. Roux, Phys. Rev. E **51**, 131 (1995).

[22] C. Bathias and J.P. Bailon, *La fatigue des matériaux et des structures*, Ed. Maloine S.A. (1980).

[23] J. Schmittbuhl, S. Roux, J.-P. Vilotte, K. J. Maloy, Phys. Rev. Lett. **74**, 1787 (1995); P. B. Thomas, M. Paczuski, preprint (1996).

[24] P. Daguier, E. Bouchaud, G. Lapasset, Europhys. Lett. **31**, 367 (1995).

[25] A. Hansen, E.L. Hinrichsen, S. Roux, Phys. Rev. Lett. **66**, 2476 (1991); S. Roux, D. François, Scripta Metall. **25**, 1092 (1991).

CONFORMAL MAP ON ROUGH BOUNDARIES APPLICATION TO FRACTURE

Stéphane ROUX, Damien VANDEMBROUCQ
Laboratoire de Physique et Mécanique des Milieux Hétérogènes, ESPCI, 10 rue Vauquelin, 75231 Paris, France.

ABSTRACT

A conformal mapping technique is presented which allows to solve efficiently harmonic and bi-harmonic problems in semi-infinite domains limited by a rough boundary. This technique is applied to obtain the statistical distribution of flux and elastic stress in the vicinity of a self-affine boundary. This computation justifies the occurence of Weibull statistics for the strength of glass fibers, with Weibull modulus depending on the roughness amplitude. Another application concerns the determination of the local mode III stress intensity factor ahead of a rough crack. Extension of the method to surface stress-corrosion, and interfacial crack propagation are discussed.

INTRODUCTION

This text presents an efficient technique to obtain the solution of two dimensional harmonic problems in the vicinity of a rough boundary, with any prescribed topography $h(x)$ provided the slope $|dh/dx|$ is less than unity — although this limitation can be overcome at the expense of a larger computation time. This technique is based on a conformal mapping technique of a semi-infinite domain $\mathcal{D}$ limited by a rough boundary onto a semi-infinite plane, $\mathcal{P}$. The conformal map can be obtained numerically from repeated one-dimensional fast Fourier transforms, and thus this technique is extremely easy to implement and requires a very low computation time to produce accurate maps. Moreover, following the first steps of the algorithm analytically, we can use this technique to obtain systematically the harmonic or bi-harmonic field perturbation from a flat boundary as a perturbative expansion on the roughness amplitude.

CONFORMAL MAPPING

The conformal map Ω associates to each point $z = x + iy$ from the semi-plane $\mathcal{P}$, $y \leq 0$, to its image $w = u + iv$ in the domain $\mathcal{D}$, $v \leq h(u)$ as shown in Figure (1). For the mapping to be conformal, Ω has to be a function of z only — independent of $\overline{z}$ — and it should be inversible over the domain of interest.

For practical purposes, we consider the boundary $h(u)$ to be periodic, and without restriction, we can choose the period to be 2π. The map is searched for as the identity plus a series of evanescent modes close to the boundary:

$$\Omega(z) = z + \sum_{n \geq 0} \omega_n e^{-inz} \tag{1}$$

The complex coefficients ω_n have to be determined so that the image of the real axis coincides with the boundary $u + ih(u)$. Considering the imaginary part of Eq.(1) on the boundary

Mat. Res. Soc. Symp. Proc. Vol. 463 © 1997 Materials Research Society

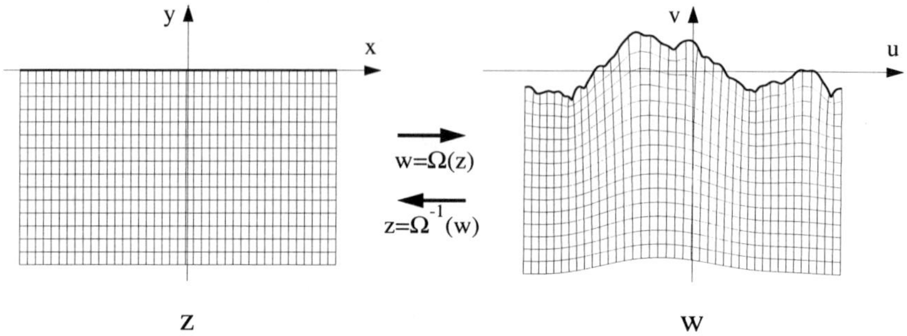

Figure 1: Schematic illustration of the geometry of the problem and of the mapping Ω.

$y = 0$, we observe that ω_n is nothing but the Fourier transform of $h(u(x))$. The difficulty is that $u(x)$ is not known. The proposed algorithm is to start from the zeroth order guess: $u(x) = x$. Using this first approximation, we can compute the coefficients ω_n as the Fourier transform of h. The real part of Eq.(1) can then be used to compute a better determination of $u(x)$. From the latter new coefficients ω_n are computed as a Fourier transform of $h(u(x))$ — i.e. h with a non-uniform sampling of points. The same procedure is repeated until the desired accuracy is reached. We refer the reader to Ref.[1] for a detailed discussion of this numerical procedure. As presented, the algorithm always converges if the slope $|dh/dx|$ does not exceed 1. If it does, under-relaxation can be used to recover convergence, at the expense of a larger computation time. Alternatively, one can contruct intermediate maps from low pass filtering of the initial profile, in such a way that each mapping problem can be solved with the previous algorithm.

STATISTICAL DISTRIBUTION OF HARMONIC FIELDS CLOSE TO SELF-AFFINE BOUNDARIES

As a direct application of the conformal map technique, we studied the statistical distribution of the flux, $f = |\nabla\phi|$, on the boundary, for a harmonic field ϕ such that $\phi = 0$ on the boundary and $\nabla\phi \to \vec{e}_v$ when $v \to -\infty$. The boundary is chosen to be self-affine, with a roughness exponent ζ. It is constructed in Fourier space as

$$\tilde{h}(k) = A\eta(k)k^{-1/2-\zeta} \tag{2}$$

with k the wavenumber, A the amplitude of the profile, and $\eta(k)$ independent random Gaussian complex numbers centered in 0 and unit variance for both real and imaginary parts. It is useful to introduce the lower cut-off of the self-affine regime as ℓ, such that the maximum wavenumber is $k_{max} = 2\pi/\ell$. Figure (2) shows the log-log plot of the histogram of flux for different amplitudes of the roughness and a roughness exponent $\zeta = 0.8$.

A first order (in A) perturbation analysis shows that the flux f has a log-normal distribution. The logarithm $g = \log(f)$ is distributed as a Gaussian, centered on 0, and of standard

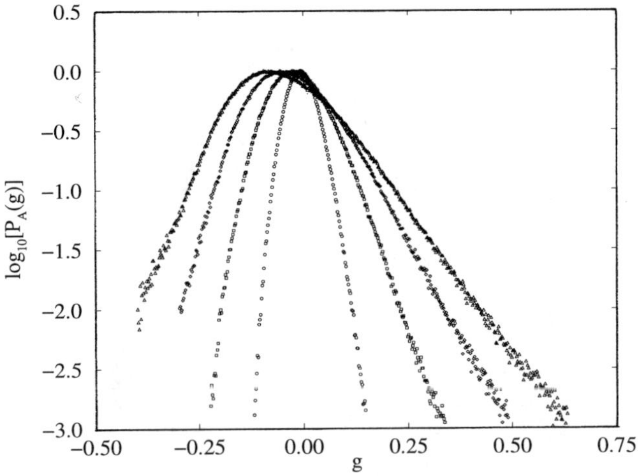

Figure 2: Log-log plot of statistical distributions of the flux, $g = \log_{10}(f)$ computed on self-affine profiles of different amplitudes with a roughness exponent $\zeta = 0.75$. Note the power-law tails of the distributions.

deviation σ_g, which assumes the following expression

$$\sigma_g^2 = \frac{\pi}{2(1-\zeta)} A^2 \left(\frac{\ell}{2\pi}\right)^{2(\zeta-1)} \tag{3}$$

A second-order analysis [3] allows to correct the large gradient behavior of the flux distribution. The statistical distribution of g can be cast in the following form

$$p(g) = \frac{1}{\sigma_g \sqrt{2\pi(1+2Kg)}} \exp\left[\frac{-1 - Kg + \sqrt{1+2Kg}}{K^2\sigma_g^2}\right] \tag{4}$$

where K is a constant, $K \approx 2$. Figure (3) shows a data collapse of distributions obtained for various roughness amplitudes when plotted as a function of $f_A(g) = (\sqrt{1+2Kg} - 1)/(K\sigma_g)$. Using the latter reduced variable, the distribution function is a Gaussian function (dotted line in Fig. (3)) independent of A up to logarithmic corrections.

The probability distribution, p, behaves for large flux f as a power-law $f^{-\tau-1}$ with an exponent

$$\tau = \frac{1}{K\sigma_g^2} \propto A^{-2}\ell^{2(1-\zeta)} \tag{5}$$

It is worth noting here that the exponent of the power-law is not directly dictated by the roughness exponent ζ, and it is a continuous function of the roughness amplitude, and of the lower cut-off of the self-affine regime. In particular as the amplitude vanishes, a diverging exponent is predicted.

Application of this result to glass fiber rupture statistics will be presented in a subsequent Section.

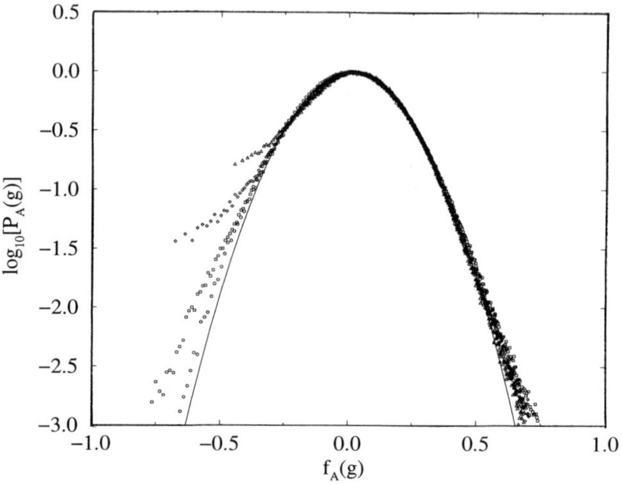

Figure 3: Collapse of different statistical distributions of the flux computed on self-affine profiles of different amplitude (same data as the previous Figure). A change of coordinate on the flux $f_A(g)$ has been made so that Eq. (4) reduces to a parabola as shown by the solid curve.

MODE III STRESS INTENSITY FACTOR

Using the above method, it is possible to address problems of anti-plane elasticity, which involve the solution of a simple harmonic field. As a particular application in this context, we have considered the effect of the crack roughness on the local stress intensity factor of a mode III crack.

The reference case we consider is a semi-infinite straight crack and we prescribe at infinity a stress field which corresponds to a pure mode III. The stress intensity factor at the crack tip is K_∞. Now we perturb the crack geometry by introducing a small amplitude roughness. We can define a local stress intensity factor K_{loc} at the crack tip. The question we wish to address is the relation between K_{loc} and K_∞.

In order to solve this problem we first apply the transformation $z' = \sqrt{w}$, so that the crack is unfolded and without roughness it would then coincide with the real axis. This puts the problem into a form similar to the previous geometry. The conformal mapping which remains to be computed makes use of the image of the crack geometry in the first transformation. In the z' plane, the latter is denoted as

$$y' = \varphi(x') \tag{6}$$

and due to its construction, φ is an odd function.

Performing a perturbation analysis in the roughness amplitude, one can show that to first order in A, these two stress intensity factors coincide. They however differ at second

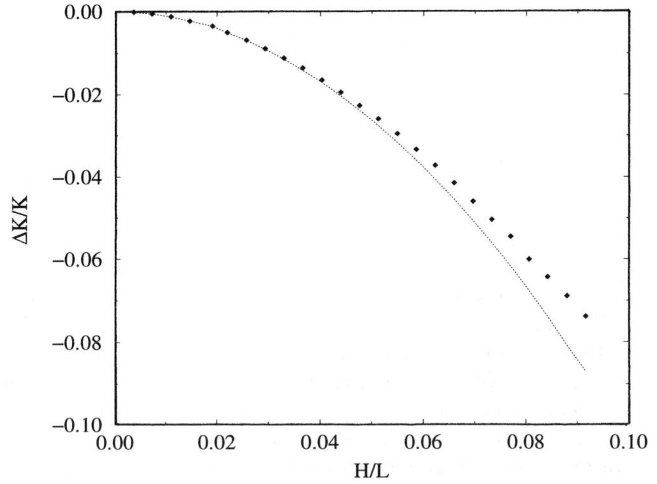

Figure 4: Relative correction to the stress intensity factor as a function of the relative roughness amplitude H/L. The numerical simulation results are shown as symbols and the perturbation analysis as a dotted line.

order. The ratio of these stress intensity factors can be written as

$$\frac{K_{loc}}{K_\infty} = 1 - \frac{\beta^2}{2} + \gamma \tag{7}$$

where β is the slope of the tangent to the crack at the tip with respect to the x-axis. This term reduces the local stress intensity factor systematically and varies quadratically with the roughness amplitude. The last term in Eq. (7), γ, is a non-local kernel whose expression to second order in amplitude is

$$\gamma = 4\int_0^\infty \int_0^k k(k' - k)\ \tilde{\varphi}(k)\tilde{\varphi}(k')\ dk\ dk', \tag{8}$$

where $\tilde{\varphi}(k)$ denotes the Fourier transform of the $\varphi(x')$. Figure (4) shows the numerically computed relative correction to the stress intensity factor, as a function of the roughness amplitude, together with the second order perturbation result.

Now we specialize our discussion to the case of a self-affine geometry since the latter is observed experimentally in a number of cases (unfortunately pure mode III cracks are rather scarse, so that we refer here to results obtained in mode I). We introduce the roughness amplitude A, lower cut-off ℓ, upper cut-off $L = 2\pi$ so as express the scaling of the correction to the stress intensity factor. Both correction terms display the same scaling so that the relative correction is proportional to

$$\langle \frac{K_\infty - K_{loc}}{K_\infty} \rangle \propto \frac{A^2}{L^2} \left(\frac{\ell}{L}\right)^{2(\zeta-1)} \tag{9}$$

The important message is that the local stress intensity factor is decreased as compared to the far-field one, implying a *strengthening* of the medium by the crack roughness.

Another interesting feature is that the far-field singular stress field is not centered exactly on the actual crack tip, but slightly off, by a distance of order A^2, for fixed spectral components $\eta(k)$ of the crack geometry. One can understand qualitatively the above results by noting that the roughness close to the crack tip has the effect of defocusing the stress field so that only part of the large scale singularity is actually carried by the crack tip.

BIHARMONIC PROBLEMS

We now present an extension of the previous computation to the case of bi-harmonic problems, $\nabla^4\Phi = 0$. The latter are naturally encountered in the context of hydrodynamics (Stokes flow), or elasticity theory (Airy potential). In the spirit of considering the effect of surface roughness at a small scale compared to the typical size of the domain or of the scale over which the loading ceases to be homogeneous, we will assume that a uniform field is applied at infinity (Couette flow for hydrodynamics, simple traction along the boundary for elasticity). We map the bi-harmonic problem on the half-plane geometry so that the potential is searched for under the following expression:

$$\Phi = \Omega(z)\overline{F(z)} + \overline{\Omega(z)}F(z) + H(z) + \overline{H(z)} \tag{10}$$

which form garantees the bi-harmonicity of Φ in the w-space.

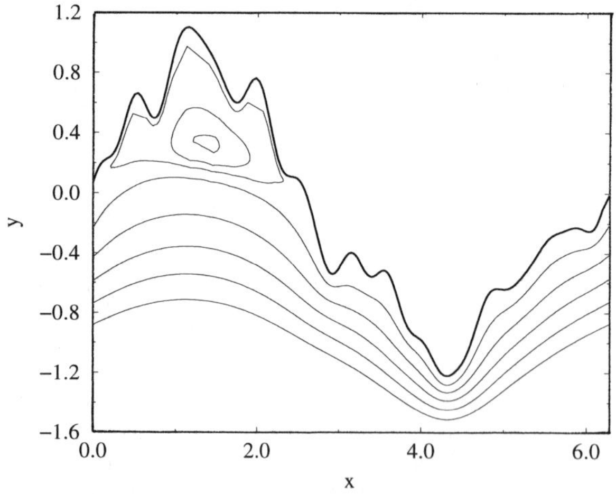

Figure 5: Contour lines of the stream function close to a rough surface. Closed contours indicate the presence of a recirculation eddy.

In the case of Couette flow at infinity, the velocity potential has the following asymptotic behavior: $\Phi \sim y^2/2$ with unit shear rate at infinity. The no-slip boundary condition along the rough surface imposes $\Phi = 0$ and $\partial_\perp \Phi = 0$. Exploiting the two latter conditions determines the two unknown functions F and H. They are both written in terms of a x periodic function, subscript p, plus an additional part determined from the far field:

$$\begin{cases} F(z) = & \dfrac{z}{8} + F_p(z) \\ H(z) = & -\dfrac{z^2}{8} - z(F_p(z) + \omega(z)/8) + H_p(z) \end{cases} \tag{11}$$

The periodic component is decomposed on a basis of evanescent modes, $F_p(z) = \sum_k f_k e^{-ikz}$, and similarly for H_p. The spectral coefficients f_k and h_k are determined from the linear set of equations

$$\begin{cases} f_k + \omega_k/8 + i \sum_n n(\overline{\omega_n} f_{k+n} + \overline{f_n}\omega_{n+k}) = 0 \\ h_k = -\sum_n (\overline{\omega_n} f_{k+n} - \overline{f_n}\omega_{n+k}) \end{cases} \tag{12}$$

An essential property is that the above system is well-conditioned, and hence the solution is obtained straighforwardly with any chosen accuracy. In particular, one can go much beyond the amplitude of the boundary roughness which gives rise to recirculation vortices. Figure (5) shows the velocity potential for a Stokes flow past a rough surface. As in the harmonic case, we can study a number of different problem with this method. In particular, we may define an equivalent no-slip straight boundary so that the velocity field matches that of a rough boundary at infinity.

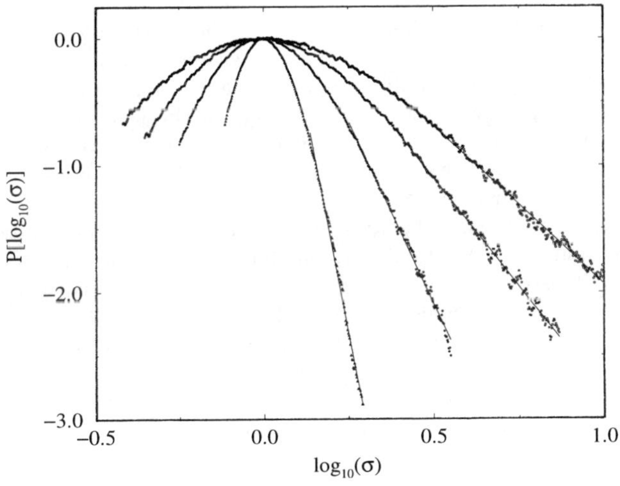

Figure 6: Log-Log plot of the tangential stress distribution along a rough self-affine boundary for different amplitudes. We have underlined the power-law tails of the distribution by fitting straight lines on the graph.

For elasticity, with a constant traction $\sigma_{uu} = \sigma_0$ exerted parallel to the boundary at infinity $v \to -\infty$, resorting to the Airy potential gives a strictly equivalent problem to the above Stokes flow with a Couette flow at infinity.

We can also study the stress field statistics along the rough boundary. The analytical approach (second order in perturbation) is much more cumbersome than for the harmonic problem, but numerically, we can easily record the statistics. Analysing the results using a similar form as in the harmonic case (see Eq. (4)) we can rescale the data on a single curve for all amplitudes using a constant $K = 0.25$. This leads to a veriation of the exponent in the power-law regime which is similar to that of Eq. (5). However, the latter form remains to be theoretically justified.

APPLICATION TO RUPTURE STATISTICS

One application of the above result can be found in the rupture statistics of fiber glass. The topography of fiber glass exhibits some roughness at a very small scale. Atomic Force Microscopy measurements have revealed that the roughness of glass displayed a self-affine character [6]. Moreover the fiber geometry allows to use freely Weibull analysis to access the failure statistics. The latter analysis is based on the weakest link assumption, stating that only the weakest element in the fiber will cause the rupture, so that failure statistics can be reduced to a pure statistical question once one can relate the local resistance to the description of the material properties, or here topography. The probability of rupture under a tension stress σ_0 of a representative element of fiber (at a scale ξ larger than the upper cut-off of the topography correlations) being written $1 - R(\sigma_0)$, the probability that a fiber of length L can sustain such a stress is written

$$P(\sigma, L) = \exp\left[-(L/\xi)R(\sigma)\right] \tag{13}$$

A very common form for the R distribution is the phenomenological one proposed by Weibull [7], namely R is often fitted by a power-law, $R \propto \sigma^m$ where m is called the *Weibull modulus*. For glass fibers, the observed modulus ranges typically from 6 to 15, from low to high quality fibers. It is important to note that in spite of the fact that this form is purely phenomenological, it describes the statistics remarkably well, and even allows to distinguish different kinds of defects. We however note that the exponential of the stress raised to such a large power does not allow for a very severe test of the power-law form.

From our previous analysis of the stress distribution at the surface, we can easily built the R distribution assuming a constant maximum tensile stress σ_c which can be supported by the material. The R distribution is nothing but the local stress distribution (σ_c/σ). Hence, the Weibull form naturally emerges for a self-affine topography. The Weibull modulus is thus expected to follow the same variation as the τ exponent from Eq. (5).

No quantitative comparison has been made yet. Qualitatively, we do obtain a decrease of the Weibull modulus as the surface gets rougher, as is observed for fibers drawn at different speeds, and for fiber ageing under hot and humid conditions where the surface corrosion increases the roughness. The next section is devoted to this problem.

SURFACE EVOLUTION UNDER STRESS CORROSION

A first order perturbation analysis shows that the tangent stress along the rough free boundary can be expressed in a simple form as

$$\frac{\sigma_{\|}}{\sigma_0} = 1 - 2\mathcal{H}\left[\frac{dh}{dx}\right] \tag{14}$$

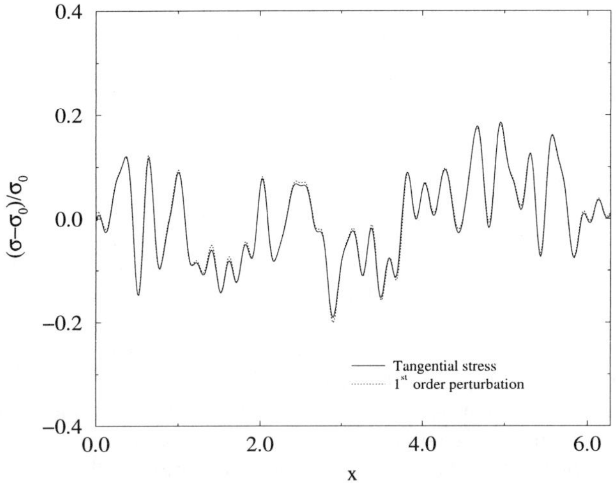

Figure 7: Plot of the tangential stress $\sigma_{\|}$ along the rough boundary (solid curve), together with the Hilbert transform of the profile slope (dotted curve).

for a tensile stress σ_0 far from the boundary. The notation $\mathcal{H}[...]$ has been introduced for the Hilbert transform (convolution with the inverse Fourier transform of $ik/|k|$). This approximation is valid for slopes typically less than 0.1. For steeper profiles, second order terms have to be taken into account. Figure (7) shows one example of the comparison between the Hilbert transform of the profile slope, and the tangential stress computed as discussed in a preceeding section. The topography of the surface is also shown on the same graph.

Materials such as glass are very sensitive to humidity. Water reacts easily with the glass surface. This process alterates the surface topography and is believed to be responsible for at least part of the ageing of glass. One important point is that the rate of reaction is dependent on the stress state of the material. The activation energy is reduced by a tensile stress[8]. On a free surface only the tangential stress matters and thus the velocity of reaction front is written — to first order in the stress — as

$$v = v_0(1 - \alpha\sigma_{\|}) \tag{15}$$

where α is proportional to an activation volume divided by kT, and v_0 is the front velocity under no stress. We introduce the simplifying assumption that the reaction front can be identified with the free surface.

Introducing an additional stochastic component η as a source term, we can write the following equation for the surface evolution

$$\frac{\partial h(x,t)}{\partial t} = v_0(1 - \alpha\sigma_0) - 2\alpha v_0 \sigma_0 \mathcal{H}\left[\frac{dh}{dx}\right] + \eta(x,h,t) \tag{16}$$

The above equation can be seen as a Langevin equation describing the front evolution. Its main characteristic as compared to classical growth models[9] is its non-local character resulting from the Hilbert transform. The constant velocity tem in the r.h.s. of the above equation can be absorbed in a moving frame of reference ($h \to h - vt$ where $v = v_0(1 - \alpha\sigma_0)$).

A comparable model has been proposed for the description of flame fronts by Sivashinsky[5], with an additional non-linear term, proportional to $(\partial h/\partial x)^2$. The latter can be argued for from Galilean invariance although it is a second-order term. An additional laplacian term, $\partial^2 h/\partial x^2$, can also be included in the r.h.s. in order to account for a finite surface tension.

This model has been studied recently [10] using a conformal mapping technique different from the above proposed one. Instead of decomposing the map on a set of evanescent waves, a set of singular fields with poles in the outside domain is used. The evolution equation then can be translated into an equation of motion of these singularities. The main output of these studies is the existence of a single cusp which is being formed, in the absence of surface tension. This cusp corresponds to the initiation of a fracture from the surface. The important and practical question is the time needed to reach such a stage, which essentially depends on the initial topography of the surface.

It is interesting to note that such a stress-corrosion model has a formal structure which is amazingly close to an interfacial crack front model introduced by Schmittbuhl et al [11]. However, the key difference is the sign of the non-local term which is such that the interfacial crack front is stable, and the front corrugation are initiated by the heterogeneity of the interface, but smoothened in the crack advance, whereas the stress-corrosion model is unstable, leading to the cusp singularity.

CONCLUSIONS

We have introduced a conformal mapping technique which offers a simple and efficient way to solve harmonic fields close to rough boundaries. This technique lies on the iterative use of one-dimensional Fast Fourier Transforms. Beyond the pure harmonic case in a semi-infinite medium, this tool has been applied to different problems and geometries: mode III stress intensity factor with a rough crack, solution of Stokes flow close to a rough boundary, stress distribution along a free rough surface. Besides an efficient way of solving problems in the vicinity of a given rough interface, this method presents two more advantages. First, the iterative algorithm naturally gives way to a perturbative expression of the mapping function and to a perturbative solution of the problem. Second, the technique is well adapted to scale invariant geometries. A simple constraint on the mapping function allows to generate automatically self-affine interfaces of given roughness exponent and makes possible at a low cost statistical studies on a large amount of samples. Recent results suggest that the stress ditribution on self-affine rough boundaries presents a power law tail. Although this behavior has still to receive a complete theoretical explantion, it could give a fundamental basis of the rupture statistics Weibull law. Furthermore, the analytical results obtained in this context open the way to a stress-corrosion model. More quantitative applications to experimental cases are under current investigation.

REFERENCES

1. D. Vandembroucq and S. Roux, *"Conformal mapping on rough boundaries I: Applications to harmonic problems"*, preprint;

2. D. Vandembroucq and S. Roux, "'*Conformal mapping on rough boundaries II: Applications to bi-harmonic problems*", preprint;
3. D. Vandembroucq and S. Roux, "*Harmonic field distributions on self-affine boundaries*", preprint;
4. D. Vandembroucq and S. Roux, "*Mode III stress intensity factor ahead of a rough crack*", to appear in J. Mech. Phys. Sol.
5. G.I. Sivashinsky, Acta Astronautica, **4**, 1177, (1977)
6. E. Guilloteau, Ph. D. dissertation, Univ. Paris-Sud, (1995)
7. W. Weibull, J. Appl. Mech. **18**, 293, (1951)
8. R. W. Davidge, "*Mechanical behaviour of ceramics*", Cambridge Univ. Press, (1980)
9. A.L. Barabási and H.E. Stanley, "*Fractal concepts in surface growth*", Cambridge Univ. Press, (1995)
10. Z. Olami, B. Galanti, O. Kuppervasser and I. Procaccia, "*Random noise and pole dynamics in unstable front propagation*", preprint
11. J. Schmittbuhl, S. Roux, J.P. Vilotte and K.J. Måløy, Phys. Rev. Lett. **74**, 10, 1787, (1995)

Part VII

Statistical Mechanics in Physics

SCALING FOR THE COALESCENCE OF MICROFRACTURES BEFORE BREAKDOWN

S. ZAPPERI[1], P. RAY[2], H.E. STANLEY[1] AND A. VESPIGNANI[3]
[1] Center for Polymer Studies and Department of Physics, Boston University, Boston, MA 02215 ,USA [2]The Institute of Mathematical Sciences, CIT Campus, Madras - 600 113, India
[3] Instituut-Lorentz, University of Leiden, P.O. Box 9506 The Netherlands.

ABSTRACT

We study the behavior of fracture in disordered systems close to the breakdown point. We simulate numerically both scalar (resistor network) and vectorial (spring network) models with threshold disorder, driven at constant current and stress rate respectively. We analyze the scaling of the susceptibility and the cluster size close to the breakdown. We observe avalanche behavior and clustering of the cracks. We find that the scaling exponents are consistent with those found close to a mean-field spinodal and present analogies between the coalescence of microfractures and the coalescence of droplets in a metastable magnetic system. Finally, we discuss different experimental conditions and some possible theoretical interpretations of the results.

INTRODUCTION

The breakdown of solids under external forces is a longstanding problem, that has practical and theoretical relevance. The way a material breaks, under the effect of an external electric field or under mechanical stress are closely related problems, due to the formal similarities in the underlying laws governing those phenomena. The first theoretical approach to fracture mechanics dates back to the twenties with the work of Griffith [1], who formulated a theory of crack formation, which is similar to the classical theory of nucleation in first-order phase transitions. Cracks grow or heal, depending on whether the external stress prevail over the resistance at surface of the crack. Similarly in bubble nucleation [2], a critical droplet will form when the change in free energy due to the bulk exceeds that of the surface. Griffith theory assume the presence a single microcrack of a particular shape surrounded by an homogeneous medium, and therefore is not appropriate in disordered systems, where cracks can start from different positions and coalescence may take place.

Spinodal nucleation [3], contrary to classical nucleation, is characterized by scaling properties and fractal droplets. The spinodal point in fact has some characteristics of a critical point in second order phase transitions. The similarity between a solid driven to the threshold of mechanical instability and spinodal nucleation has been discussed in the past. Rundle and Klein [4], using a Landau-Ginzburg analysis of a single crack, showed that the crack growth obeyed scaling laws expected for spinodal nucleation. Selinger et al. [5] have shown by numerical simulations and mean-field theory of thermally activated fracture that the breakdown has the characteristics of a spinodal point.

In this paper we concentrate on disordered media and we disregard the effect thermal fluctuations. The system is driven by an increasing external load to the point of global failure. It has been experimentally observed that the response, detected by acoustic emission (AE) measurements, to an increasing external stress takes place in bursts or avalanches distributed over a wide range of scale. Examples of this are found in foam glasses [6], fiber matrix composites [7], concretes [8], hydrogen precipitation [9] and volcanic rocks [10]. We

Mat. Res. Soc. Symp. Proc. Vol. 463 © 1997 Materials Research Society

observe a similar behavior for two dimensional discrete models. We show that the scaling behavior close to the breakdown is in quantitative agreement with the mean-field theory and it is suggestive of a first-order transition. The values of the scaling exponents are consistent with those found close to a spinodal point in *thermally* driven homogeneous systems.

THE MODELS

We study here two models, the random fuse model [11] for electric breakdown and a spring network model [12] for fracture. In the random fuse model [11] each bond of a two dimensional lattice is occupied by a fuse of conductivity $\sigma = 1$, which burns when the current flowing in it exceeds a quenched random threshold. We consider a rotated square lattice with periodic boundary conditions in one direction. We impose a constant external current on the two other edges of the lattice. The currents in each bond are computed solving the kirchhoff equations. This step corresponds to the minimization of the total energy dissipated in the lattice

$$E(\{\sigma\}) \equiv \frac{1}{2}\sum_i \sigma_i (\Delta V)_i^2, \tag{1}$$

where $(\Delta V)_i$ is the voltage drop in the bond i. We employ a multigrid relaxation algorithm with precision $\epsilon = 10^{-10}$. When all the currents are below the threshold we increase the current until the next bond reaches the threshold. The process is continued until a path of broken bonds spans the lattice and no current flows anymore. We chose a uniform distribution of thresholds, $D \in [0,2]$.

The second model is an elastic network [12] which has central and rotationally invariant bond-bending forces. The potential energy is

$$E = \frac{a}{2}\sum_i (\delta r_i)^2 \sigma_i + \frac{b}{2}\sum_{<i,j>} (\delta\theta_{ij})^2 \sigma_i \sigma_j \tag{2}$$

where δr_i is the change in the length of the bond i and $\delta\theta_{ij}$ is the change in the angle between neighboring bonds i and j. The constant σ_i is equal to one if the bond is present and it is zero otherwise. A slowly increasing external stress is applied on all the edges and the lattice dynamics is obtained by numerically solving the equations of motion for each spring. Bonds break when stretched beyond a randomly chosen threshold.

SIMULATION RESULTS

The response of the model to the increase of the external force takes place in widely distributed avalanches. The average size of the avalanches (i.e. the number of broken bonds) increases when the global failure is approached. We were able to show [13] using mean-field theory that the average avalanche size $\langle m \rangle$ diverges at the breakdown as

$$\langle m \rangle \sim (f_c - f)^{-\gamma} \qquad \gamma = 1/2. \tag{3}$$

where f is the stress or the current per unit length imposed on the lattice. We note that the same scaling law is expected close to a spinodal point, in the case of thermally driven first-order transitions. The macroscopic quantities of the system (i.e. elasticity) have a finite jump at the breakdown, indicative of a first-order transition.

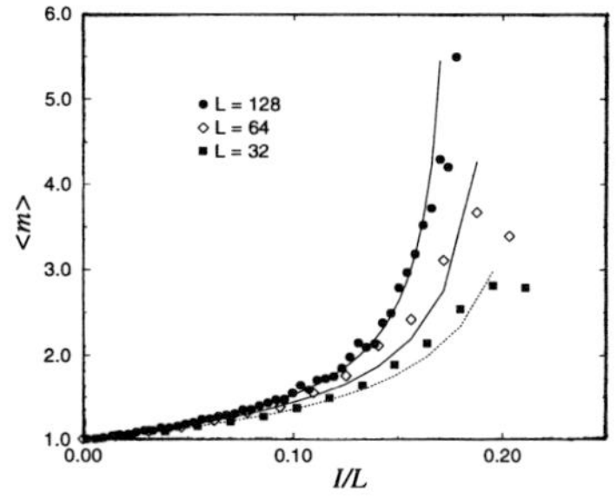

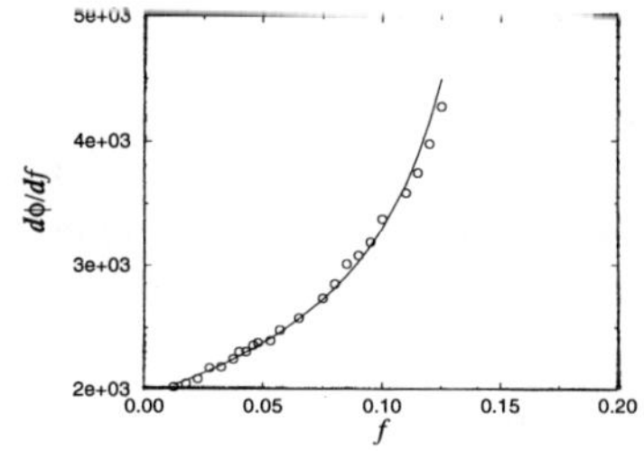

Figure 1: a) The average avalanche size $\langle m \rangle$ is plotted as a function of $f = I/L$, the fit is done using the mean-field value $\gamma = 1/2$. b) The "susceptibility" of the spring network with the mean field fit ($\gamma = 1/2$). The parameter ϕ is the fraction of bond that are not broken. The average avalanche size $\langle m \rangle$ is proportional to $d\phi/df$.

We confirm the validity of mean-field scaling by computer simulations of two dimensional models. For both models mean-field theory is obeyed remarkably well (see Fig. 1a and Fig. 1b).

The reason for the observed mean-field behavior is probably due to the long-range nature of elastic interactions. The formation of cracks in those models takes place by the coalescence of several microcracks. This is confirmed by the behavior of the average crack size which *does not diverge* at the breakdown (see Fig 2).

DISCUSSION AND CONCLUSIONS

The breakdown of driven disordered media is described by scaling law which are reminiscent of those found close to a spinodal point. It appears that the behavior of a driven disordered system is similar to that of a thermally driven homogeneous system. This analogy is not too strict since the concept of metastability and spinodal are not well defined in the first case.

Despite several experimental investigations of avalanche dynamics in fractures [6, 7, 8,

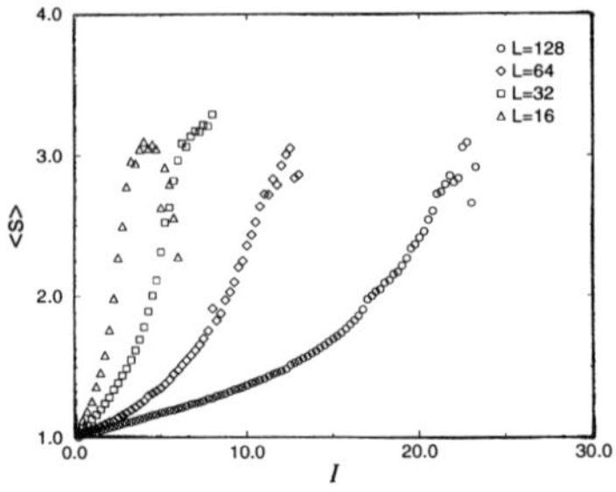

Figure 2: The average crack size for the fuse model as a function of the current for different systems sizes. The crack size does not seem to diverge at the breakdown.

9, 10], there is not a clear theoretical interpretation of the results. We believe that different experimental conditions can all give rise to similar scaling behavior, but the underlying physical mechanisms could be quite different. We can distinguish the following experimental setups:

1. A solid driven by a constant stress rate can be described in the framework discussed in this paper. The system responds to the increase of the external load by AE bursts of increasing size [7], diverging at the point of global failure. It would be interesting to check if the scaling exponents agree with the mean-field theory.

2. A solid subject to a constant load breaks because of thermal fluctuations. The AE is due to the formation of "droplets" and should be power law distributed close to the limit of stability (spinodal). Scaling exponents consistent with those of spinodal nucleation were observed in a recent experiment on cellular glass [6]. To confirm this interpretation it would be necessary to study the scaling for different values of the applied load.

3. A solid in a perfectly plastic state could respond to the increase of the external strain by a *stationary* AE signal. In this case one can interpret the results as a manifestation of self-organized criticality. Such a behavior was shown in numerical models [14, 15], but to our knowledge it has not yet been observed in experiments.

We believe that extensive and systematic experiments along these lines can resolve these longstanding problems.

ACKNOWLEDGMENTS

We wish to thank P. Cizeau, W. Klein, F. Sciortino and H. Strauven for useful discussions and remarks. The Center for Polymer studies is supported by NSF.

REFERENCES

[1] A. A. Griffith, Phil. Trans. Roy. Soc. London A **221**, 163 (1920).

[2] J. D. Gunton, M. San Miguel and P. S. Sahini, in Phase Transitions and Critical Phenomena, Vol. 8, edited by C. Domb and J. L. Lebowitz (Academic, London, 1983).

[3] C. Unger and W. Klein, Phys. Rev. B **29**, 2698 (1984); ibid.**31**, 6127 (1985); for a review, see L. Monette, Int. J. of Mod. Phys B **8**, 1417 (1994).

[4] J. B. Rundle and W. Klein, Phys. Rev. Lett. **63**, 171 (1989)

[5] R. L. B. Selinger, Z.-G. Wang, W. M. Gelbart and A. Ben-Saul, Phys. Rev. A **43**, 4396 (1991).

[6] H. Strauven, G. Claes, G. Heylen, G. Crevecoer and C. Maes, in 22nd European Conference on Acoustic Emission Testing Proceedings (Aberdeen, 1996)

[7] J.-C. Anifrani, C. Le Floc'h, D. Sornette and B. Souillard, J. de Phys. I **5**, 631 (1995).

[8] A. Petri, G. Paparo, A. Vespignani, A. Alippi and M. Costantini, Phys. Rev. Lett. **73**, 3423 (1994).

[9] G. Cannelli, R. Cantelli and F. Cordero, Phys. Rev. Lett. **70**, 3923 (1993).

[10] P. Diodati, F. Marchesoni and S. Piazza, Phys. Rev. Lett. **67**, 2239 (1991).

[11] L. de Arcangelis, S. Redner and H. J. Herrmann, J. Phys. Lett. (Paris) **46**, L585 (1985); P. Duxbury, P. D. Beale and P. L. Leath, Phys. Rev. Lett. **57**, 1052 (1986).

[12] P. Ray and G. Date, Physica A **229**, 26 (1996).

[13] S. Zapperi, P. Ray, H. E. Stanley and A. Vespignani, unpublished.

[14] S. Zapperi, A. Vespignani and H. E. Stanley in Fracture-Instability Dynamics, Scaling and Ductile/Brittle Behavior, edited by R. B. Selinger, J. Mecholsky, E. R. Fuller, Jr. and A. Carlsson (Mat. Res. Soc. Proc. **409**, Pittsburgh, 1996) pp. 355-358.

[15] H. J. Tillemans and H. J. Herrmann, Physica A **217**, 261 (1995).

A MODEL OF CRACK PROPAGATION IN A 2D HETEROGENEOUS MATERIAL

P. DAGUIER, E. BOUCHAUD, G. LAPASSET
O.N.E.R.A. (OM), 29 Avenue de la Division Leclerc,
B.P. 72, 92322 Châtillon Cedex, FRANCE

ABSTRACT

A continuous numerical model of pure mode I crack propagation in a bidimensional heterogeneous material is presented. This model describes the propagation of a macrocrack into a two-phases brittle material containing a finite density of second phase precipitates. The morphology of cracks produced for various mechanical and microstructural conditions is analysed. It is shown that the simulated cracks are self-affine with a roughness index $\simeq 0.6$ independent of the microstructure. Relevant length scales, on the contrary, strongly depend on the microstructural parameters, and indicate an optimum density leading to a maximum fracture toughness.

INTRODUCTION

It is now well established that fracture surfaces can be considered as self-affine fractal objects [1]. Although many experiments performed on metallic alloys [2, 3], rocks [4, 5], glass [6] or numerical simulations [7] report a roughness exponent ζ (related to the fractal dimension of the surface) which is independant of the material ($\zeta \simeq 0.8$ for three dimensional systems, and $\zeta \simeq 0.7$ for two dimensional ones [8, 9, 10]), the universality of this index is still controversial [11, 12].

It has been argued recently that, contrarily to the roughness exponent, relevant length scales are strongly correlated to the microstructural parameters, and hence to the fracture toughness. As a matter of fact, by rewritting the Griffith criterion for a self-affine crack, Bouchaud *et al.* [13] argued that the fracture toughness is related to the spikiness h_{max}/ξ of the fracture surface, where ξ is the self-affine correlation length and h_{max} is the typical heigth on the fracture surface outside the fractal regime. The higher the ratio h_{max}/ξ, the steepest the local slopes on the considered surface or profile.

The central point of this paper is to deal with the influence of the microstructural parameters on the morphology of the simulated cracks and hence on the fracture toughness. So, a continuous numerical model of pure mode I crack propagation in a bidimensional heterogeneous material is built up. The principles of the simulation are presented first, followed by an analysis of the morphology of the simulated cracks.

PRINCIPLES OF THE SIMULATION

Description of the model

The propagation of a macrocrack lying along the x-axis under a purely Mode I loading condition is considered. The crack, after running over a distance l_o, enters into a box of dimensions (x_l,y_l), with $x_l, y_l \ll l_o$. This box contains N sites (the positions of which are randomly distributed with a uniform probability law) where microcracks may nucleate. A cleavage type criterion is chosen for the nucleation of microcracks: if the stress produced by the macrocrack at site i overcomes a critical stress $\sigma_C(i)$, then a microcrack of length s_o nucleates. The threshold cleavage stresses $\sigma_C(i)$ are random variables, distributed with a uniform probability law. The stress acting on site i is given by:

Mat. Res. Soc. Symp. Proc. Vol. 463 © 1997 Materials Research Society

$$\sigma_i = \frac{k_I}{\sqrt{\pi d_i}} \tag{1}$$

where d_i is the distance between the macrocrack tip and site i. k_I is the opening stress intensity factor (SIF) at the tip of the macrocrack in the direction of crack propagation. The computation of k_I takes into account both the macrocrack deflection and the presence of the microcracks.

Once opened, the cleavage microcracks may progress towards the main crack tip if the local SIF overpasses a threshold K_I^{th} for propagation into the matrix. In order to take into account the presence of heterogeneities in a "mean field" type of approximation, each microcrack follows a directed random walk to the main crack tip.

If the distance between the tips of the macrocrack and microcrack i is smaller than a distance R_{plast0}, they coalesce instantaneously. The back tip of the microcrack then becomes the new main crack tip. After the microcraks' nucleation / growth stage, the modified SIFs of the main crack tip, due to the presence of microcracks, are computed assuming that the main crack is approximately straigth and under mode I and that the interactions between the microcracks are negligible compared to the interaction with the macro-crack.

The computation of the new SIFs at the main crack tip is based on first order analytical results due to Rose [14]. The modification of the SIFs due to the presence of microcracks is mainly governed by the ratio of the length $2s_i$ of microcrack number i to its distance to the main crak tip r_i, and the position and orientation of the ith microcrack, defined by two angles θ_i and α_i (see Fig. 1):

$$\begin{cases} K_I = K_{I,0}\ \left[1 + \sum_{i=1}^{N_{open}} \mathcal{F}(\frac{s_i}{r_i}, \theta_i, \alpha_i)\right] \\ K_{II} = K_{I,0}\ \sum_{i=1}^{N_{open}} \mathcal{G}(\frac{s_i}{r_i}, \theta_i, \alpha_i) \end{cases} \tag{2}$$

where $K_{I,0}$ is the remote SIF, $\mathcal{F}$ and $\mathcal{G}$ are first order pertubative terms, and N_{open} is the number of microcracks.

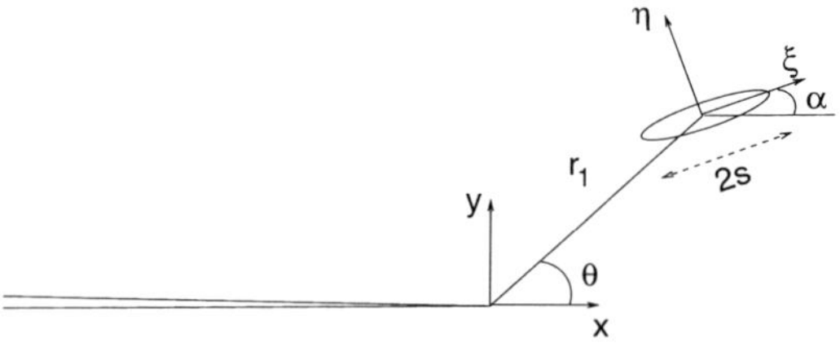

Figure 1: **Interaction between the macrocrack and the ith microcrack.**

The appearence of a local non-zero K_{II} produces crack deflection. The crack will propagate in the direction where , where θ is the kink angle. Within the Cotterell-Rice approximation [15], this two criteria give the same result:

$$\frac{K_{II}}{K_I} = -tg(\theta/2)\frac{\cos^2(\theta/2)}{3\cos^2(\theta/2) - 2} \tag{3}$$

As the main crack propagates into the box, more microcracks are produced or grown. The damage produced generates a decrease of the elastic moduli ν (Poisson modulus) and μ (shear modulus) of the material following (for $f \ll 1$) :

$$\left\{ \begin{array}{l} \frac{\overline{\nu}}{\nu} \simeq 1 - \pi f \\ \frac{\mu}{\overline{\mu}} \simeq 1 + \frac{\pi f}{1-\nu} \end{array} \right. \tag{4}$$

where $f = \sum_{i=1}^{N_{open}} \frac{s_i^2}{x_l y_l}$ is a microcrack density parameter and $\overline{\nu}$ and $\overline{\mu}$ are the elastic moduli of the damaged material.

For a circular damage zone, the modified SIF has been computed by Hutchinson [16]:

$$\frac{K_{I,damaged}}{K_{I,undamaged}} = 1 - \frac{5}{8}\frac{1}{1-\nu}(\frac{\mu}{\overline{\mu}} - 1) + \frac{3}{4}\frac{1}{1-\nu}(\overline{\nu}\frac{\mu}{\overline{\mu}} - \nu) \tag{5}$$

This approximation is used to take into account the shielding effect due to the nucleation and growth of the microcracks.

Parameters of the model

The influence of the following parameters on the morphology of the cracks is studied:
- $\rho = \frac{N}{x_l y_l}\frac{\pi s_o^2}{4}$, the surface fraction of precipitates;
- s_o, the initial size of the microcracks;
- σ^{ext}, the external stress.

while:
- $v_o = 0.1$, the microscopic dimensionless speed of the macrocrack and of the microcracks;
- K_I^{th}, the propagation threshold into the matrix;
- $R_{plast0} = 0.05$, the "plastic zone size";
- $\sigma_C(i)$ is the critical stress above which microcrack i is created (random variable, uniformely distributed)

are kept constant.

For each condition, severals cracks are produced which correspond to various samples of the same microstructure.

Crack morphology parameters

Both the roughness exponent and the spikiness of the simulated cracks are measured in order to have their morphology characterized. The roughness exponent ζ is determined with the Hurst method [17]:

$$z_{max}(r) \;=< max(z(r'))_{r_o<r'<r_o+r} - min(z(r'))_{r_o<r'<r_o+r} >_{r_o} \tag{6}$$

$$with\ z_{max}(r) \propto \left\{ \begin{array}{ll} r^{\zeta} & (r \lll \xi) \\ h_{max} & (r \gg \xi) \end{array} \right.$$

where r is the width of the window. $z_{max}(r)$ is the difference between the maximun and the minimun heights z within this window, averaged over all possible origins r_o of the window belonging to the profile. In practice, the saturation value h_{max} is taken to be equal to $z_{max}(x_l)$, and ξ is defined as the intersection of the power law regime valid at small length scales with the saturation plateau $z_{max}(r) = h_{max}$ (see Fig. 2).

It has been argued by Bouchaud *et al.* [13] that the spikiness ratio h_{max}/ξ is related to the fracture toughness through the relation:

$$K_{IC} \simeq 2\sqrt{\gamma\, E}(\frac{h_{max}}{\xi})^{\frac{1}{2}} \tag{7}$$

This result is in qualitative agreement with the common observation that "shallow" fracture surfaces are associated to more brittle materials (rocks, glass, ...) while spikier ones are observed in the case of toughter materials (metallic alloys for example). It will be used in the following this parameter and the microstructure is performed.

RESULTS

The simulations are performed on a personnal computer PENTIUM 90 Mhz. The cracks simulated as explained above are analysed in this section. The influence of the microstructural parameters on the morphology of the simulated cracks is considered first.

Effect of the microstructural parameters

To analyse the effect of density, simulations are performed for ρ, ranging from 8% to 78%.

As it can be seen in Fig. 2, the craks are self-affine and the roughness exponent ζ seems independant of ρ. Small variations of the measured exponent are due to the small size of the scaling domain.

On the contrary, the spikiness evolves with ρ in a non monotonic way and a maximum is observed for $\rho \simeq 58\%$.

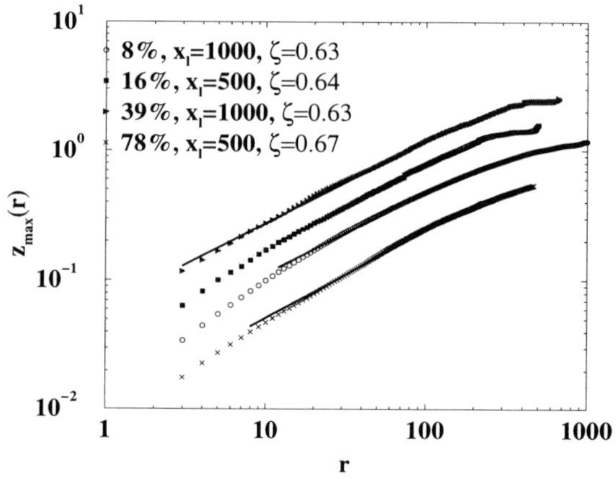

Figure 2: **Evolution of $z_{max}(r)$ for various values of ρ. The measured roughness index is close to 0.63 in each case.**

s_o is varied between 0.05 and 0.5. No significant variation of ζ is observed either, while the fracture toughness monotonically increases with s_o.

Finally, because there are two mechanisms for crack propagation – pure deflection or microcrack coalescence –, the macro crack speed increases beyond v_o when the external load is increased. When the external load is increased, the number of microcracks increases as well, and so the probability of coalescence. Hence, the average crack speed increases also, and saturates when the external load is able to open microcracks in all the brittle precipitates.

DISCUSSION

Bidimensional cracks propagating in mode I through a heterogeneous material have been simulated with a local model and their morphology has been analysed. The cracks are shown to be self-affine, with a roughness exponent $\zeta \simeq 0.63$, in agreement with the

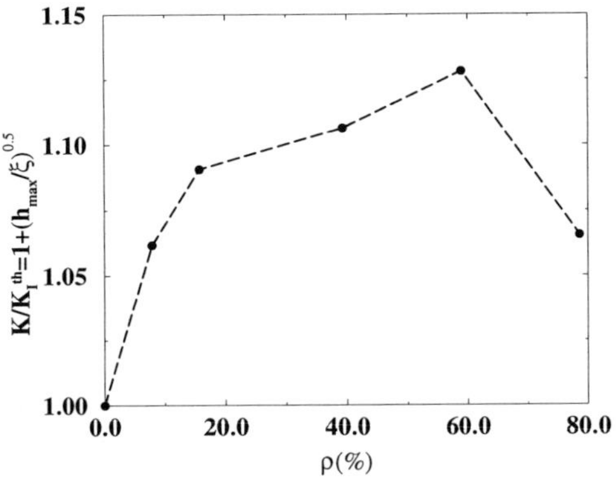

Figure 3: **Evolution of the fracture toughness K of the material with the site density ρ. The maximum fracture toughness is obtained for $\rho \sim 58\%$.**

results of the model of Hansen *et al* [8], and with experimental observations on paper sheets [9] or on wood [10]. The value of ζ is independant of the density and size of precipitates (see Fig. 2).

On the contrary, the relevant length scales h_{max} and ξ and their ratio h_{max}/ξ vary with the microstructural parameters. An optimal fracture toughness should be obtained for a surface fraction of secondary phases of 58%. The very existence of an optimum is due to shielding: by screening the stress field due to the presence of the macrocrack, damage around the tip decreases the effective probability of opening/growing microcracks, which cannot exceed a certain value.

On the other hand, the evolution of the spikiness reflects the well-known result that fracture toughness increases with the inclusions size.

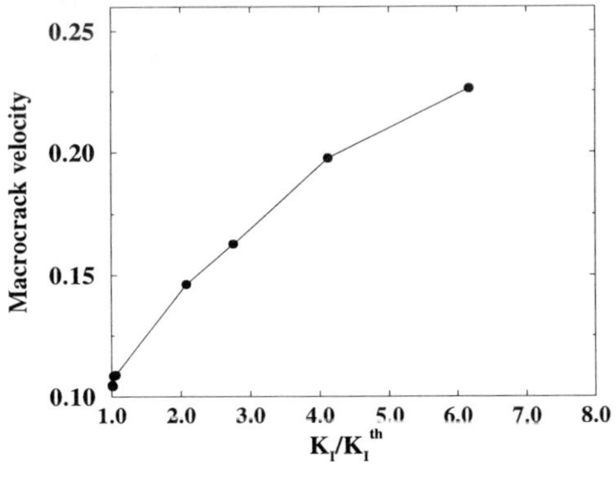

Figure 4: **Evolution of the macrocrak speed with the external applied stress**

CONCLUSION

A continuous numerical model of pure mode I macrocrack propagation into a two phases bidimensional material is presented. The morphological parameters of self-affine cracks produced for several mechanical and microstructural conditions are analysed. An optimum of the fracture toughness with the surface fraction of secondary phases is predicted by this model.

ACKNOWLEDGMENTS The authors are indebted to J.M. Dorvaux for his computational advices, and to J.-P. Bouchaud for interesting discussions.

References

[1] B.B. Mandelbrot, Phys. Scr. **32**, 257 (1985)

[2] E. Bouchaud, G. Lapasset and J. Planès, Europhys. Lett. **13**, 73 (1990)

[3] P. Daguier, S. Hénaux, E. Bouchaud and F. Creuzet, Phys. Rev. E **53**, 5637 (1995)

[4] J. Schmittbuhl, S. Gentier and S. Roux, Geophys. Lett. **20**, 639 (1993)

[5] K.J. Maloy, A. Hansen, E.L. Hinrichsen and S. Roux, Phys. Rev. Lett. **68**, 213 (1992)

[6] E. Guilloteau, H. Charrue, F. Creuzet, Europhys. Lett. **34**, 549 (1996)

[7] A. Nakano, R. Kalia and P. Vashishta, Phys. Rev. Lett., **73**, 2336 (1994)

[8] A. Hansen, E.L. Hinrichsen and S. Roux, Phys. Rev. Lett., **66**, 2476 (1991)

[9] J. Kertesz, V. Horvath and F. Weber, Fractals, **1**, 67 (1993)

[10] T. Engoy, K.J. Maloy, A. Hansen and S. Roux, Phys. Rev. Lett., **73**, 834 (1994)

[11] B.B. Mandelbrot, D.E. Passoja and A.J. Paullay, Nature (London) **308**, 721 (1984)

[12] J.J. Mecholsky, D.E. Passoja and K.S. Feinberg-Ringel, J. Am. Ceram. Soc. **72**, 60 (1989)

[13] E. Bouchaud, J.-P. Bouchaud, Phys. Rev. B **50**, 17752 (1994)

[14] L.R.F Rose, Int. Journal of fracture **31**, 233 (1986)

[15] B. Cotterell and J.R. Rice, Int. journal of fracture **16**, 155 (1980)

[16] J.W. Hutchinson, Acta Metall. **6**, 1605 (1987)

[17] J. Schmittbuhl, J.P. Vilotte and S. Roux, Phys. Rev. E **51**, 31 (1995)

PATTERN EVOLUTION OF GRANULAR MEDIA ROTATED IN A DRUM MIXER

K. M. Hill, J. Kakalios
School of Physics and Astronomy, The University of Minnesota, Minneapolis, MN 55455 USA
A. Caprihan
The Lovelace Institute, 2425 Ridgecrest Drive Southeast, Albuquerque, NM 87108

ABSTRACT

A homogeneous mixture of different types of granular material will often segregate when rotated in a drum mixer. In the traditional axial segregation effect, the mixture of two different sizes of granular media will appear from the surface to separate into bands of relatively pure single concentrations along the axis of rotation. The initial pattern is not stable, but evolves in time with continued rotation through metastable states of fewer and fewer bands. We describe two experimental methods for measuring this evolution which provide a more complete picture of the dynamics involved in the pattern progression. The use of a CCD camera in conjunction with digital analysis techniques provides a consistent and precise measure of the state of the surface as a function of time. Nuclear Magnetic Resonance Imaging (MRI) techniques are used to noninvasively measure the segregation beneath the surface. These methods indicate that the underlying mechanisms of the pattern evolution originate in the bulk of the material beneath the avalanching surface.

INTRODUCTION

Mixtures of granular materials such as powders and sand tend to segregate by particle property, which is a significant problem for many areas of particle processing, such as pharmaceutical and agricultural industries. (See, for example, Ref. 1-2). In pharmaceutical industries, for example, different powders must be well-mixed within a length scale of a single pill diameter. Imperfect mixing remains a deterrent to consistent efficient production, as granular materials can segregate in any one of the various handling stages, even in situations where one is trying to mix different components. When a mixture of two different sizes of granular materials is rotated in a horizontal cylinder, the mixture will first radially segregate, with the smaller component becoming more highly concentrated near the center of rotation. Further segregation into what appears from the surface to be bands of relatively pure single concentration along the rotational axis then occurs, as illustrated in fig. 1a and 1b [3-11]. Figure 1a shows a 50/50 mixture of 0.75 mm (dark) and 3.0 mm (clear) glass spheres in a 2 ft. long cylinder. After 5 minutes of rotation at 15 rpm, segregation of the large and small beads into sharp bands is observed (fig. 1b). When a mixture which has segregated into bands continues to rotate for an extended period of time, the original pattern of bands is not stable but evolves as bands move and merge as shown in fig. 1c and 1d [8,10,11]. This may have strong implications for the understanding of axial segregation, as theories which appear to successfully model the axial segregation effect using surface dynamics do not include consideration of this band evolution. (See, for example ref. 11 and references therein.)

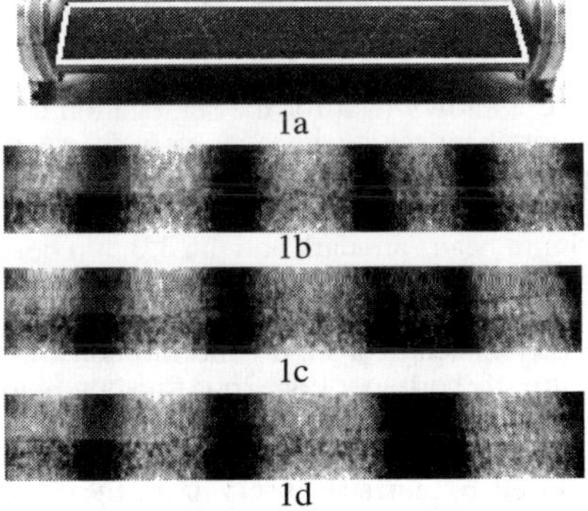

Fig.1 Banding patterns observed in granular media. (Figs. 1b,c,d are images of the top surface of the mixture as indicated by the white box in fig.1a.) A mixture of spheres only differing in size (1a) segregates into bands along the axis of rotation (1b). When rotated for an extended period of time bands merge two at a time (1c and 1d).

Mat. Res. Soc. Symp. Proc. Vol. 463 © 1997 Materials Research Society

The band merging events appear to involve subsurface interactions. When two bands merge, they do not, in general, simply move toward each other until they meet. Rather, in the case of the merging of two small bead bands, for example, it appears from the surface that the region between the small bead bands becomes increasingly more concentrated with small beads, while the small bead concentration within the bands themselves appear to decrease slightly, until the concentration of all three regions is roughly uniform. Finally, a new single band of small beads "sharpens". That is, the band itself narrows while the region within it becomes more highly concentrated with small beads. The pattern is then stable again for a period of time with only minor axial movement of the bands until the next relatively sudden merging event occurs.

Studies of the surface segregation using a CCD camera and digital image analysis techniques provide a quantitative evaluation of the surface segregation. Results of these studies presented in the next section indicate that there is indeed some local remixing that is associated with band merging events and suggest that interactions beneath the surface are at least in part responsible for the band evolution. The use of Nuclear Magnetic Resonance Imaging (MRI) provides a method for non-invasively examining concentrations and other details beneath the surface of granular materials [12-15].

DIGITAL SURFACE ANALYSIS

The experimental apparatus for the surface segregation studies is similar in part to that in Ref. [11]. The drum mixer used was a two-foot long Plexiglas cylinder, five inches in diameter, rotated about its axis with a 1/17 hp motor (providing rotational speeds of 0 - 15 rpm). For the results reported here, the cylinder was one-third filled with the same mixture shown in the images in Figure 1: a 50/50 mixture of .75 mm (dark) and 3.0 mm (clear, appearing light) glass spheres. The results from this mixture are typical of results from other bead mixtures of similar sizes but different bead diameter ratios. While the mixture was rotated in the cylinder, digital images were taken of the surface of the beads as in figs. 1b-d at regular time intervals with a Cohu 4915 CCD camera and Scion LG-3 frame grabber operated with the public domain NIH Image software on a Power Macintosh. The regions near the endplates were not included in this analysis as, unlike the majority of the axial bands farther from the endplates, the development of the bands near the ends of the cylinder appear to depend largely on endplate-bead interactions and evolve differently from the rest of the pattern [11].

Figure 2 shows data for the segregation of the mixture described above where segregation was monitored by taking images every 30 seconds for the first 20 minutes of rotation at 15 rpm. Figure 2a shows the average intensity as a function of distance along the cylinder axis for the mixed state (similar to the area outlined in figure 1a) after only 30 seconds of rotation. Since the 0.75 mm beads are dark and the 3.0 mm beads are clear, the intensity in the images of figure 1 at any given point roughly corresponded to the concentration of large beads near the surface: thus a high intensity in the graphs of figure 2a-d corresponds to high concentration of large beads; a low intensity indicates a high concentration of small beads, and an intermediate intensity reflects a mixture of the two. Fig. 2b corresponds to the image in Figure 1b when the mixture had been rotated for 5 minutes and the axial bands had formed. Fig. 2c and 2d correspond to the images in 1c and 1d respectively, and show some of the intensity data during the merging event. As suggested by surface observation, these traces indicate that the band merging event does not simply involve two bands moving together. Instead, all the bands involved in the merging become less distinct until the final band sharpens.

The mean and the standard deviation of the mean for all the images in this experiment are shown in figure 2e. While the mean value (dashed line) is roughly constant in time, indicating no loss of beads of either type and no loss of information beneath the surface, the standard deviation is closely related to the state of segregation. When the mixture was still fairly uniform in distribution as indicated by "a" in fig. 2e, the pixel value at every point in the surface is fairly close to the mean value, and the standard deviation of the mean is low. If the bands are sharply defined, the pixel values in images taken of the surface should be far from the mean value. Thus the sharp rise in the standard deviation in the graph corresponds to the initial formation of the bands, until the bands finally sharpen, indicated by "b" in fig. 2e. The dip in the standard deviation as the bands are merging support the subjective surface observation that there is local remixing during a band merging event. When the band number is finally reduced, there are fewer band-band interfaces and thus fewer regions closer to the mean value of the traces, which is presumably why the next semi-stable pattern (indicated by "d" in fig. 2e) has a higher standard deviation than the previous one.

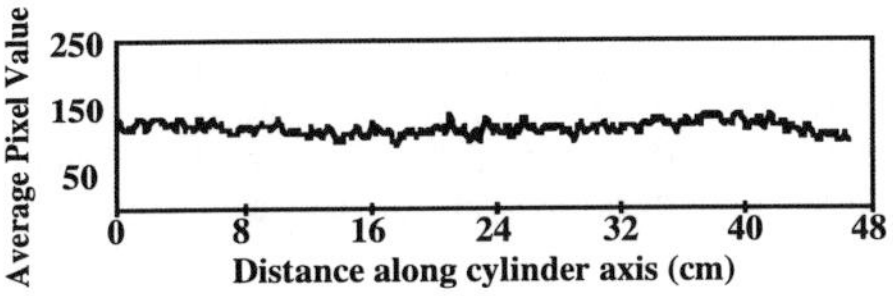

2a

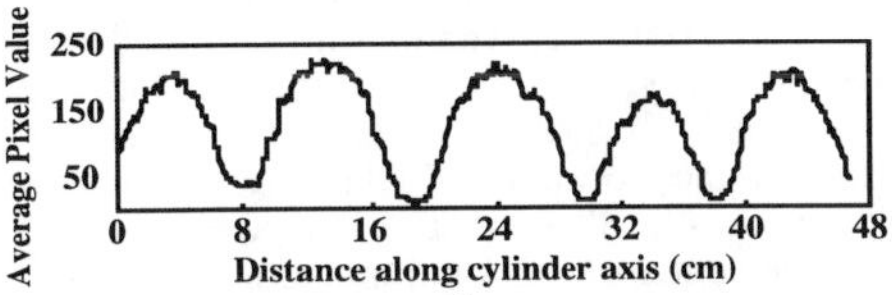

2b

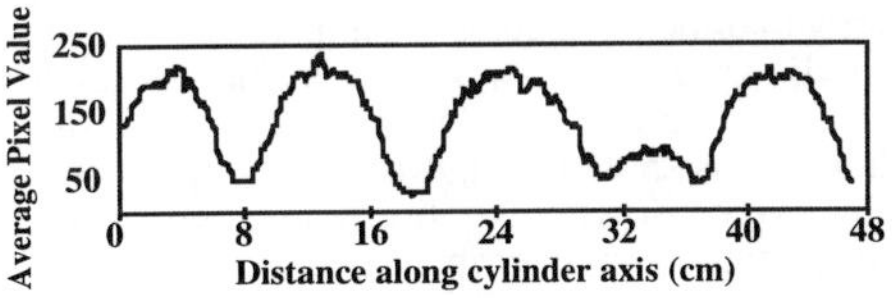

2c

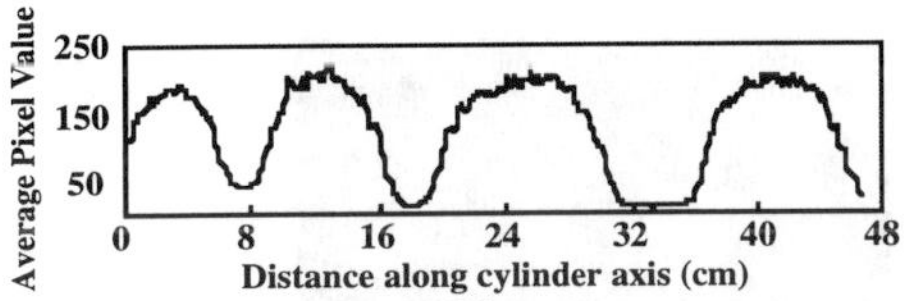

2d

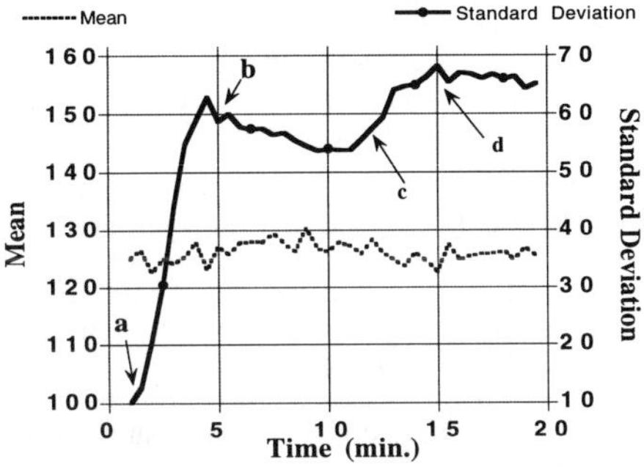

2e

Fig. 2: Results from digital image analysis for the mixture shown in figure 1. Figures 2a-d shows the average intensity as a function of distance along cylinder axis for images taken of mixture surface run shown in fig. 1a-d. Figure 2e is a graph of the mean value and the standard deviation of the mean for similar traces taken from images during the entire experiment The approximate locations on the graph for images and graphs 1a-d and 2a-d respectively are indicated.

NUCLEAR MAGNETIC RESONANCE IMAGING EXPERIMENTS

In order to investigate the local remixing during band merging events Nuclear Magnetic Resonance Imaging (MRI) was used. MRI allows for non-invasive observation of the evolution of the pattern within the bulk as it moves from one stage to the next. Nakagawa and coworkers first demonstrated that liquid state MRI techniques may be used to image arbitrary planes within the bulk of granular media for material containing some sort of fluid [12]. They have used these techniques to investigate the dynamics of granular flow in a partially filled rotating drum as well as the migration of the radially segregated core of a mixture with continued rotation.

In a recent report [15] we described the use of MRI to obtain a full three-dimensional picture of axial segregation by imaging planes both perpendicular and parallel to the cylinder axis at successive stages of the development of the bands. The experimental setup used for the band development studies is similar to that above. We use a mixture of one MRI sensitive component and one that is not MRI active, so that a concentration image will directly show the distribution of the MRI active species. The MRI-sensitive beads are 1 mm diameter pharmaceutical spherical pills while the inert beads are 3 mm plastic spheres. The spheres were rotated in a 24 cm long, 7.5 cm diameter acrylic cylinder. Fluid-filled spheres were taped around the outside of the cylinder at 5 cm intervals to provide reference points in the images. An NMR imager/spectrometer (Nalorac Cryogenics Corp.) was used at the Lovelace Institutes with a 1.9 Tesla Oxford superconducting magnet with a bore diameter of 31 cm. The magnet bore is horizontal, so the cylinder may be placed into the magnet without disturbing its contents. Digital image analysis was performed on a Power Macintosh 7100/80 computer using the public domain NIH Image program. The partially-filled cylinder was first rotated outside the magnet and any segregation was noted. Rotation was then stopped and the cylinder placed within the bore of the magnet where the material was imaged using standard stationary NMR imaging techniques. This procedure was repeated for extended rotation periods. In this way, the evolution of the radial and axial segregation was serially monitored. The MRI experiment images a 5 mm thick slice either perpendicular to the axis at any point along the length of the cylinder as illustrated in fig. 3a or parallel to and containing the axis of the cylinder as illustrated in fig. 3c. The smaller MRI sensitive spheres appear as bright regions in the images and dark regions correspond to high concentrations of the larger non-MRI sensitive spheres.

After the mixture had been rotated for approximately 5.5 minutes at 30 rpm in a 7.5 cm diameter 24 cm long cylinder, the mixture appeared both radially and axially segregated on the outside as indicated in the sketch in figure 3a. Images taken beneath the surface at this point indicate an interdependence between radial and axial segregation. Figure 3a illustrates the state of the mixture as observed from the outside, as well as the approximate location of the images taken perpendicular to the cylinder axis shown in fig. 3b. The images in 3b serve two purposes. They indicate that the high concentration of small spheres near the center of the rotational axis, the radial segregation, exists both within the bands of the smaller spheres and the larger spheres. These images also assure that the pattern is roughly cylindrically symmetric around the axis of rotation at any given location, and thus the images containing the cylinder axis (as in 3d) are sufficient for a first order representation of the state of the mixture below the surface. Figure 3c shows the location of the image taken down the length of the cylinder axis shown in figure 3d. In figure 3d, it is verified that the "radial mode" of small beads appears to run down the entire length, widening to the surface in the regions where axial bands of small spheres are seen from above.

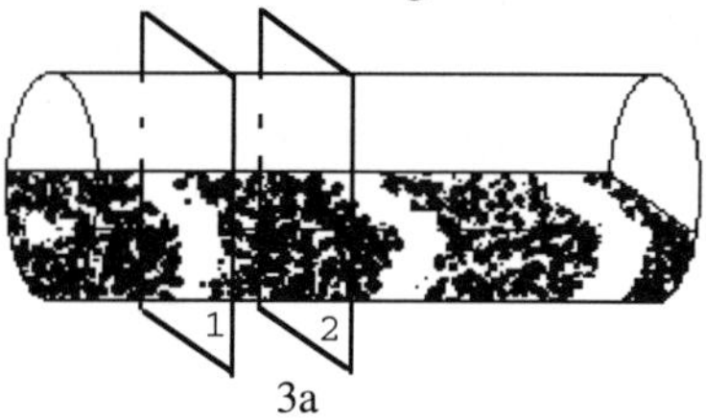

3a

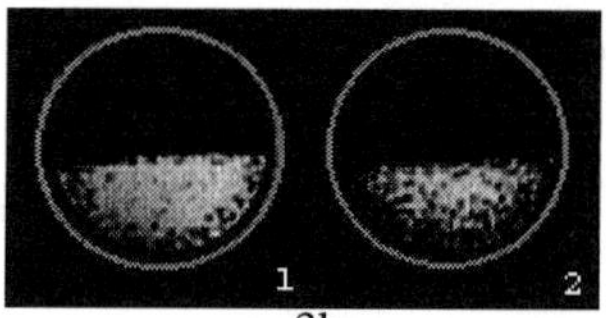

3b

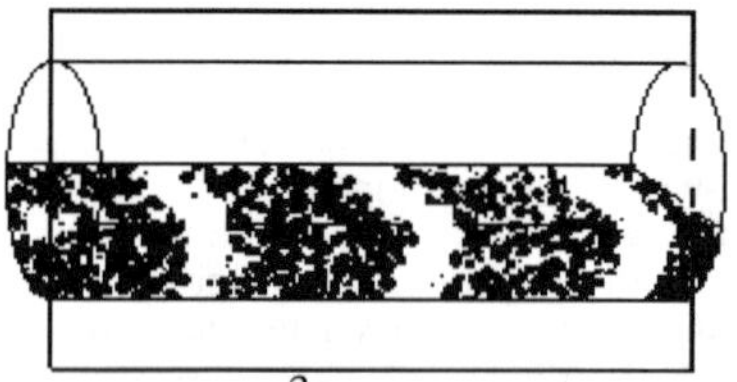

3c

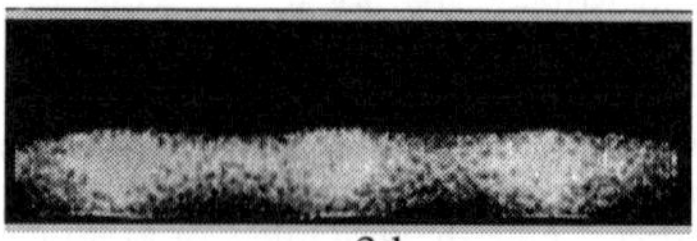

3d

Fig. 3 MRI data for a mixture of pharmaceutical spheres (1 mm) and plastic spheres (3 mm) rotated in a 24 cm long cylinder for 5.5 min. at 30 rpm. From outside the cylinder, the mixture appears to have segregated radially and axially. Fig. 3a illustrates the state of the system as viewed from outside after rotation and the approximate locations of the planes from where the images shown in 3b were taken: within small sphere band (1), and within the large sphere band (2). Fig. 3c indicates the location of the image taken parallel to the axis shown in figure 3d. The brighter regions represent higher concentrations of small spheres.

Figure 4 illustrates the pattern development of this granular material system. Figure 4a-4c are the subsurface images taken using MRI as two of the bands merged. Figure 4d-4f are sketches illustrating the state of the surface extrapolated from MRI data taken near the surface. These images are all vertical and parallel to the cylinder axis in the position indicated in figure 3c. Figure 4a is the subsurface image taken after approximately 15.5 minutes of rotation after the mixture had radially segregated and axially segregated into three bands of small beads and four bands of large beads (as shown in figure 4d). After approximately eight minutes of further rotation, the bands had begun to merge, mostly beneath the surface as illustrated in figures 4b and 4e. The radial mode of small beads increased significantly in the region between the small beads which were merging, and that region became more highly concentrated with small beads. The bands as seen on the surface had not moved much closer together. Finally, after ten more minutes of rotation the bands had completely merged as illustrated by figs. 4c and 4f.

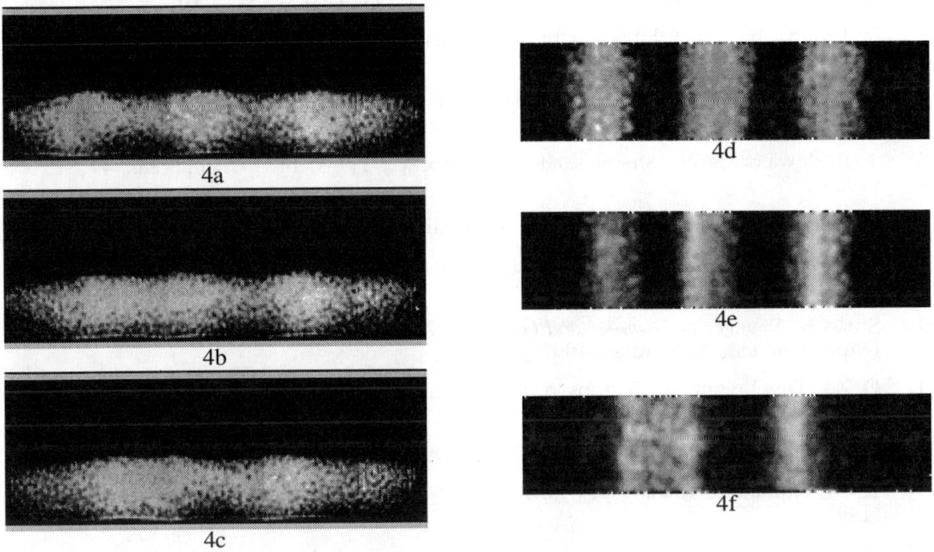

Fig. 4 Subsurface MRI data (4a-4c) and corresponding surface illustrations (4d-4f) for a mixture of pharmaceutical spheres (1 mm) and plastic spheres (3 mm) rotated for an extended period of time in a 24 cm cylinder at 30 rpm. The images were taken while a band merging event was occurring. Figs. 4a and 4d correspond to the state of the mixture after 15.5 minutes of rotation. Figs. 4b and 4e correspond to the system after 23 minutes of rotation and 4c and 4f to the state of the system after 33 minutes of rotation. Again, brighter regions represent higher concentrations of 1 mm spheres

CONCLUSIONS

The evolution of the axial banding pattern in a system of granular material rotated in a drum has been studied. Analysis of digital images taken of the surface segregation in conjunction with sub-surface measurements using MRI techniques provide important information about the details of the evolution of the surface pattern. The data from the surface segregation shows that as a band merging event occurs there is some local mixing, indicating the existence of interactions beneath the surface which are driving the pattern evolution. MRI data support this conjecture as these images reveal a relationship between the radial and axial segregation and movement in the radial mode when the axial bands are merging.

ACKNOWLEDGMENTS

This research was supported, in part, by the Minnesota Graduate School Grant-in-Aid program, NSF-CTS-9501437, the U.S. Department of Energy, Pittsburgh Energy Technology Center via grant #DE-FG22-94PC94248, and the U.S. Department of Energy, Office of Basic Energy Sciences via grant #DE-FG03-93ER14316. We gratefully acknowledge the help and hospitality of Steve Altobelli, Eiichi Fukushima, Masami Nakagawa, Allen Waggoner, and the Lovelace Institutes, where the MRI data were taken.

REFERENCES

[1] M. H. Cooke, D. J. Stephens, and J. Bridgwater, Powder Technol. **15**, 1 (1976).

[2] L. T. Fan, Y.-M. Chen, and F. S. Lai, Powder Technol. **61**, 255 (1990).

[3] Y. Oyama, Bull. Inst. Phys. Chem. Res. (Tokyo), Rep. **5**, 600 (1939), Y. Oyama and K. Ayaki, Kagaku Kikai **20**, 6 (1956).

[4] M. B. Donald and B. Roseman, British Chem. Eng. **7**, 749 (1962); Brit. Chem. Eng. **7**, 823 (1962).

[5] J. Bridgwater, N.W. Sharpe, and D. C. Stocker, Trans. Inst. Chem. Eng. **47**, T114 (1969).

[6] S. Das Gupta, D. V. Khakhar, and S. K. Bhatia, Chem. Eng. Sci. **46**, 1513 (1991).

[7] S. Fauve, C. Laroche, and S. Douady, in *Physics of Granular Media*, edited by Daniel Bideau and John Dodds (Nova Science, Commack, NY, 1991), p.277.

[8] Stuart B. Savage, in *Disorder and Granular Media*, edited by D. Bideau and A. Hansen (North-Holland, Amsterdam, 1993), p.255.

[9] O. Zik, Dov Levine, S. G. Lipson, S. Shtrikman, and J. Stavans, Phys. Rev. Lett. **73**, 644 (1994).

[10] M. Nakagawa, Chem. Eng. Sci, **49**, 2540 (1994).

[11] K. M. Hill and J. Kakalios, Phys. Rev. E **49**, R3610 (1994); Phys. Rev. E **52**, 4393 (1995).

[12] M. Nakagawa, S.A. Altobelli, A. Caprihan, E. Fukushima, and E.-K. Jeong, Experiments in Fluids **16**, 54 (1993); S.A. Altobelli, A. Caprihan, E. Fukushima, M. Nakagawa, R.C. Givler, Multiphase Flow Studies by NMR with Application to Granular Flows in Joint DOE/NSF Workshop on Flow of Particulates and Fluids, 34-40 (1991).

[13] E.E. Ehrichs, H.M. Jaeger, G.S. Karczmar, J.B. Knight, V.Y. Kuperman, and S.R. Nagel, Science **267**, 1632 (1995).

[14] G. Metcalfe and M. Shattuck, Radial and Axial Transport of Solids in a Slowly Rotating Tube in 12th Australasian Fluid Mechanics Conference, (1995).

[15] K. M. Hill, A. Caprihan, and J. Kakalios, submitted to Physical Review Letters; K.M. Hill, A. Caprihan, J. Kakalios, Magnetic Resonance Imaging of granular media segregated in a rotating drum in Fifth World Congress of the American Institute of Chemical Engineers, (1996).

TEMPERATURE AND CHEMISTRY EFFECTS IN POROUS-MEDIA ELECTROKINETICS

DAVID B. PENGRA* AND PO-ZEN WONG
Department of Physics and Astronomy, University of Massachusetts,
Amherst, MA 01003, U.S.A.

ABSTRACT

Electrokinetic phenomena in brine-saturated porous media, such as electroosmosis (fluid-flow induced by applied electric fields) and streaming current (the complementary process) depend on the density of ions adsorbed on the pore surface and the characteristic thickness of the diffuse space-charge layer λ. These, in turn, depend on brine chemistry, ambient temperature and possibly other parameters. We report on a series of measurements of natural rock and synthetic glass-bead samples: for one sample group, we varied the temperature over 0–50 ° C; for another, we changed the brine cation species. We find that the electrokinetic coefficients depend only weakly on temperature; this is shown to follow from the expected trends in λ, η, and σ. The chemistry dependence follows qualitatively but not quantitatively the predictions of the Debye-Hückel approximation.

INTRODUCTION

In brine-saturated porous material, such as porous rock, the flow of fluid current is accompanied by a weak electric current situated near the pore surface. This *streaming current* arises from the *electrochemical double-layer* that forms whenever a fluid electrolyte comes into contact with a solid [1]. The layer may form from two types of processes: either preferential solvation of certain ions in the solid, or preferential adsorption of ions from the solution. In either case, a tightly-bound charged layer develops at the solid surface while ions of the opposite charge form a diffuse mobile space-charge layer in the fluid. In the model due to Stern, [1, see Kortüm] tightly-bound ions at the solid surface are separated from the closest ions in the fluid by a minimum distance d given by twice the mean radius of the two (possibly hydrated) charge species. Beyond d, the charge density of the diffuse layer falls off with a characteristic length λ. Within the Debye-Hückel approximation, which treats the charges as point particles,

$$\lambda = \left(\frac{\varepsilon k_B T}{\sum_i n_i q_i^2} \right)^{1/2}, \tag{1}$$

where ε is the dielectric constant of the brine, k_B is Boltzmann's constant, T is the absolute temperature, and n_i and q_i are the number-density and charge of the ith ion in the brine. For 0.1 M NaCl solution, $\lambda \approx 10$ Å. The electrostatic potential at the hydrodyanmic slp-plane is known as the ζ-potential. In the limit of $q\zeta \ll k_B T$, the charge density in the fluid and the static potential fall of exponentially. It can be shown that ζ is related to the areal charge density Σ_S by $\varepsilon\zeta = \Sigma_S \lambda$. The flow of fluid in the pores drags along the charge in the diffuse layer. This gives rise to a streaming current proportional to $\Sigma_S \lambda$ and inversely proportional to the fluid viscosity η [1]. Conversely, upon the application of an electric field, the charge flow in the diffuse layer drags the fluid to flow. This results in an *electroosmotic* current. The phenomena of these coupled currents is known as *electrokinetics*, and dates to the experiments of Saxén [2]. These effects were not well studied historically because they are very weak and difficult to measure. With modern techniques and

Mat. Res. Soc. Symp. Proc. Vol. 463 © 1997 Materials Research Society

instrumentation, we have shown recently that high precision measurements can be made and they are extremely useful probes of porous-media properties [3, 4, 5].

In the techniques we developed, we do not measure electrokinetic currents directly, but measure steady-state potentials that arise from them: the *streaming potential* and *electroosmotic pressure*. We place a brine saturated sample between two closed reservoirs of brine, suitably equipped with electrodes and attendant instrumentation [4, 5]. We then measure the streaming potential coefficient K_S, defined by

$$K_S \equiv -\Delta V_s / \Delta P_a \ , \tag{2}$$

where ΔP_a is an applied pressure differential across the sample, and ΔV_s is the voltage generated by the displacement of charge coming from the streaming current. We also measure the electroosmotic pressure coefficient K_E, as defined by

$$K_E \equiv -\Delta P_e / \Delta V_a \ , \tag{3}$$

where ΔV_a is an applied voltage across the sample, and ΔP_a is the back-pressure resulting from the displacement of fluid by electroosmosis. We have shown previously that the hydraulic permeability k of a brine-saturated sample can be obtained by determining K_S, K_E, and the electric conductivity σ_r of the sample [3]. This result is independent of the details of the double-layer. It is a general consequence of Onsager's relations in nonequilibrium thermodynamics [6], and may be shown to follow from conditions even less restrictive than used in Onsager's original proof [7]. We have also proposed that K_S and K_E can be used to define and *effective pore radius* and *effective ζ-potential* for porous media [3].

While nonequilibrium thermodynamics is useful for relating K_S and K_E to other bulk properties of the porous media, their actual values are determined by the microscopic details of the brine/pore interface, such as the local surface geometry, chemistry of the ions, and other fluid properties. A starting point is to consider the predictions of the simple Debye-Hückle theory for a cylindrical pore with radius $R \gg \lambda$ and compare them with experiments. The most important predictions of the theory are

$$K_S = \varepsilon\zeta / \eta\sigma_w \approx \Sigma_S \lambda / \eta\sigma_r \ , \tag{4}$$

and

$$K_E = 8\varepsilon\zeta / R^2 \approx \Sigma_S \lambda / k \ , \tag{5}$$

where we have identified the permeability k with $R^2/8$ and replaced the brine conductivity σ_w with the brine-saturated media's conductivity σ_r in the approximation. The approximate equalities are valid in the $q\zeta \ll k_B T$ limit. With the exception of k, each of the parameters in these equations is either temperature dependent, chemistry dependent, or both. In this paper, we report measurements of K_S and K_E for a selection of samples to investigate the effects of changing the temperature and the cation species. As we show below, the simple theory captures the physics of the system to a surprising degree despite the enormous complexities of real porous media such as rocks.

TEMPERATURE EFFECTS

For the temperature dependence study, we built a temperature controlled insulated box for our experimental cell (see [4, 5] for more details). Within the box, four platinum resistance thermometers continuously monitored the temperature uniformity and stability, which was within 0.01°. We selected four samples from among a suite of twelve samples we have studied previously [4, 5]: a Fontainebleau sandstone (k = 2500 mD), a Berea sandstone (k = 680 mD), an Indiana limestone (k = 5 mD), and a 100 μm fused-glass-bead sample (k = 620 mD). They were saturated with 0.1M

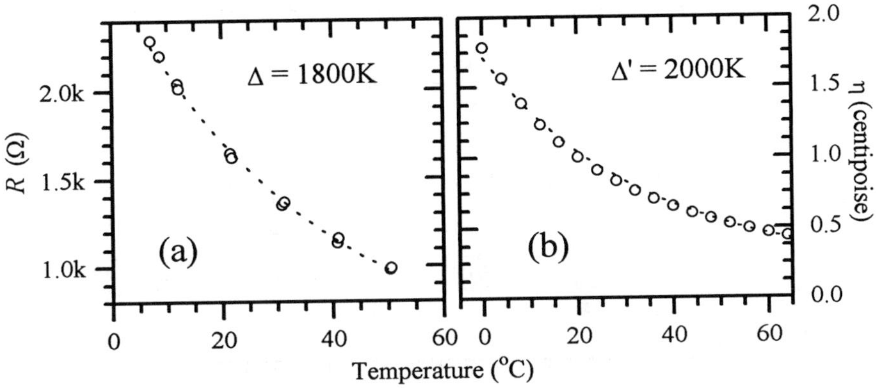

Figure 1: (a) Berea sandstone electrical resistance R and (b) water viscosity η as a function of temperature. The fit in each is to the thermally activated form $Ae^{\Delta/T}$. Viscosity values are taken from the CRC handbook [8].

NaCl brine. The sample resistance R, and the electrokinetic coefficients K_S and K_E were measured at roughly 10° intervals from 0–50°C. Fig. 1 shows the temperature dependence of the resistance R of the Berea sandstone along with the data for water viscosity taken from Ref. [8]. Figure 2 shows the temperature dependence for K_S (a) and K_E (b). These data are representative of the behavior found for all four samples: R decreases with increasing T, while K_S and K_E are nearly constant.

The behavior of R is readily understood because the conductivity σ_r of a brine-saturated rock comes mainly from the conductivity of the of the brine σ_w which is well known to be thermally activated: $\sigma_w \propto e^{-\Delta/T}$. This is due to the fact that the ions move by exchanging positions with water molecules, and that involves overcoming an energy barrier $k_B\Delta$. Thus we expect

$$R = R_0 e^{\Delta/T} \; . \tag{6}$$

Fitting the data in Fig. 1(a) and those for other samples to Eq. (6) yields $\Delta \approx 1800$ K, which corresponds to an energy barrier of about 0.15 eV. We note that fitting the viscosity data in Fig. 1(b) to the same activated form ($\eta = \eta_0 e^{\Delta'/T}$) yields a similar value for Δ' (≈ 2000 K). This shows that the movement of water molecules also takes place by the exchange mechanism with a similar energy barrier.

To understand why K_E is insensitive to temperature change, note that in Eq. (5), only the numerator should depend on temperature, as k depends only on the pore geometry. The surface charge density Σ_S should depend very weakly on temperature, as the binding energies of the tightly-bound layer are typically much higher than k_BT. Where temperature might be important is in the screening length λ. However, in the numerator of Eq. (1), the dielectric constant of water ε is known to roughly follow a $1/T$ Curie-law behavior near room temperature, so εk_BT is nearly temperature independent. Using the empirical data of $\varepsilon(T)$ for pure water from Ref. [8] would cause a change by no more than two percent over the temperature range studied. The denominator in Eq. (1) gives the density of unpaired ions in the solution. For low concentrations of soluble salts, such as the 0.1M NaCl used here, this number is a constant. Thus λ is essentially constant and, according to Eq. (5), the same is true for K_E. That is indeed what we observed.

For K_S, we note that Eq. (4) has the same numerator as Eq. (5), so we only need to consider the behavior of the denominator $\eta\sigma_r$. It is easy to see that this term is also nearly temperature independent. As noted above, charged ions and water molecules move by the same activation

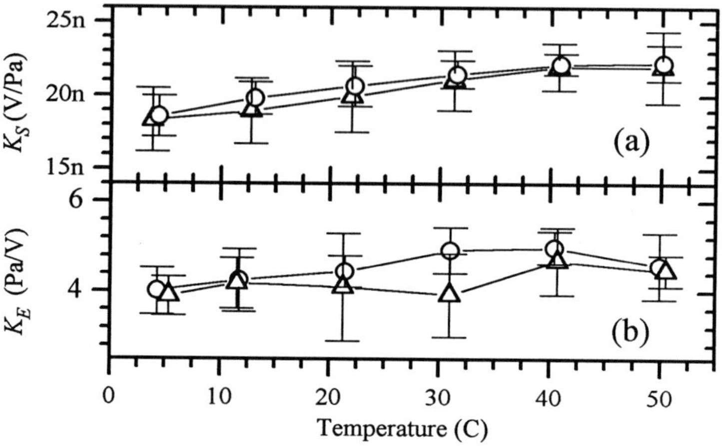

Figure 2: Variation of K_S (a) and K_E (b) as a function of temperature. The double data points represent separate measurements, taken in order of increasing ($\triangle$) and then decreasing ($\bigcirc$) temperature.

mechanism with very similar energy barriers. From the discussion following Eq. (6), we expect $\eta\sigma_r \propto e^{(\Delta'-\Delta)/T} \approx e^{200\mathrm{K}/T}$. As T changes from 273 K to 323 K, this factor decreases from 2.08 to 1.86. This would cause K_S to increase by about 10%, consistent with the data in Fig. 2. Instead of using these figures for Δ and Δ' obtained from empirical fits, a simple model can be used to explain why $\eta\sigma_r$ is temperature independent. If we think of the ion species i as (possibly hydrated) spheres of radius r_i and charge q_i, under the influence of applied field E, the ions will reach a steady-state speed v_i when the viscous force F_V from Stoke's law balances the electric force F_E:

$$F_E + F_V = q_i E - 6\pi r_i \eta v_i = 0 \ . \qquad (7)$$

Upon summing the contributions from the different ion species, the conductivity of the brine σ_w is given by

$$\sigma_w = \frac{1}{E}\sum_i n_i q_i v_i = \sum_i \frac{n_i q_i^2}{6\pi\eta r_i} \ . \qquad (8)$$

Hence the product $\eta\sigma_w$ depends on the parameters n_i, q_i and r_i, each of which is a constant. This is an example of Walden's rule [9].

While the above arguments explain why K_S and K_E are insensitive to changes in temperature, we do note that the two "dirtier" samples we studied—Indiana limestone and Berea sandstone—showed slightly larger increase in K_S and K_E than the two "cleaner" samples—glass beads and Fontainebleau sandstone. One possible explanation is that Σ_S may actually increase slightly with temperature. The reason is that, for rocks, the mobile ions in the diffuse layer comes from the hydration of a fraction of the counterions originally bound to the pore surface. With increased temperature more of these counterions will go into the hydrated state, leaving a larger Σ_S of the opposite sign on the solid surface. Dirtier samples have more surface counterions, so this desorption effect is likely to be stronger. The magnitude of this effect will clearly depend on the adsorption energies of different ions on different surfaces, and how they compare with the hydration energy. For the same sample and the same ions, the temperature dependence of K_S is slightly larger than K_E because of the $\eta\sigma_r$ factor as explained above.

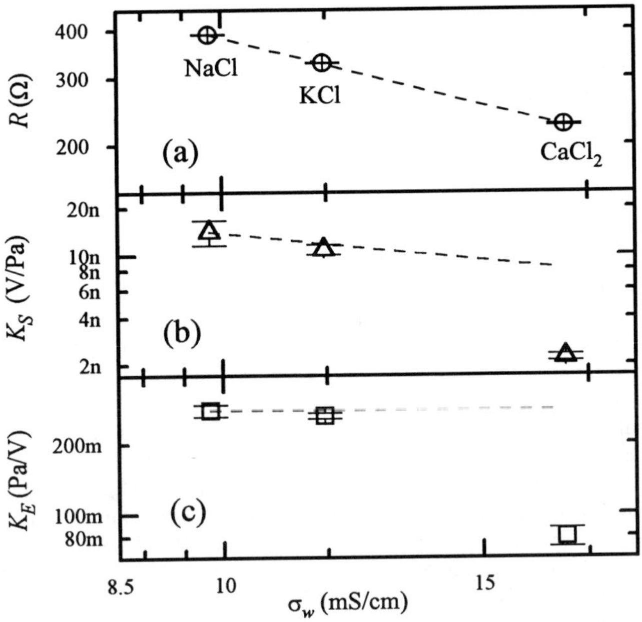

Figure 3: Dependence of resistance R (a) and coefficients K_S (b) and K_E (c) for a 200 μm sample on cation species versus conductivity of three 0.1M brines: NaCl, KCl, and $CaCl_2$. The sample is 200 μm glass bead. The lines show the expectations from only a change in σ_w.

BRINE CHEMISTRY EFFECTS

In previous work, we have studied the effects of changing the conductivity of the brine using different concentrations of NaCl [4, 5]. For the twelve samples studied, we found that $K_S \propto \sigma_w^{-\alpha}$ and $K_E \propto \sigma_w^{-\beta}$, where α and β were sample dependent, but $\alpha > \beta$ always. The Debye-Hückel theory predicts $\alpha = 3/2$ and $\beta = 1/2$. Deviations from this prediction are substantial in some samples and the cause of this is not undertsood. It may have to do with microscopic structural changes in the double layer when the brine concentration is varied.

To study the effects of changing the cation species, we selected a different group of four samples from our original twelve: a Berea sandstone ($k = 34$ mD), a Whitestone calcite ($k = 8$ mD), a 50 μm fused-glass-bead sample ($k = 68$ mD) and a 200 μm fused-glass-bead sample ($k = 5000$ mD). Three different types of brine were used: 0.1M solutions of NaCl, KCl, and $CaCl_2$. In turn, each sample was saturated, allowed to equilibrate to the new brine, and then measured at room temperature (23 ± 2°C).

Figure 3 shows the effects of changing the cation type on R, K_E and K_S for the 200 μm glass bead sample. We have plotted the results against brine conductivity σ_w, because, even though the molar concentrations are the same, the brine conductivities are not. This is due to two effects. First, the cation mobilities vary from species to species because the effective size of the ions vary; differences in hydration energies attract different numbers of loosely bound water molecules, thus varying r_i. Second, calcium is doubly-valent in solution, so it carries a larger charge and there are more chlorine ions in the brine. By plotting the results against σ_w, we can see if the change in cation has any other affect than that due to the change in conductivity alone.

In Fig. 3(a), the dotted line has a slope equal to -1, indicating a direct proportionality between the sample resistivity and the brine conductivity. That the points lie close to this line indicate that the sample was well equilibrated. In Figs. 3(b) and (c), the dotted lines have slopes of -1 and 0 which represent the denominators in Eqs. (4) and (5). Interestingly, K_E and K_S show the same deviations and it is logical to attribute them to the common numerator $\Sigma_s\lambda$. The large drop in the response with $CaCl_2$ can be qualitatively understood as two separate effects. First, the doubly charged Ca-ions are more tightly bound to the surface, so a much smaller fraction of them would become hydrated. This reduces Σ_s substantially. Second, we note that, for the same molarity, Eq. (1) predicts that λ for $CaCl_2$ would be smaller than that for NaCl or KCl by a factor of $\sqrt{3}$. Comparing NaCl to KCl, we note that they have the same λ according to Eq. (1), so the difference can only be attributed to the areal charge density Σ_s. We note that the surface binding energy U_b and the hydration energy U_h should both be weaker for K^+ ions than for Na^+ ions due to its larger size. We expect $\Sigma_s = \Sigma_0 e^{-\Delta U/k_BT}$ where $\Delta U \equiv U_h - U_b > 0$ and Σ_0 is the total density of exchangeable ions on the surface. The fact that K_E and K_S are similar for Na^+ and K^+ suggests that ΔU for the two ions are close to each other. Among the four samples we studied, the sandstone and glass bead samples show the same trends as Fig. (3). Only the Whitestone calcite shows a different behavior: K_E and K_S for $CaCl_2$ are enhanced above the dotted lines that represent the conductivity trends. This may be due to the fact the the surface interaction (U_b) for calcium based material is very different from that of silicon based materials. Although these conclusions are tentative, it is clear that the details of surface chemistry play an important role in determining the electrokinetic effects. This, in turn, makes the measurements valuable as tools for studying solid-liquid interfaces.

ACKNOWLEDGMENTS

We have benefited from experimental assistance and discussions with X. Li and P. Johnson. This work is supported by the Gas Research Institute under Contract 5090-260-1953, and the National Science Foundation under Grant DMR-9404672.

*Present address: Department of Physics and Astronomy, Ohio Wesleyan University, Delaware, OH 43015.

REFERENCES

1. G. Kortüm, Treatise on Electrochemistry, 2nd revised English edition (Elsevier, Amsterdam, 1965).
2. U. Saxén, Ann. Physik und Chemie **47**, 46 (1892).
3. S. X. Li, D. B. Pengra, and P.-z. Wong, Phys. Rev. E **51**, 5748 (1995).
4. P.-z. Wong and D. B. Pengra, in *Access in Nanoporous Materials*, T. J. Pinnavaia and M. F. Thorpe, eds., pp. 295–317 (Plenum, New York, 1995).
5. D. B. Pengra and P.-z. Wong, in *Disordered Materials and Interfaces—Fractals, Structure and Dynamics*, Mat. Res. Soc. Symp. Proc. **407**, H. Z. Cummins, D. J. Durian, D. L. Johnson and H. E. Stanley, eds., pp. 3–14 (MRS, Pittsburg, 1996).
6. See, for example, S. R. deGroot and P. Mazur, Non-Equilibrium Thermodynamics (North-Holland, Amsterdam, 1962).
7. S. Pride, Phys. Rev. B **50**, 15678 (1994).
8. CRC Handbook of Chemistry and Physics, R. C. Weast, ed. (CRC Press, Florida).
9. J. O'M. Bockris and A. K. N. Reddy, *Modern Electrochemistry* (Plenum, New York 1973).

AC RESPONSE OF HETEROGENEOUS MATERIALS: A NUMERICAL STUDY

A. A. RODRIGUEZ, J. VALBUENA
Intevep S.A., Apartado 76343, Caracas 1070A, Venezuela.

ABSTRACT

The ac and dc electrical properties of composite materials are studied using hierarchical lattices. First we show that the hierarchical model can correctly account for the main scaling properties of critical percolative structures. Then we study the effect of potential disorder by assuming that the microscopic conductances are distributed according to a power law distribution function. We find that in the limit of strong disorder, the predictions are in qualitative agreement with reported experimental measurements.

INTRODUCTION

This paper concerns the ac and dc electrical properties of a model of heterogeneous materials such as water-saturated rock, granular metallic films, etc. The electrical properties of such materials vary continuously when the volume fraction of the conducting phase changes. It is well known that when the fraction of conducting material (p) approaches the percolation threshold (p_c), the effective dielectric constant and electric conductivity behave in the following way:
for $\omega \to 0$ (dc)

$$\sigma_e(p) \sim (p - p_c)^t, \quad \epsilon_e(p) \sim |p - p_c|^{-s}, \tag{1}$$

while in the frequency range $\omega_1 \leq \omega \leq \omega_0$:

$$\sigma_e \sim \omega^x, \quad \epsilon_e \sim \omega^{-y} \tag{2}$$

where ω_1 and ω_0 are the macroscopic and microscopic cutoff frequencies of the system which will be defined later in this paper. The above exponents have been determined numerically [1]: $t = s \approx 1.3$, $x = y = 1/2$ for $d = 2$, and $t \approx 2.0$, $s \approx 0.7$, $x \approx 0.74$ and $y = 0.26$. For a recent review on this subject see [2].

On the experimental side [3], measurements generally indicate that x is close to 1 while y is close to 0. The difference between the dispersion exponent found experimentally and the predictions of percolation theory has been recently attributed to the presence of mechanisms of conduction, such as hopping and assisted tunneling, that are generally neglected in the numerical simulations [4]. In this paper we show that the deviations may be attributed to hopping conduction, but also to disorder of geometric origin.

Here, we use hierarchical lattices to model the heterogeneous material. Hierarchical lattices are self-similar structures generated recursively by replacing every bond by a chosen motif. The advantage of this kind of lattices is that they not only simplify the calculations but also allow us to change the effective dimension with ease. Figure

Mat. Res. Soc. Symp. Proc. Vol. 463 © 1997 Materials Research Society

1a depicts a few steps of the generation of a hierarchical lattice of effective dimension $d_e = 2$, figure 1b shows the motif used here two generate a lattice with $d_e = 3$.

Two Kinds of disorder are studied: structural and potential. In the first case, each bond of the lattice is assumed to be either metallic with conductivity g, or dielectric with capacitance C. For potential disorder, the microscopic conductances are sampled from a power law distribution function.

The macroscopic admitance (complex conductance) of the sample is given by: $Y_e = L^{d-2}\sigma_e^*$ where $\sigma_e^* = \sigma_e + i\omega\epsilon_e$ is the complex macroscopic conductivity, L is the size of the sample and d is the dimension. As we will show later, σ_e and ϵ_e may or may not depend on the scale, depending on whether the size of the system is above a disorder length of the system or not.

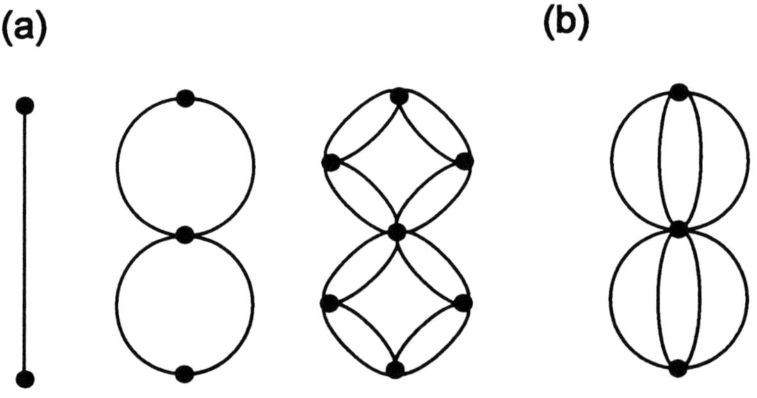

Fig. 1. (a) Recursive construction of a hierarchical lattice of $d_e = 2$ and (b) motif used to generate a hierarchical lattice with $d_e = 3$.

STRUCTURAL DISORDER

For structural disorder, we assume that each bond is either metallic or dielectric, with admitances $Y = g = 1/R$ and $Y = i\omega C$, respectively. When the fraction of metallic bonds is near the percolation threshold, a characteristic length $\xi \sim (p - p_c)^{-\nu}$ of geometrical origin emerges, which is the usual percolation correlation length. When the size of the system is below this scale, the effective transport coefficients depend on L, while for $L > \xi$, the transport coefficients assume their macroscopic values.

Figure 2 and 3 show the results for the macroscopic electric conductivity and dielectric constant of the network as functions of frequency, for $L = 16$ to 1028 at $p \approx pc$, and $d_e = 2$ and 3, respectively. The number of realizations were 1000 for $d_e = 2$ and 20 for $d_e = 3$. We observe from these figures, that there exists dispersive behavior of the form: $\sigma_e \sim \omega^x$ and $\epsilon_e \sim \omega^{-y}$, in the range $\omega_1 \leq \omega \leq \omega_0$, where ω_1 is the macroscopic cutoff frequency of the sample given by $1/R_eC_e$, where R_e and C_e are the effective resistance and capacitance of the sample, respectively, and ω_0 is the microscopic cutoff frequency, given by $1/RC$. It is found that $x + y = 0.99 \pm 0.02$, in agreement with the expected dimensional relation $x + y = 1$.

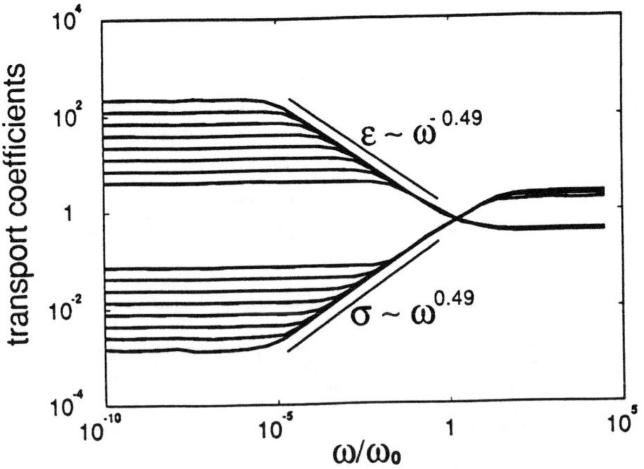

Fig. 2. Results for the effective dielectric constant and electric conductivity for $d_e = 2$ at the percolation threshold.

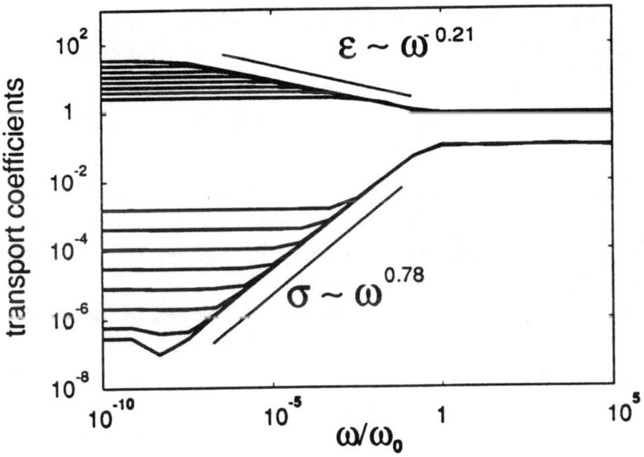

Fig. 3. Results for the effective dielectric constant and electric conductivity for $d_e = 3$ at the percolation threshold.

For frequencies ω lower than the macroscopic frequency cutoff of the network, the system is effectively at dc. In this range of frequencies, the dielectric constant and electric conductivity of the sample depend on the size of the system according to:

$\epsilon_e \sim L^{s/\nu}$ and $\sigma_e \sim L^{-t/\nu}$, respectively. The values for the exponents s, t, x and y obtained in this work are presented in table 1 for both two and three dimensions. Observe that they are in good agreement with the reported results obtained using regular lattices [2].

We have also observed that when the dielectric phase is at the percolation threshold there is also a dispersive behavior in the same range of frequencies observed before. The difference is that, under these conditions, $\sigma_e \sim \omega^{-y}$ and $\epsilon_e \sim \omega^{x}$. This behavior is expected at least for high frequencies, where the dielectric phase is far better conductor than the metallic phase, resulting in an inverted percolation problem. We also observe that, in the infinite frequency limit ($\omega > \omega_0$), $\sigma_e \sim L^{s/\nu}$ and $\epsilon_e \sim L^{-t/\nu}$.

d	x	y	s	t	ν
2	0.49 ± 0.01	0.49 ± 0.01	1.30 ± 0.05	1.4 ± 0.05	1.63
3	0.78 ± 0.02	0.21 ± 0.02	0.64 ± 0.05	2.32 ± 0.05	1.23

Table 1. Dispersion and scaling exponents for $d_e = 2$ and $d_e = 3$ obtained in this work.

POTENTIAL DISORDER

For potential disorder we assume that each bond of the lattice has a capacitor and a resistor, connected in series. We have studied the two-dimensional case in which the distribution function for the capacitances is a Dirac delta function, while the conductances are sampled from a power-law distribution function:

$$P(g) = |\mu| g^{\mu-1}. \tag{3}$$

In this case, a finite disorder length $\xi_D = \xi_0 \mu^{-\nu}$ will emerge [5]. Various limits of the above distribution are interesting: (a) The case $0 < \mu < 1$, which is related to continuum percolation and (b) the limit $\mu \to 0$, which is related to the case of disordered semiconductors in the hopping regime [6].

Figure 4 shows results for σ_e and ϵ_e as functions of frequency for values of L from 2 to 1028 at $\mu = 0.03$. In this case we observe that there exists a dispersive behavior in the frequency range defined by $\omega_1 \leq \omega \leq \omega_0$, where ω_1 is the macroscopic cutoff frequency defined in the previous section, and ω_0 is the maximum microscopic frequency of the system. At least for high frequencies this dispersive relation seems to adjust well to a power law behavior, deviations occur when the frequency approaches ω_1 and only for systems of size $L \approx \xi_D$. Our most important result is the fact that strong heterogeneity (i.e. $\mu \to 0$) seems to drive the dispersion exponents to the experimental values of $x \to 1$ and $y \to 0$. The strongly heterogeneous limit, in our view, makes at least two possible scenarios: 1) μ small but finite which is related to narrow necks in the structure of conducting structures and 2) $\mu = 0$ which is related to hopping conduction.

In the dc regime it is observed that the macroscopic transport coefficients, σ_e and ϵ_e, depend on L when the size of the system is below the disorder length. Above that, σ_e and ϵ_e reach their macroscopic value.

It is well known that for highly disordered materials, the conductivity exponent t depends on μ as $t(\mu) = (d-2)\nu + 1/\mu$. Preliminary results show that the superconductivity exponent s depends weakly on μ. This result is consistent with the fact that x tends to 1 when μ goes to zero since $x = t/(s+t)$, this implies that, in the case that s depends on μ, this dependence must be weaker than $1/\mu$.

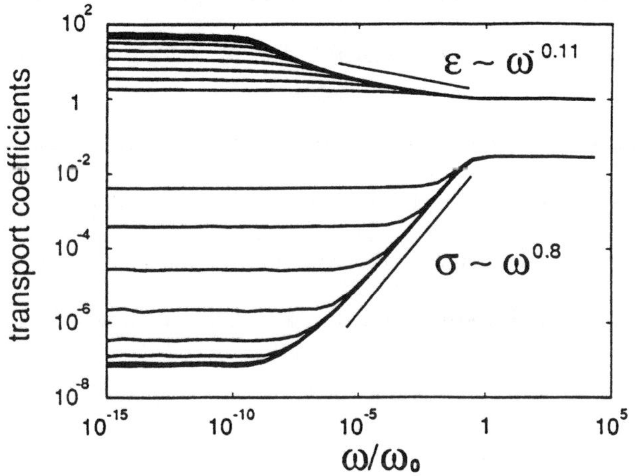

Fig. 4. Results for the dielectric constant and electric conductivity for the potential disordered case with $\mu = 0.03$.

CONCLUSIONS

In this work we have used hierarchical lattices to model the ac response of heterogeneous materials. In the case of structural disorder we have found dispersive behaviors for both the effective dielectric constant and electric conductivity, that are in agreement with previously reported results on percolative structures. The conductivity and superconductivity exponents obtained here are also in agreement with the reported values for two and three dimensional systems.

The most important result of this work is the fact that the exponents x and y depend on the strength of the disorder of the material. We have found, for the two dimensional case, that x and y approach to 1 and 0, respectively, when the disorder parameter μ goes to zero. This results support the theory of [4] which suggests that the deviations of experimental dispersive exponents from numerical calculations may be due to the existence of hopping conduction, but also suggests that such deviations may also be due to heterogeneities of geometrical origin, such as the presence of narrow necks in the conducting material.

ACKNOWLEGEMENTS

The authors thank Intevep, S.A. for permission to publish this paper. We thank E. Medina and V. Alvarado for helpful discussions and for carefully reading the manuscript.

REFERENCES

1. R. B. Laibowitz and Y. Gefen., Phys. Rev. Lett. **53**, pp. 380-383 (1984).

2. J. P Clerc, G. Giraud, J. M. Laugier and J. M. Luck., Advances in Physics **39**, pp. 191-308 (1990).

3. D. J. Bergman and D. Stroud, Solid State Phys. **46**, 147-157 (1992).

4. A. K. Sarychev and F. Brouers, Phys. Rev. Lett. **73**, pp. 2895-2898 (1994).

5. R. F. Angulo and E. Medina, J. Stat. Phys. **75**, pp. 135-150 (1994).

6.V. Ambegaokar, B. I. Halperin and J. S. Langer, Phys. Rev. B, **4**, pp. 2612 (1971).

Influence of sheet electron beam irradiation (SEBI) on water wettability of hydroxy apatites for artificial bone

J. KAWANO*, H. IZUMI*, K. OGURI**, A. TONEGAWA**, T. KAWAI***, M. OCHI****, and Y. NISHI*
*Department of Materials Science, Tokai University, 1117 Kitakaname, Hiratsuka, Kanagawa 259-12 Japan.
**Department of Physics.
***Chemical Technology Research Section, Chemical Polymer Bio Technology Laboratory, Kobe Steel Limited.
****Iwasaki Electric Group Company.

ABSTRACT

An influence of sheet electron beam irradiation (SEBI) on the wettability is investigated of the hydroxy apatite (HAP) [$Ca_{10}(PO_4)_6(OH)_2$]. The wettability is one of the important factors to control bio-compatibility. The SEBI is homogeneously performed by an electrocurtain processor. The temperature of the sample is below 323 K just after the irradiation. The wettability is evaluated by measuring the wet angle θ in a drop of water. The SEBI increases the wettability. Based on rate process, the influence of SEBI on wettability is discussed. Using the SEBI, we can precisely control the surface condition of HAP.

INTRODUCTION

Based on biological application, it is important the bio-compatibility of biological materials. The wettability from the biological point of view, it is important to know a wettability of hydroxy apatite (HAP) of artificial bone. The wettability is one of dominant factor to control the growth rate of HAP. The rapid growth rate likely shortens the term of cure.

On the other hand, an argon ion irradiation modifies surface properties [1,2]. However, the irradiation retains the argon atoms in the sample. Since a sheet electron beam irradiation (SEBI) doesn't retain the impurity atoms, the reproducible values can be obtained. Therefore, an influence of SEBI on the wettability is investigated for HAP.

EXPERIMENTAL PROCEDURE

The SEBI [3,4] was homogeneously performed by an electrocurtain processor. Figure 1 shows the schematic diagram of the apparatus. The acceleration potential and the irradiating current were 175 kV and 4 mA, respectively. The SEBI treatment was not continuously performed. In order to control the temperature of the sample surface, the conveyer speed was kept constant at 0.17 m/s. The temperature of the sample was below 323 K just after the irradiation. The repeated performance increased the dose of irradiation. Although the SEBI generalized in vacuum. The irradiated specimen was under a nitrogen with atmospheric pressure in the SEBI process zone. The oxygen concentration was less than 400 ppm in nitrogen atmosphere. The amount of irradiation dose was proportional to the SEBI time (t_i). The dose rate was 36.15 Mrad/s.The structure of the sample was monitored by x-ray diffraction (Rotafrex RU-200B, Rigaku Denki, Tokyo).The diffraction was performed under a step scanning method.

Mat. Res. Soc. Symp. Proc. Vol. 463 © 1997 Materials Research Society

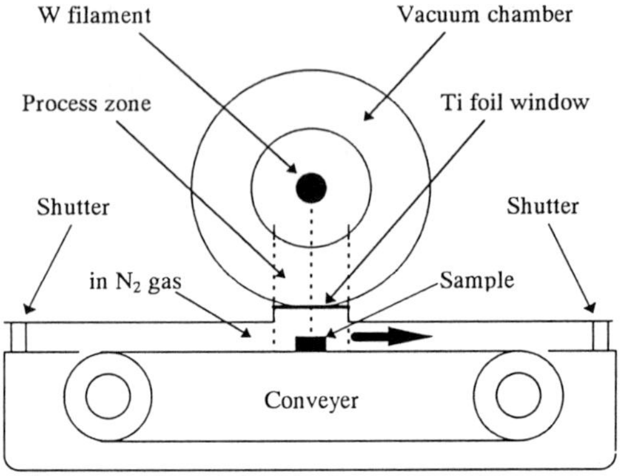

Fig. 1. Schematic diagram of electrocurtain processor.

RESULT AND DISCUSSION

Structure changes

Figure 2 shows the x-ray diffraction patterns of the samples. HAP and β-tricalcium phosphate (β-TCP) phases are found. (a) and (b) are before and after SEBI for 4.5 s, respectively. The SEBI decreases the peaks angles of both phases. Namely, the SEBI increases the lattice constant. We deduce that the SEBI increases the attractive binding force between atoms. Thus, we have studied the x-ray photoelectron spectrum (XPS) analysis, as shown in Fig. 3.

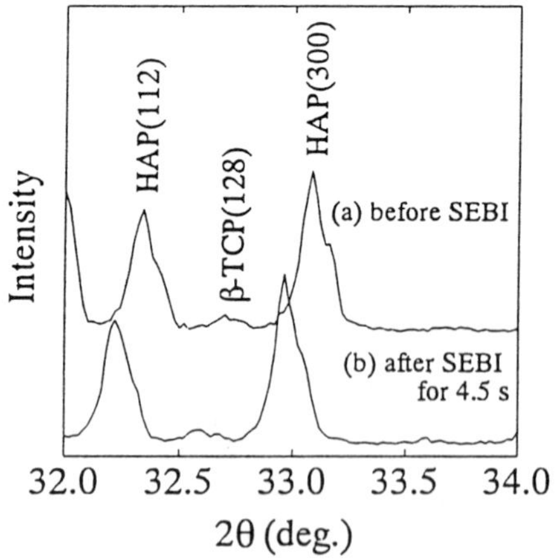

Fig. 2. X-ray diffraction patterns of HAP before and after SEBI.

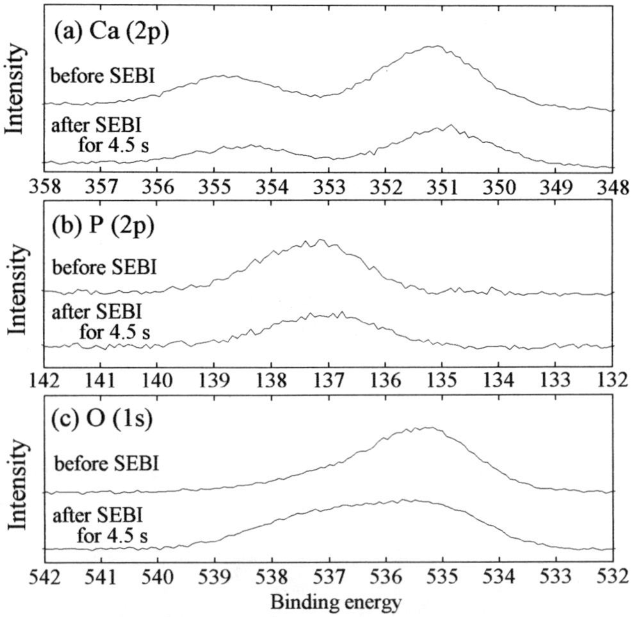

Fig. 3 X-ray photoelectron spectra before and after SEBI. (a) Ca (2p), (b) P (2p), (c) O (1s)

The SEBI increases the electron binding energy of oxygen atoms and slightly decreases the energies of calcium and phosphate atoms. If the SEBI affects the atom distribution (see Fig. 4), the polarization probably occurs. Since the polarization parts of water molecule attracts the polarized surface of HAP, the SEBI enhances the wettability.

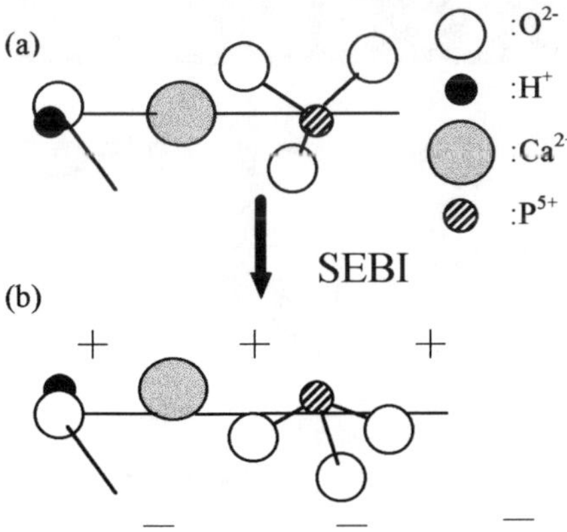

Fig. 4. Schematic diagrams of hydroxy apatite molecule.

(a) and (b) are before and after SEBI for 4.5 s, respectively.

(b) is for polarized condition.

Wettability

The wettability is evaluated by measuring the contact angle θ in a drop of water. The θ value is measured by taking a photograph of the drop of water (see Fig. 5). The mass of the drop is approximately 0.001 g. Figure 5 shows the θ value change against the SEBI. The SEBI decreases the θ value. If the θ value shows a low value, the wettability is high. The wettability is often evaluated by the interfacial energy (W_a) at constant temperature.

$$W_a = \sigma_0 (1+\cos\theta) \qquad (1)$$

Here, σ_0 is interfacial energy of water between liquid and gas. The reduced interfacial energy R(x) depends on the term (1+cosθ) of the contact angle.

$$R(x) = W_a / \sigma_0 = (1+\cos\theta) \qquad (2)$$

Figure 6 shows the change in reduced interfacial energy R(x) versus t_i. The R(0) is about 1.391 for the HAP sample before irradiation. The SEBI enhances the wettability, as shown in Fig. 6. The longer the t_i, the larger the reduced interfacial energy R(x) becomes.

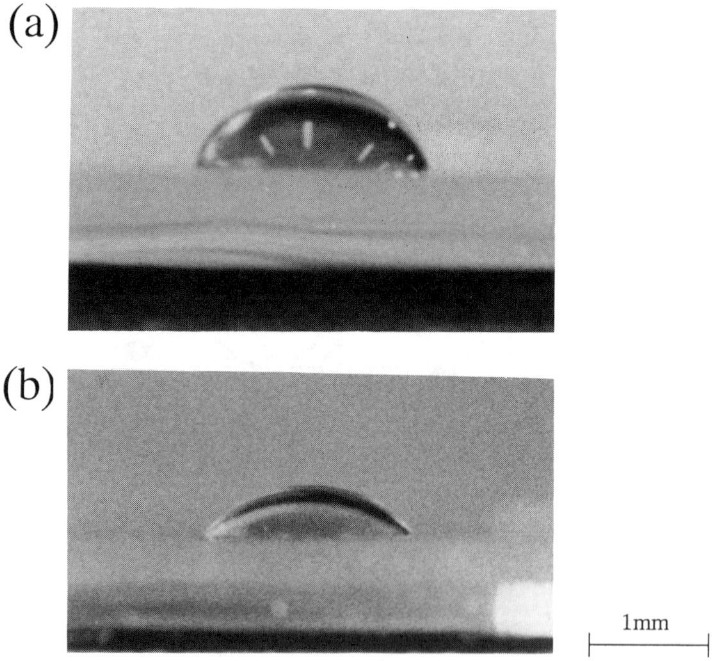

Fig. 5. Photographs of sessile drop of water on apatite ceramics.
(a) before SEBI, (b) after SEBI for 4.5 s .

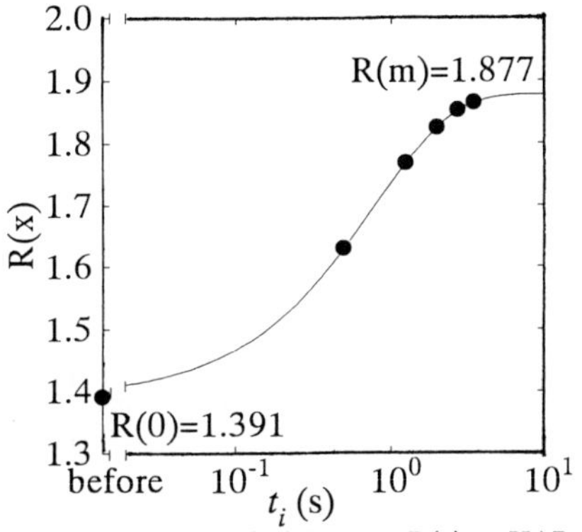

Fig. 6. Change in reduced interfacial energy R(x) on HAP versus SEBI time (t_i).

Rate process

The rate process is a useful equation to describe the phenomena [5-7]. If it is applied here to the SEBI process, the change X in reduced interfacial energy R(x) is assumed to be expressed by the following equations in relation to the t_i:

$$X = 1 - \exp(-k t_i^{\,n}) \qquad (3)$$

Here, k and n are constants. X is assumed to be expressed by

$$X = [(R(x) - R(0)) / (R(m) - R(0))] \qquad (4)$$

where R(m) and R(0) are the R(x) values after the infinite irradiation time and before the irradiation, respectively. The values are calculated in a self-consistent way. R(x) of the irradiation sample approaches R(m). It is the saturated value. When the correlation coefficient, F, of equation (3) is maximum, as shown in Fig. 7, the saturated reduced wettability value R(m) is 1.877. From these results, X is expressed by the following equation:

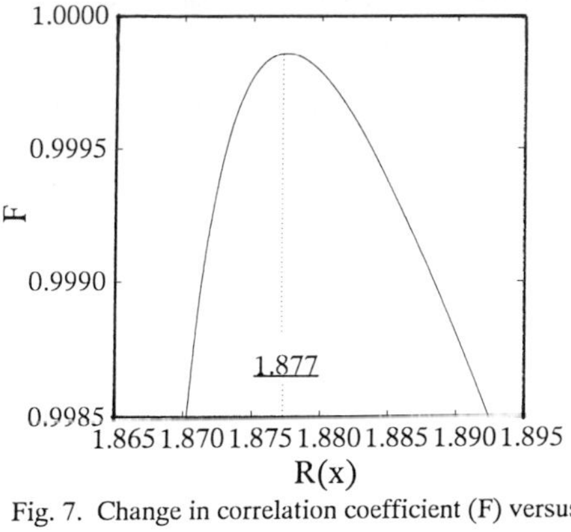

Fig. 7. Change in correlation coefficient (F) versus R(x).

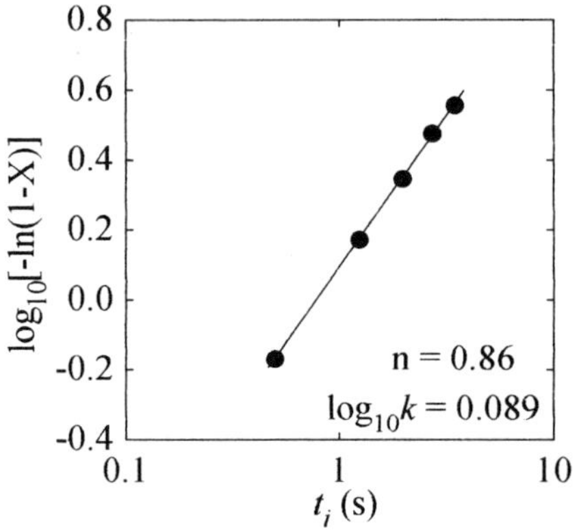

Fig. 8 Variation of $\log_{10}$[-ln(1-X)] versus t_i.

$$\log_{10}[\,-\ln(1-X)\,] = n \log_{10} t_i + \log_{10} k \qquad (5)$$

The linear plot (see Fig. 8) of equation (5) confirms the assumption of the rate process (see equation (3)). The values of n and $\log_{10}$k are 0.86 and -0.089, respectively.

We conclude that we can precisely control the surface active condition of HAP by using the SEBI. It's useful for patients to shorten the cure team in a hospital.

CONCLUSION

In summary, the SEBI enhanced the water wettability of HAP. The wettability wae evaluated by the reduced interfacial energy R(x) between HAP and water. Since the change in R(x) is applied by the rate process, we can precisely control the surface condition of HAP.

REFERENCES

1. N. Inoue, M. Kikuchi, T. Manabe, T. Shima and Y. Nishi, Nucl. Instrum. Methods B **59/60**, p. 1,328 (1991).

2. N. Inoue, M. Kikuchi, T. Shima, M. Iwase and Y. Nishi, Phys. Lett. A **157**, p. 299 (1991).

3. Y. Nishi, S. Takagi, K. Yasuda and K. Itoh, J. Appl. Phys. **70**, p. 367 (1991).

4. Y. Nishi, S. Takagi, K. Yasuda, K. Itoh and M. Kawakami, Nucl. Instrum. Methods B **59/60**, p. 1,422 (1991).

5. W. A. Johnson and R. F. Mehl, Trans. AIME **35**, p. 416 (1939).

6. Y. Nishi and H. Harano, J. Appl. Phys. **63**, p. 1,141 (1988).

7. Y. Nishi, H. Harano, T. Fukunaga and K. Suzuki, Phys. Rev. B **37**, p. 2,855 (1988).

RANDOM WALKS FOR MAGNETIC DECAY IN POROUS MEDIA

WEICHENG CAI*, THOMAS C. HALSEY*, JOHN H. HARDENBERGH*, MICHAEL LEIBIG**

*Exxon Research and Engineering Co., Annandale, NJ 08801
**Institute for Theoretical Physics, University of California, Santa Barbara, CA 93106

ABSTRACT

We propose a Monte Carlo method to simulate the magnetization decay for NMR in porous media. In this method, the diffusive spins in the fluid phase are modeled by 3-d random walkers, whose absorption at the solid surface simulates surface-mediated decay. Our method is 20 times faster than direct relaxation of the diffusion equation when a variable-step size implementation is used. We demonstrate our method by computing the surface relaxivity ρ for a Fontainebleau sandstone, whose structure was determined by X-ray tomography.

INTRODUCTION

Nuclear Magnetic Resonance (NMR) has emerged as one of the most useful probes of the structure of porous media [1]. However, the forward modelling of magnetization decay for these systems is complicated by the difficulty of solving partial differential equations in the complex geometry of the interconnected pore space in these materials. In this contribution, we present a fast Monte Carlo simulation method, which is particularly well-suited for this type of problem.

Consider a porous material whose pore space is filled with a liquid. The liquid can be magnetized parallel to an external field by an applied pulse. One can then study the relaxation this moment ("T_1"-relaxation), or one can rotate the magnetization into the plane perpendicular to a constant field with an rf pulse, and then study the relaxation of this transverse, precessing magnetization ("T_2" relaxation) [2]. There are a number of mechanisms by which this magnetization can relax. Inhomogeneous processes in the fluid can lead to the "bulk" relaxation of the magnetization; field gradients in the sample can also lead to relaxation. We are primarily concerned in this study with surface-mediated relaxation, in which microscopic magnetic moments relax only in the neighborhood of the surface of the pore space [3]. Unlike bulk relaxation processes, this type of relaxation can yield information about the microscopic structure of the pore space. For surface-mediated relaxation, the equations determining T_1 and T_2 relaxation are identical, provided the constant background precession is removed in the latter case.

The basic physics is described by [4]

$$\begin{aligned} \frac{\partial m(\mathbf{x},t)}{\partial t} &= D\nabla^2 m(\mathbf{x},t) \\ m(\mathbf{x},t=0) &= M_0/V \\ D\hat{\mathbf{n}}\cdot\nabla m(\mathbf{x},t)|_{\mathbf{x}\in S} &= -\rho m(\mathbf{x},t)|_{\mathbf{x}\in S} \end{aligned} \tag{1}$$

where $m(\mathbf{x},t)$ is the local magnetization density in the porous volume V bounded by the surface S. M_0 is the total initial magnetization, D the diffusion constant, and ρ the surface relaxivity. The simplest observable quantity is the total magnetization $M(t) = \int_V m(\mathbf{x},t)d^3x$.

Mat. Res. Soc. Symp. Proc. Vol. 463 © 1997 Materials Research Society

Note that the elementary diffusion constant D will presumably be the same as for the self-diffusion of the molecules of the fluid in bulk, and is thus typically well-known. By contrast, the surface relaxivity ρ is a complex function of the interaction between the fluid near the surface and the surface itself. Thus its value will vary with rock type. Typically, values in the range $5\mu\text{m/s} < \rho < 30\mu\text{m/s}$ are obtained, albeit by highly indirect means [5]. One advantage of our approach is that the surface relaxivity ρ can be left as a fitting parameter to reconcile simulation with experimental results, allowing fairly direct determination of this important quantity.

The problem described by Eq. (1) is analogous to the relaxation of random walkers, which are partially absorbed at the pore walls, in the same pore space [6]. Consider N_0 such walkers, of diffusivity D', whose distribution is described by the local number density $u(\mathbf{x}, t)$. When walkers contact the pore surface, there is a probability of σ that the walkers are absorbed when they contact the wall, and a probability of $1-\sigma$ that they are reflected (in practice, moved a small distance a from the wall) and continue their walk. In the continuum limit, $u(\mathbf{x}, t)$ is governed by the equations

$$\begin{aligned}\frac{\partial u(\mathbf{x}, t)}{\partial t} =& D'\nabla^2 u(\mathbf{x}, t), \\ u(\mathbf{x}, t=0) =& N_0/V.\end{aligned} \tag{2}$$

A straightforward calculation shows that the boundary condition at the pore wall is

$$D'\hat{\mathbf{n}} \cdot \nabla u(\mathbf{x}, t)|_{\mathbf{x}\in S} = -\frac{1}{a}\frac{\sigma}{1-\sigma} u(\mathbf{x}, t)|_{\mathbf{x}\in S}, \tag{3}$$

provided that a is much smaller than any surface length scale. Comparing to Eq. (1) for $m(\mathbf{x}, t)$, we see that the systems are identical if

$$\frac{\rho}{D} = \frac{a\sigma}{(1-\sigma)}\frac{1}{D'}. \tag{4}$$

This relation provides us with a convenient way to determine ρ by comparing experimental results to simulation results obtained with varying values of σ. To obtain a fast method, we need only speed up the Monte Carlo process by 1) constructing our simulation so that we can obtain results for different values of σ simply by varying σ at the end of the simulation, with a fixed data set, 2) using variable-step size random walks, and 3) working in the Laplace transform variable s instead of in time t. This is a generalization of the method of ref. 6. In the next section we describe our Monte Carlo process in more detail, and in the final section we demonstrate it by computing ρ for a Fontainebleau sandstone.

IMPLEMENTATION OF MONTE CARLO METHOD

The quantity analogous to total magnetization $M(t)$ is $U(t) = \int_V u(\mathbf{x}, t) d^3x$, the total number of walkers still inside the pore space at time t. This can be given in terms of a set of functions $\{P_n(t)\}$, which refer to the motion of random walkers inside the identical pore space with completely reflecting boundary conditions; $P_n(t)$ is the probability that such a walk makes its n'th contact with the wall before time t. Then

$$U(t) = N_0 \left[1 - \sigma \sum_{n=0}^{\infty} (1-\sigma)^n P_{n+1}(t)\right]. \tag{5}$$

We can calculate $P_n(t)$ in the simulation by averaging over all possible paths of walk as well as all possible initial positions of walkers. Note that although the magnetization which we are simulating decays in time, our simulation procedure utilizes "immortal" random walkers.

To accelerate the simulation process, we adopt a variable-step size algorithm [7]. In such an algorithm, the walker step size is fixed to be just smaller than the closest distance between the walker's current position and the pore wall. Of course, the step size has some minimum value, Δ. If the walker is closer than Δ to the wall, it still takes a step of size Δ, which creates the possibility that the step may cross the wall. Steps in which the walker crosses the wall are taken back, and count as contacts with the wall.

It is convenient to choose the step sizes always to be integral multiples of Δ. Formally, we write $P_n(t)$ as

$$P_n(t) = \left\langle \int_0^t dt_m p(t_m, r_m) \cdots \int_0^{t-t_m-\cdots-t_2} dt_1 p(t_1, r_1) \right\rangle_n . \tag{6}$$

Here $p(t_i, r_i)$ is the probability density that a walker setting off from the center of the sphere with radius r_i reaches its surface at the first time at time t_i. The angle brackets $\langle\,\rangle_n$ represent the average on all possible paths of walk and all possible initial positions of walkers, with the restriction that the walk must make n contacts with the wall of the pore space. It is convenient to avoid the convolutions by working with the Laplace transforms,

$$\begin{aligned} \tilde{P}_n(s) =& \frac{1}{s} \langle \tilde{p}(s, r_m) \cdots \tilde{p}(s, r_1) \rangle_n \\ \tilde{U}(s) =& N_0 (1/s - \sigma \sum_{n=0}^{\infty} (1-\sigma)^n \tilde{P}_{n+1}(s)), \end{aligned} \tag{7}$$

where $\tilde{p}(s, r) = \int_0^\infty e^{-st} p(t, r)$, and similarly for the other Laplace transforms.

The function $p(t, r)$ is related to the local probability density of a walker $\phi(\mathbf{x}, t)$ in a sphere of radius r through

$$\begin{aligned} p(t, r) =& -\frac{d}{dt} \int d^3x \phi(\mathbf{x}, t) \\ =& -D \int d^3x \nabla^2 \phi(\mathbf{x}, t) \\ =& -(4\pi r^2 D) \hat{\mathbf{r}} \cdot \nabla \phi(\mathbf{x}, t)|_{|\mathbf{x}|=r}, \end{aligned} \tag{8}$$

where $\phi(\mathbf{x}, t)$ obeys the boundary conditions

$$\begin{aligned} \frac{\partial \phi(\mathbf{x}, t)}{\partial t} =& D \nabla^2 \phi(\mathbf{x}, t) \\ \phi(\mathbf{x}, t=0) =& \delta(\mathbf{x}) \\ \phi(|\mathbf{x}| = r, t) =& 0, \end{aligned} \tag{9}$$

corresponding to a step which starts at the center of the sphere at time $t = 0$ and arrives at the sphere surface at time t. Equations (8-9) can be solved analytically, and the result for $p(t,r)$ is

$$p(t,r) = -\frac{8\pi r D}{(4\pi Dt)^{3/2}} \sum_{k=0}^{\infty} \left(1 - \frac{(2k+1)^2 r^2}{2Dt}\right) \exp\left(-\frac{(2k+1)^2 r^2}{4Dt}\right), \qquad (10)$$

and the Laplace transform of $p(t,r)$ is $\tilde{p}(s,r) = (r\sqrt{s/D})/\sinh(r\sqrt{s/D})$.

The Monte Carlo procedure is now straightforward to design. First, we acquire a digital representation of a pore space. For materials of geological or petrophysical interest, this representation may be obtained by X-ray tomography of an appropriate sample [8]. We segment the tomographic volume image to produce a binary representation of the pore space, with each voxel assigned either to the pore space or to the rock matrix. Then we make a "6-neighbor distance map" for the pore space, determining in each elementary voxel of the digital representation the distance to the nearest pore wall. Now variable-step size walks may be conducted, using the Euclidean distance map to determine local step sizes. Note that the walk need not be confined to any lattice, even though the representation of the pore space is digital, we used continuous random walks.

The output of each history of an individual walk is then a sequence of step sizes, as well as information about the sequence of contacts between the walk and the pore walls. Averaging over many walks, Equation (7) then allows the determination of $\tilde{P}_n(s)$. We repeat that in our simulation, the walks are never absorbed–they are allowed to continue up to some fixed, arbitrary number of contacts with the pore walls.

One technical problem with this procedure lies in the correspondence between the surface absorption σ and the surface relaxivity ρ of the original magnetic relaxation problem. The correspondence given in Eq. (4) above requires the use of a microscopic parameter a, which gives the distance from the surface at which a walk is restarted after contact is made. However, in the random walk algorithm, the step leading to contact is taken back, so while the walker restarts a distance of order Δ, the elementary step-size, from the wall, in practice the exact magnitude of this distance will vary. We deal with this problem by

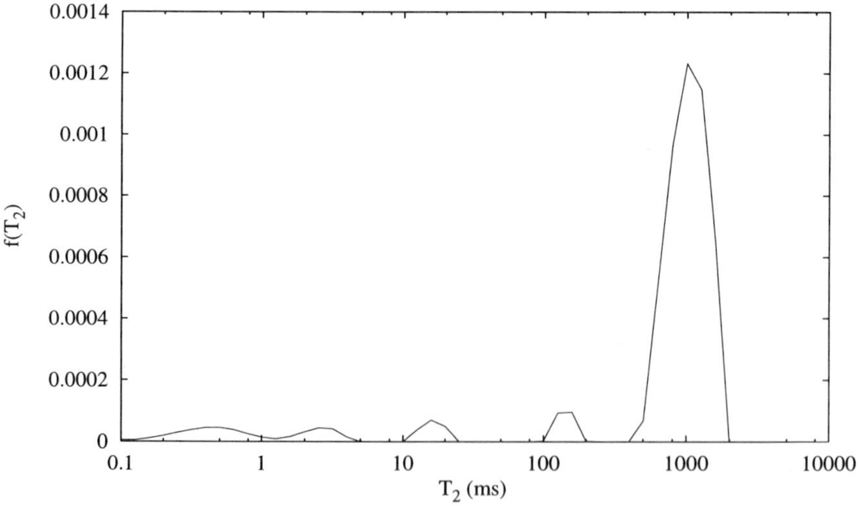

Figure 1: T_2 distribution from the experimental data for a Fontainebleau sandstone. $f(T_2)$ is normalized by $\int dT_2 f(T_2) = 1$.

calibrating our algorithm against a digitized sphere, for which we know the exact solution to the relaxation problem. This calibration yields the result $a = 0.64\Delta$, which allows us to uniquely compare a random walk simulation with a particular value of Δ to a relaxation problem whose fundamental length scale is D/ρ.

APPLICATION TO FONTAINEBLEAU SANDSTONE

We tested our method by determining the surface relaxivity ρ for a Fontainebleau sandstone, by comparing simulation results in a representative region of the pore space (obtained by X-ray tomography) with experimental results [9]. NMR relaxation spectra were obtained using the commercially available Resonance Instruments Maran NMR spectrometer. The sample was saturated with water; we then applied a magnetic pulse, and measured the magnetization decay $M(t)$. The results are plotted as a T_2 distribution, shown in Fig. 1. This distribution is defined implicitly by

$$M(t) = \int_0^\infty dT_2 f(T_2) \exp(-t/T_2). \tag{11}$$

The T_2 distribution is the spectrum of the magnetic relaxation, which characterizes the pore-size distribution of the medium [3].

We wish to compare this experimental result with simulation results. The pore structure of the sample in a small, presumably representative region can be obtained by X-ray tomography. The tomography has the resolution of $6.9\mu m$ which was defined as one voxel; we used a spatial volume of $298 \times 298 \times 298$ voxel. We chose Δ to be the side of one pixel. Since the sample had about 0.12 pore volume fraction, the total pore volume for the walkers in simulations was about $3 \times 10^6 (\Delta^3)$.

For this pore geometry, we performed the Monte Carlo simulation and computed $\tilde{U}(s)$ for a variety of values of σ. We found that the simulation data for $1 - s\tilde{U}(s)/N_0$ was well

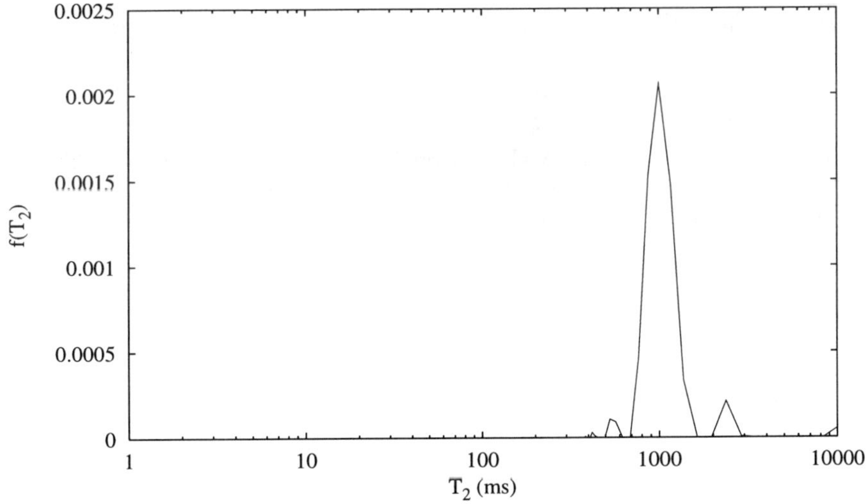

Figure 2: T_2 distribution of the simulation for $\sigma = 0.03$. We have also added the effect of the bulk relaxation time $T_{2B} \approx 3$ sec to this data, to simplify comparison with the experimental result in Fig. 1.

fit by the function $(1/2)[1-\tanh(as+b)]$ where $a(\sigma)$ and $b(\sigma)$ are two fitting parameters. We then calculated $U(t)$ by performing the inverse Laplace transform of the fitting function and plotted the results as a T_2 distribution, shown in Fig. 2. In this latter figure we have included the effect of the bulk magnetic relaxation, which modifies the surface T_{2S} according to the formula $1/T_2 = 1/T_{2B} + 1/T_{2S}$, where T_{2B} is the bulk relaxation time, ≈ 3 sec for water [10].

To fit the T_2 distribution from our simulation with that from the experiment, we then matched the main peak position by varying σ. Our result is $\sigma = 0.03$. From this, we determined $\rho \approx 15\mu$m/s where we used $D \approx 2.3\mu^2/ms$ [10]. This result is in good agreement with results obtained by the Schlumberger group using an entirely different method [5].

Acknowledgements

We would like to thank John Dunsmuir, who provided the X-ray tomography data for the Fontainebleau sandstone.

References

1. For a recent review of applications in the petroleum industry, see R. Kleinberg, Industr. Phys. **2** (2), 18 (1996).
2. A. Abragam, Principles of Nuclear Magnetism, (Oxford U.P., New York, 1994) Chapter III.
3. K.R. Brownstein and C.E. Tarr, Phys. Rev. A **19**, 2446 (1979).
4. H.C. Torrey, Phys. Rev. **104**, 563 (1956); M.H. Cohen & K.S. Mendelson, J. Appl. Phys. **53**, 1127(1982).
5. M.D. Hürlimann, K.G. Helmer, L.L. Latour, and C.H. Sotak, J. Mag. Res. A **111**, 169 (1994).
6. M. Leibig, J. Phys. A: Math. Gen. **26**, 3349 (1993).
7. P. Meakin, J. Phys. A **18**, L661 (1985).
8. B.P. Flannery, H.W. Deckman, W.G. Roberge, and K.L. D'Amico, Science **237**, 1439 (1987).
9. The X-ray tomography data for the sample was provided by J. Dunsmuir at Exxon Research and Engineering Co.
10. "NMR Short Course," SPWLA 36th Annual Symposium (in Paris) 1995.

ANOMALOUS RELAXATION IN DISORDERED MEDIA

MARIELA ARAUJO †, ORLANDO GONZALEZ ‡
† Reservoir Department. INTEVEP S.A, Apartado 76343. Caracas 1070–A Venezuela.
‡ Departamento de Física, Universidad Central de Venezuela. Caracas, Venezuela.

ABSTRACT

Several random walk-algorithms are used to model relaxation processes on disordered structures. Substrates are regular lattices in d = 1,2 with disorder variables attached to each lattice site. For the simulations, we consider diffusing particles obeying different rules according to the transport process being studied. Relaxation description is done in the time domain by calculating the characteristic function F(k,t), and in the frequency domain through the study of the complex susceptibility $\chi(k,\omega)$. Two types of relaxation mechanisms are seen both in simulations and in Nuclear Magnetic Resonance (NMR) experiments done on sandstones. These are expressed as stretched exponential forms of the relaxation function, and the Cole-Cole form of the imaginary part of the complex susceptibility. Longitudinal and transversal NMR measurements were done on fully saturated samples at 90 MHz. NMR relaxation data can be qualitatively understood using the random walk models proposed here.

INTRODUCTION

Relaxation phenomena occurring in many systems are very complex and not described by the classical exponential form in the time domain, or the Debye function in the frequency domain. Frequently relaxation data is fitted to empirical laws such as stretched exponentials, power laws, Cole-Cole form among others [1–3]. These relaxation processes are called anomalous.

Anomalous relaxation results from many body effects within a single system making it very hard to study from first principles. This is why there has been some effort to reproduce many of its features in terms of a one-body picture by applying geometrical constraints to the movement of the relaxing particles [4–7]. We have explored the "limits" of this approach by using random walks on disordered structures, such as regular lattices with disorder variables attached to each site, and percolation clusters at criticality. The effect of disorder on such structures is to temporarily trap the particles, thus reducing their diffusion. We use these models to analyze Nuclear Magnetic Resonance (NMR) data of water saturated sandstones trying to identify different relaxation mechanisms on those complex systems. The model is able to give essential features of the relaxation process.

Quantities related to the relaxation process, such as the relaxation function and the complex susceptibility, are calculated for the simulation model. Also, the relaxation response of water saturated sedimentary sandstones is used to calculate the complex susceptibility, and finally we made a comparison between the experimental data and results from the simulations of the random walk model.

Mat. Res. Soc. Symp. Proc. Vol. 463 © 1997 Materials Research Society

MODEL AND SIMULATIONS

A particle released from a chosen site on a disordered structure, cannot move freely, since spatial irregularities will set restrictions on its motion. Frequently, particles are trapped (sometimes temporarily) in poorly connected regions, thus their diffusive transport is effectively reduced. In the model, a localized signal is allowed to relax on a disordered substrate. The "signal" is made up of a swarm of walkers which are allowed to diffuse in such medium. Two different substrates are considered. One is a regular lattice with random variables taken from a power-law distribution of the form $|\mu|\xi^{\mu-1}$ with $0 < \mu < 1$. The second substrate is a percolation cluster at criticality.

In the case of a regular lattice, a walker is released from the origin, and allowed to move if a random number (drawn from a uniform distribution) is smaller than the value of the disorder variable attached to its site. It moves to one of its nearest neighbors with equal probability. We distinguish between the "quenched" case, where the disorder variables are frozen on the lattice, and the "annealed", where disordered variables are updated each time a walker attempts to move. A sampling over many disorder configurations is done, typically for 10^4 realizations. Lattices in d = 1 and 2, are used for the simulations of particles' motion up to 2^{20} time steps with system sizes big enough to avoid situations where the walkers reach the boundaries.

The transport on the percolation cluster is simulated similarly. A site is randomly chosen on each percolating system, and a random walker is released from it. At each unit of time, the walker is allowed to move on the structure, with equal probability to a nearest neighbor site if it belongs to the structure. Percolation clusters generated on a 400 x 400 lattice were used averaging the transport related quantities over 20 different cluster realizations. For the two cases, the walker position is recorded as a function of time.

With this information we calculate the relaxation function, $F(k,t)$, which is the characteristic function of the random variable $\vec{r}(t) - \vec{r}(0)$, i.e.,

$$F(k,t) = < e^{i\vec{k}.(\vec{r}(t)-\vec{r}(0))} >,$$

where $\vec{k}$ is the wave vector (for the 2D lattice it is chosen as $\vec{k} = (1,1)$ i.e., along the main diagonal of the lattice), $\vec{r}$ is the particle's position vector, and the angular brackets denote the sample average, and the complex susceptibility

$$\chi(k,\omega) = 1 + i\omega \int_{0}^{\infty} e^{i\omega t} F(k,t)dt.$$

These quantities are compared with their corresponding counterparts calculated for transport on a regular non-disordered structure.

NMR MEASUREMENTS

NMR has proven to be a valuable technique for understanding porous media [8]. Information about the geometrical attributes of the sample, such as porosity, is easily obtained from the NMR signal, whereas the relaxation times provide a measure of the characteristic length scales in the porous medium.

Samples of sedimentary rocks were cut from original rock nuclei, cleaned and completely water saturated with distilled water. The NMR longitudinal and transversal relaxation times were measured with a Bruker spectrometer at 90 MHz. For the transversal relaxation measurements, the Carr-Purcell-Meiboom-Gill (CPMG) pulse sequence was used, and for the longitudinal measurements it was used the inversion-recovey pulse sequence [9,10].

RESULTS

The relaxation function was calculated using the position of diffusing particles, recorded as a function of time, for two types of substrates: regular lattices with disorder variables and a percolation aggregate. We find that for regular lattices, when the disorder variables are quenched on the lattice, F(k,t) is drastically affected by the disorder "intensity", measured by exponent μ (which describes the power-law distribution from which the disorder variables are drawn).$\gamma = (1+\mu)/\mu$ for 1D lattices and $\gamma = 2/\mu$ for 2D lattices respectively. The response was less sensitive for the annealed case. We also find that the relaxation time, τ ,(calculated as the value of time at which $F(k,t)$ decays to a factor $1/e$) follows a power-law dependence with the wave number value, i.e., $\tau \sim k^{-\gamma}$. Exponent γ being a function of μ for quenched disorder. γ increases as μ decreases. For the annealed case γ = 2 as for regular transport (normal relaxation). Figure 1, shows a log-log plot of τ as a function of k for different μ values and 2D lattices with quenched and annealed variables. A similar behavior was observed on the percolation cluster. For this case the value of exponent γ is 3.06 ± 0.09.

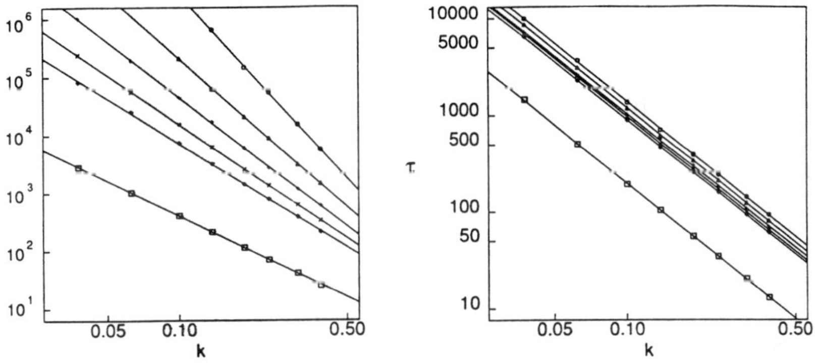

Figure 1. Relaxation time as a function of wave number magnitude for 2D regular lattices with quenched (left) and annealed disorder variables (right). Different lines are (from top to bottom) for μ = 0.4, 0.5, 0.6, 0.7, 0.8. Bottom curve is for a 2D non-disordered lattice.

For the two types of substrates, one with potential disorder (regular lattice with disorder variables attached to each site) and the other with structural (percolation cluster), we find that the product of the exponents $\theta \times \gamma/2 = 1$, here θ is the exponent that describes the time dependence of the mean-squared displacement $\langle r^2(t) \rangle \sim t^{\theta}$ holds regardless of the disorder strength (in the case of regular lattices) and system's dimensionality.

Regarding the complex susceptibility, we have computed its imaginary part for relaxation in regular lattices and the percolation cluster. As a function of frequency, the curves $\chi''(k,\omega)$ increase monotonically up to a maximum at a particular frequency ω_{max}; beyond that value, the curves decrease monotonically. Around the maximum the curves have the Cole-Cole form

$$\chi(\omega) = \frac{1}{1+(-i\omega\tau)^{\alpha}} \qquad 0 < \alpha < 1 .$$

For small and high frequencies linear regimes are observed on a log-log plot. Figure 2 shows a plot of the imaginary part of this function for potential and annealed disorder. The functional Cole-Cole form of $\chi''(k,\omega)$ for various values of α is shown in Figure 3. The slopes of the lines are different for the two frequency regimes suggesting different relaxation mechanisms as a function of frequency. This linear behavior is associated with stretched exponential relaxation. The value of ω_{max} depends on the dimensionality of the system and the particular type of disorder.

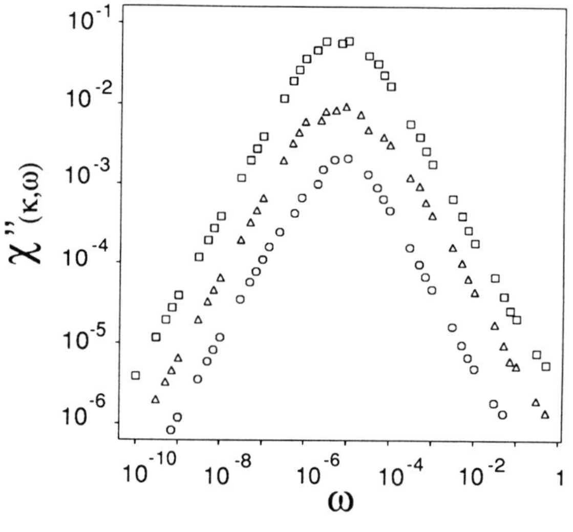

Figure 2. Imaginary part of complex susceptibility $\chi''(k,\omega)$ as a function of frequency for 2D lattices with: quenched disorder, μ = 0.8, k = $3\pi/250$ (□) ; annealed disorder, μ = 0.5, k = $3\pi/250$ (o), and percolation cluster at criticality for k= $3\pi/125$ (△).

For the NMR relaxation measurements of water in sandstones we find substantial nonexponential time dependence. The stretched exponential behavior is generally followed, although some departure for very small times is observed. This is characteristic of samples with different relaxation times acting independently in different parts of the samples. Figure 4 shows typical relaxation curves for the rock samples studied. These curves were fitted using non-linear regression with a stretched exponential form $F(t) = \exp\left[-(t/\tau)^{\beta}\right]$. The exponent β is found to monotonically increase with the width of the distribution of pore sizes, which can be considered as a disorder indicator. β values are in the range $0.51 \le \beta \le 0.89$. Using these relaxation curves, we calculated the imaginary part of the complex susceptibility and found that as a function of frequency this function has qualitatively the same behavior as the one seen in the simulations, i.e., there is a maximum at some characteristic frequency, and the linear regime observed in the simulations in a double log plot is only observed in the low frequency regime.

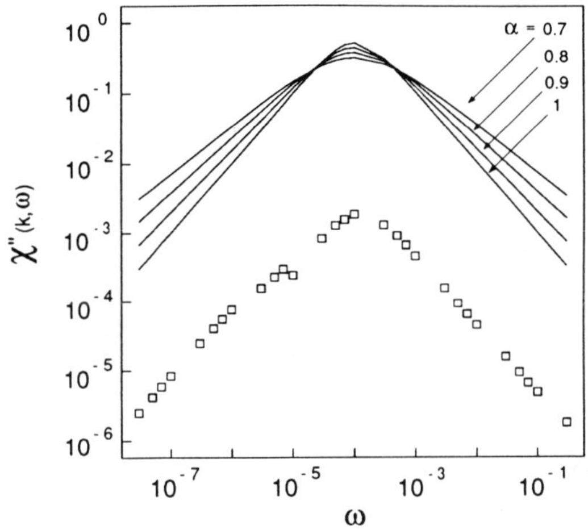

Figure 3. Cole-Cole form of the complex susceptibility for α = 0.7, 0.8, 0.9 and 1, and a typical simulation on a 2D regular lattice with quenched disorder for μ = 0.8 and k = $3\pi/250$.

CONCLUSIONS

In summary, we have used a simple model based on the motion of random walkers on a disordered structure to describe anomalous relaxation. The model is able to reproduce qualitative features commonly observed in experiments such as NMR longitudinal and transversal relaxation in porous rocks. This approach indicates that a one-body picture is able to reproduce the essential features of the relaxation process.

We find that the relaxation function in time domain, for a fixed wave vector magnitude, depends strongly on the value of the disorder parameter μ for quenched variables on a regular lattice, whereas for the annealed case, the results are clearly described by the current relaxation theory. A power-law dependence of the relaxation time with the wave number allows to verify that the product of exponents $\theta \times \gamma/2 = 1$ holds for both potential and structural disorder and is independent of dimensionality.

The imaginary part of the complex susceptibility makes possible the identification of different relaxation mechanisms. As observed for relaxation processes on the two types of substrates, around the maximum of this function the relaxation is of Cole-Cole type, whereas for small and high frequencies the relaxation is stretched exponential.

For small wave numbers, as the NMR longitudinal and transversal relaxation data clearly shows, the simple model presented here describes qualitatively magnetic relaxation processes observed in fully saturated consolidated porous rocks.

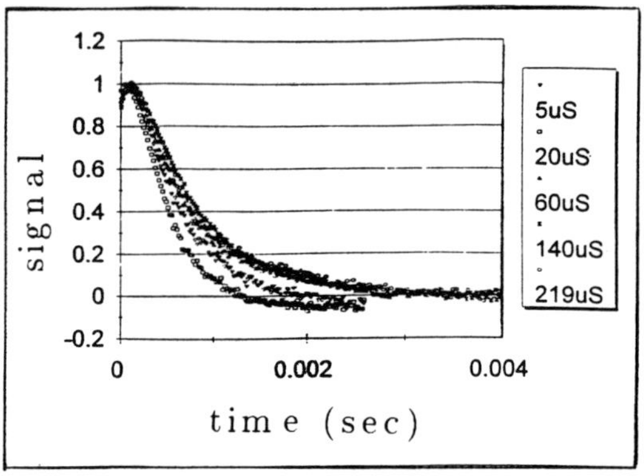

Figure 4. Typical relaxation curves of fully saturated rock samples measured by NMR at 90 MHz.

ACKNOWLEDGMENTS

The authors would like to thank Intevep S.A. for permission to publish this paper.

REFERENCES

1. K. Binder and A.P. Young, Rev. Mod. Phys. **58**, 801 (1986); R. Kohlraush, Ann. Phys. (Leipzig) **12**, 393 (1847) ; K.S. Cole and R.H. Cole, J. Chem Phys. **9**, 639 (1941).
2. J.P. Hansen, Phys. World **4**, 32 (1992)
3. K.H. Fisher and J.A. Hertz, Spin Glasses (Cambridge Univ. Press, Cambridge, 1991)
4. S Fujiwara and F. Yonezawa, Phys. Rev. E **51**, 2277 (1995); Phys. Rev. Lett. **74**, 4229 (1995).
5. S. Gomi and F. Yonezawa, Phys. Rev. Lett. **74**, 4125 (1995).
6. M. Araujo and O. González, in Mat. Res. Soc. Symp. Proc. Vol. **407**, 63–68. Edited by H. Cummins, D. Durian, D. Johnson and H.E. Stanley (1996).
7. O. González and M. Araujo, Phys. Rev. E., in press.
8. W.P. Halperin, F. D'Orazio, S. Bhattacharja and J.C. Tarczon, in Molecular Dynamics in Restricted Geometries, edited by J. Klafter and J.M. Drake, Wiley InterScience, 1989.
9. T.C. Farrar and E.D. Becker, Pulse and Fourier Transform NMR. New York. Academic Press. 1971.
10. NMR measurements were done at the Center of Magnetic Resonance of Universidad Central de Venezuela.

MONTE CARLO SIMULATIONS OF PHASE TRANSITIONS IN A TWO-DIMENSIONAL RANDOM-BOND POTTS MODEL

R. PAREDES, J. VALBUENA
INTEVEP S.A, Apartado 76343. Caracas 1070–A Venezuela.

ABSTRACT

Motivated by recent experiments on phase behavior of systems confined in porous media, we have studied the effect of quenched bond randomness on the nature of the phase transition in the two dimensional Potts model. To model the effects of the porous matrix we chose the couplings of the q state Potts Hamiltonian from the distribution $P(J_{ij}) = p\delta(J_{ij} - J) + (1-p)\delta(J_{ij})$. For a range of p values, away from the percolation threshold, the transition temperature follows the mean field prediction $T_c(p) = T_c(1)p$. Furthermore, we observed that the strong first order transition, that appears in the pure case for $q = 10$, changes two a second order transition. It is also clear from our simulations that the second order transition of the $q = 3$ pure case changes to a second order transition of a different universality class. A finite size scaling analysis suggests that in both cases the critical exponents, in the presence of disorder, fall into the universality class of the two dimensional pure Ising model. This result agrees with theoretical calculations recently published [1].

INTRODUCTION

During the past few years a great deal of effort has been devoted to the study of the effect of quenched randomness on phase transitions. Bond and field randomness can produce drastic effects in a phase transition. For example, critical exponents change in systems with positive heat capacity exponent if bond randomness is introduced [2]. Additionally it has been argued that the introduction of quenched bond randomness has a drastic effect on all temperature driven first-order phase transitions [3]. There is some experimental evidence that disorder produced by porous media affect drastically phase transitions of systems inside them. An extensive study of the isotropic to nematic phase transition of nCB liquid crystals in aerogel shows that the transition temperature is lowered compared to the pure situation and that the order of the transition changes from first to second [4–7]. The same effect is observed in ^{3}He-^{4}He mixtures in aerogels [8]. Using renormalization group calculations [9] and Monte Carlo simulations [10] Falicov et al. explained the latter experiment introducing bond randomnes in a lattice model of ^{3}He-^{4}He mixtures.

The q-state Potts model is a simple model that exhibits first and second order phase transitions depending on a simple parameter q. Chen et al. made an extensive Monte Carlo study on the random-bond Potts model in 2D [11]. They used a double delta distribution with two different bond strengths, occuring with equal probability. The ratio between the two strength values was finite. They found a second order phase transition instead of the first order found in the pure case. Additionally they found that this second order transition belongs to the same univerasality class as the Ising model. In 3D, Uzelac et al., also performed Monte Carlo simulations, but instead of randomizing the bonds they eliminated sites according to a random dilution model emulating the structure of aerogels and site percolation [12]. They found, in both cases, that it is

Mat. Res. Soc. Symp. Proc. Vol. 463 © 1997 Materials Research Society

neccesary to have a finite amount of disorder to observe a change in the order of the transition. Additionally they computed the phase diagram where it was found that the transition temperature is proportional to the concentration of Potts sites.

In this paper we describe the effects of random bond dilution on the $q = 3$ and $q = 10$ Potts model in 2D. The $q = 3$ and $q = 10$ exhibit respectively second and first order transitions for the pure case. In both cases we study the behavior of the temperature transition with disorder. The specific heat exponent of the $q = 3$ Potts model is positive and its critical exponents must change with the introduction of disorder [2]. It is interesting to study both cases because recently it was predicted that, with disorder, both systems should belong to the same universality class [1].

In the next section we discuss the random-bond Potts model. Then we discuss the methodology followed by us. We end with a discussion and conclusions.

THE RANDOM-BOND POTTS MODEL

The q-state Potts model is described by the Hamiltonian

$$\beta H = \sum_{<i,j>} J_{ij} \delta_{s_i s_j}, \tag{1}$$

where $\beta = 1/k_B T$. The spin s can take the values $1, 2, \ldots, q$ and δ is the Kroneker delta function. The sum runs over all nearest-neighbor bonds in the system and J_{ij} is the strength of the interaction betwen s_i and s_j. In a pure system, J_{ij} is constant for all bonds. The q-state Potts model is a simple generalization of the Ising model which has the advantage of exhibiting first order phase transitions for $q > q_c(d)$ and second order transitions for $q \leq q_c(d)$. In 2D ($d = 2$), q_c is equal to 4.

The random-bond Potts model is described by the above Hamiltonian with couplings J_{ij} randomly selected from the distribution

$$P(J_{ij}) = p\delta(J_{ij} - J) + (1 - p)\delta(J_{ij}). \tag{2}$$

METHODS

We performed extensive simulations of $L \times L$ lattices ($16 \leq L \leq 128$) with periodic boundary conditions using the Swendsen-Wang multiple spin flip method [13]. Histogram techniques were used to determine several thermodynamic quantities over a range of βJ [14]. Between 5×10^5 and 4×10^6 Monte Carlo steps were performed and up to 25 different bond samples were used in the configurational average over randomness for bulk properties.

We considered the probability distribution of energies defined by

$$P_L(E) = \frac{N_L(E) exp(-\beta J E)}{Z_L(-\beta J E)}, \tag{3}$$

where E is the energy in units of J, $Z_L(\beta J E)$ is the partition function, and $N_L(E)$ is the number of configurations of the system with energy E and size L. It has been shown that a finite size scaling analysis of $P_L(E)$ can be very efficient in detecting first order transitions even when the system size is smaller than the correlation length [15]. When the transition is of first order,

$P_L(E)$ exhibits two maxima which are of the same magnitude for $T = T_c$. Denoting E_M the energy of one of these maxima and E_m the energy of the minimun in between, the quantity

$$\Delta F_L = \ln \frac{P_L(E_M)}{P_L(E_m)} \tag{4}$$

is known to represent the interface free energy. If the transition is first order, ΔF_L should scale as L^d while it should go to zero otherwise [15].

We also calculate the specific heat which involves higher moments of the energy distribution:

$$C_L(T) = \left(\frac{\beta J}{L}\right)^2 \left(\langle E^2\rangle_L - \langle E\rangle_L^2\right), \tag{5}$$

where the moments are given by

$$\langle E^n\rangle_L = \sum_E E^n P_L(E). \tag{6}$$

According to finite size scaling theory, the maximum in the specific heat scales with system size as

$$C_{max} \propto L^{\alpha/\nu}. \tag{7}$$

For a first order transition, the specific heat grows as L^d, so a measure of the scaling behavior of C_{max} will also provide additional confirmation of the order of the transition.

RESULTS

The phase diagram for the random-bond Potts model, taking specific heat maxima to indicate critical temperatures, is shown in Fig. 1. For a range of p values, close to $p = 1$, the transition temperature follows the mean field prediction $T_c(p) = T_c(1)p$[12]. This result is also observed using Monte Carlo simulation for the $q = 3$ and $q = 4$ Potts model in 3D [12] and using real space renormalization group in 2D and 3D for several q values [16].

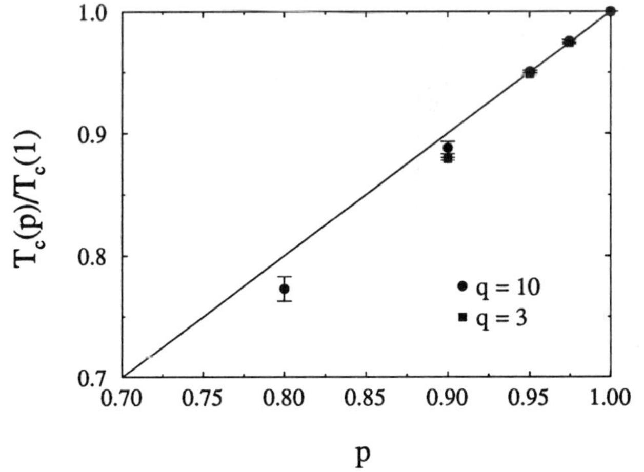

Figure 1 Phase diagram of the random-bond Potts model.

We are interested in what happens to the order of the transition for the $q = 10$ case when quenched disorder is included. We calculated the probability distribution of the energy for $p = 0.9$ to obtain the interface free energy. The resultant free-energy barrier $\Delta F(L)$ is plotted as a funtion of inverse lattice size in Fig 2. Clearly the trend for large L is towards a zero free-energy barrier which indicates that the transition has indeed changed from first to second. Note that for small L, $\Delta F(L)$ can be quite large indicating the presence of strong finite size effects. In Fig. 3 the probability distribution of the energy ($P_L(E)$) is plotted for several sizes. We note that the double peak structure, typical of the first order transition, is maintained until large L.

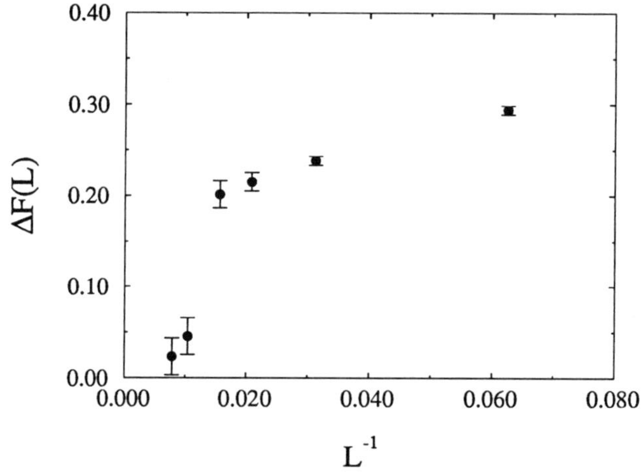

Figure 2 Free energy barrier, $\Delta F(L)$, versus inverse lattice size ($p = 0.9$).

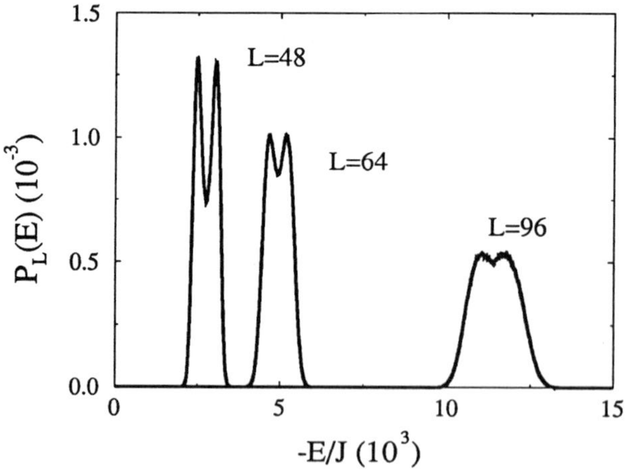

Figure 3 Probability distribution of energy for several lattice sizes. The double peak structure tends to dissapear for larger L.

Using Monte Carlo simulations in 2D, Chen et al. [10] have recovered the Ising exponents for the second order phase transition, induced by random impurities, for the $q = 8$ Potts model. On the other hand, critical behavior of the specific heat in a 2D Ising model has a logarithmic size dependence, i.e.,

$$C_{max} = a + b \log L. \tag{8}$$

To see if our data follows this behavior for the $q = 3$ and the $q = 10$ random-bond Potts model, we plotted in fig. 4 the maxima of the specific heat versus $\log L$. Both figures show the $p = 0.9$ case. In figure 4.a the pure case ($p = 1$) is also plotted. It is clear, for $q = 3$, that the pure case does not follow the logarithmic behavior but the disordered case does. This suggests a change in the universality class of the system. Note that for large L the tendency to a linear behavior is also clear for $q = 10$ (fig 4.b). We omit the smallest values of L because finite size effects are strong. We also fit these data with a power law but the fit was not as good. At this moment, from our data, we can conclude that our specific heat results for the $q = 3$ and $q = 10$ random bond Potts model do not exclude an Ising type behavior as was predicted theoretically by Kardar and collaborators [1]. To improve these results it is neccesary to go to stronger disorder and try to reduce finite size effects.

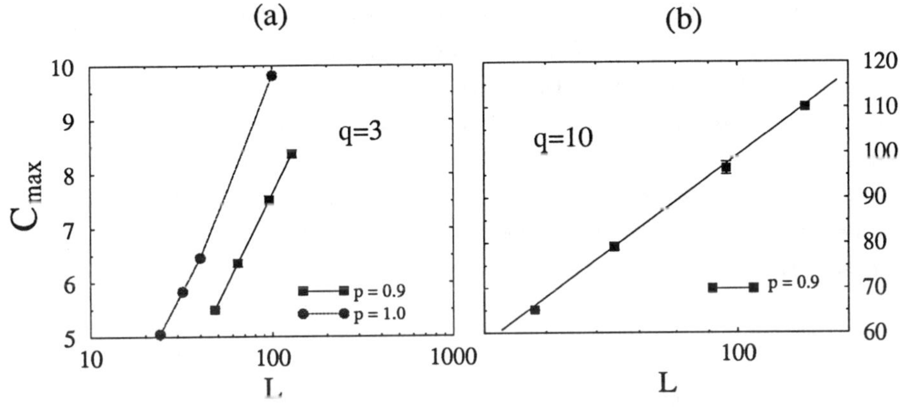

Figure 4 Specific heat maxima versus lattice size L. (a) $q = 3$ Potts model for the pure ($p = 1$) and the disorder ($p = 0.9$) cases. (b) $q = 10$ Potts model for the disorder case ($p = 0.9$).

CONCLUSIONS

From the results above it is clear that introducing bond randomness, in the $q = 3$ and $q = 10$ Potts model, changes the nature of the transition. The presence of disorder also decreases the transition temperature as has been seen in liquid crystal systems. The dependence with p found is linear close to $p = 1$. It is similar to previous results using renormalization calculations [16] and Monte Carlo in 3D[12]. For $q = 10$ the order of the transition changes from first to second. In $q = 3$ there is a change of universality class. Futhermore, we found evidence that in both cases the specific heat maxima follow a logarithmic behavior which suggests that when bond randomness is introduced the q-state Potts model changes to the universality class of the Ising model.

ACKNOWLEDGMENTS

The authors would like to thank Intevep S.A. for permission to publish this paper. Useful conversations with E. Medina are gratefully acknowledged.

REFERENCES

1. M. Kardar, A. Stella G. Sartoni and B. Derrida, Phys. Rev. E **52**, R1269 (1995).
2. A. B. Harris, J. Phys. C **7**, 1671 (1974).
3. A. N. Berker, J. Appl. Phys. **70**, 5941 (1991).
4. X-I. Wu, W. I. Goldburg, M. X. Liu and J. Z. Xue, Phys. Rev. Lett. **69**, 470 (1992).
5. T. Bellini, N. A. Clark, C. D. Muzny, L. Wu, C. W. Garland, D. W. Schaefer and B. J. Olivier, Phys. Rev. Lett. **69**, 788 (1992).
6. G. S. Iannacchione and D. Finotello, Phys. Rev. Lett. **69**, 2094 (1992).
7. N. A. Clark, T. Bellini, R. M. Malzbender, B. N. Thomas, A. G. Rappaport, C. D. Muzny, D. W. Schaefer and L. Hrubesh, Phys. Rev. Lett. **71**, 3505 (1993).
8. S. B. Kim, J. Ma and M. H. W. Chan, Phys. Rev. Lett. **71**, 2268 (1993)
9. A. Falicov and A. N. Berker, Phys. Rev. Lett. **76**, 4380 (1996).
10. A. Falicov and A. N. Berker, Phys. Rev. Lett. **74**, 426 (1995).
11. S. Chen, A. M. Ferremberg and D. P. Landau, Phys. Rev. Lett. **60**, 1213 (1992).
12. K. Uzelac, A. Hasmy and R. Jullien, Phys. Rev.Lett **74**, 422 (1995).
13. R. H. Swendsen and J.-S. Wang, Phys. Rev. Lett **58**, 86 (1987).
14. A. M. Ferremberg and R. H. Swendsen, Phys. Rev. Lett. **63**, 1195 (1989).
15. J. Lee and J. M. Kosterlitz, Phys. Rev. B **43**, 3265 (1991).
16. R. Paredes and A. N. Berker, (1996) unpublished

COLLOIDAL AGGREGATION WITH A DRIFT: A COMPUTER SIMULATION

Agustín E. González* and François Leyvraz**
*Instituto de Física, Universidad Nacional Autónoma de México, Apartado Postal 20-364, 01000 México, D.F., Mexico
**Instituto de Física, Laboratorio de Cuernavaca, Universidad Nacional Autónoma de México, Apartado Postal 139-B, 62191 Cuernavaca, Morelos, Mexico

ABSTRACT

Preliminary results of an aggregation model that takes into account both the Brownian motion as well as the gravitational drift experienced by the colloidal particles and clusters is presented. It is shown that for high strengths of the drift the system crosses over to a regime different from diffusion-limited colloid aggregation, for which there is an increase of the fractal dimension, a speeding up of the aggregation rate and a widening of the cluster size distribution, becoming algebraically decaying with an exponent τ.

INTRODUCTION

With a few exceptions [1, 2, 3], most of the studies on colloidal aggregation [4, 5] neglect the effect of sedimentation due to gravity on the resulting cluster formation. However, the combined action of Brownian motion and the external gravitational field may lead to a different behavior on the structural and kinetic quantities describing the aggregates. Although for theory and simulations it may be an advantage to single out the aggregation under Brownian motion only, experimentalists need to resort to a number of tricks to eliminate the action of the gravitational field on the aggregating system, like flipping the sample very often during the aggregation, considering a suspension fluid whose density matches closely that of the colloidal particles, or even considering a space lab experiment. It seems therefore to be necessary to advance our theoretical and simulational knowledge of colloidal aggregation under the action of a gravitational drift.

Currently, our basic knowledge [6] of diffusion-limited colloid aggregation (DLCA), the limiting regime of colloidal aggregation *with no external field* that occurs when the sticking probability after encounters is one, is as follows: (a) The fractal dimension d_f^o of the clusters in the dilute limit is close to 1.80 and has a square root type of increase with concentration, $d_f = d_f^o + a\phi^\beta$, where $a \approx 0.91$ and $\beta \approx 0.51$, ϕ being the volume fraction of the colloidal particles. (b) The number-average $(S_n(t))$ and weight-average $((S_w(t))$ cluster sizes grow linearly with time in the dilute limit: $S_n(t) \sim t^{z'}$ and $S_w(t) \sim t^z$, with $z = z' = 1$. However, the exponents z and z' increase from their zero-concentration value, again as a square root type of growth. This behavior of the average cluster sizes is in opposition to the reaction-limited colloid aggregation (RLCA) case, that occurs when the sticking probability tends to zero, for which an exponential increase of the averages is obtained. (c) The cluster size distribution $N_s(t)$, that gives the number of clusters of size s at time t, is bell shaped in the late stages of the aggregation. This is also in opposition to the RLCA case, for which an algebraic decay $N_s(t) \sim s^{-\tau}$ ($\tau \approx 1.5$) [7] is seen, again in the late stages of the aggregation.

Much of this basic knowledge has been obtained through computer simulations, which have assisted researchers by proportioning a great deal of information about the structural and kinetic properties of the formed clusters. After a successful development of an aggregation model [8, 9], capable of providing the correct fractal dimensions, the simulations of

Mat. Res. Soc. Symp. Proc. Vol. 463 © 1997 Materials Research Society

colloidal aggregation have advanced up to the point as to furnish the correct kinetics and scaling of the cluster size distribution [10], as well as the scaling of the structure factor [11, 12]. In this paper we present preliminary results of a colloid aggregation model that takes into account both the Brownian motion experienced by the particles and clusters, as well as the sedimentation due to a gravity field also undergone by the same particles and clusters. In the next section we write down the basic sedimentation equations leading to the aggregation model developed here. Then, in the section that follows we present and discuss the results of the model for the fractal dimension, the average cluster sizes and the cluster size distribution function. It is worth mentioning here that the deposition of big clusters on the bottom of the container as well as the depletion of them on top are not considered here; in our simulations we have periodic boundary conditions not only on the x and y directions but also on the z direction. This would describe the aggregation well in the bulk of a tall container. We also need to mention that all our simulations were stopped just before gelation, always working in the sol phase. We hope to relax those two restrictions in the future.

THE MODEL

The sedimentation velocity v_s experienced by a cluster of N spherical particles of radius a and mass m_o is

$$v_s = \frac{m_o(1-\rho/\rho_o)gN}{f} = \frac{m_o(1-\rho/\rho_o)g}{k_BT}DN \quad , \tag{1}$$

where ρ_o is the density of the particles, ρ is that of the suspension fluid, f is the friction coefficient of the cluster, $D(\sim N^{-1/d_f})$ its diffusion coefficient and T is the temperature. Let t_o be the time for which the cluster diffuses a particle diameter ($2a$), that is, $t_o = 2a^2/D$. During the same time, the cluster drifts a distance

$$d_s = v_s t_o = KNd \quad , \tag{2}$$

where $K \equiv m_o(1-\rho/\rho_o)ga/k_BT$ and d is the diameter of the particles. It is found that K is of the order of unity if the particles are 1 μ in diameter, ρ_o is of the order of 1 gm/cm^3, T is room temperature and $1-\rho/\rho_o$ is not too small. However, if the diameter is 0.1 μ such quantity is of the order of 10^{-4}, while if the diameter is 10 μ, K goes up to 10^4. Therefore, 1 μ marks the transition size between diffusive and drifting behavior for *individual* particles, with density different from that of the medium.

We will simulate the physical system on a simple cubic lattice with periodic boundary conditions and with lattice spacing equal to d. Each of the cells of the lattice can be occupied by a colloidal particle or can be empty, in which case is considered as occupied by the suspension fluid. As the aggregation proceeds, we deal with a collection of clusters made of nearest neighbor lattice cells that are diffusing randomly and sedimenting downwards.

Let us define $\Delta t_{dif} \equiv d^2/2D_{max}$ as the time taken by the most mobile cluster of the sample (the smallest one) to diffuse one lattice spacing. Also, let $\Delta t_{drif} \equiv d/v_s^{max}$ be the time taken by the largest cluster to sediment downwards one lattice spacing. The algorithm goes as follows:

a) If $\Delta t_{drif} < \Delta t_{dif}$ then

1. We pick a cluster in a cyclic way.
2. The time is increased by $\Delta t_{drif}/N_c$, where N_c is the number of clusters at that time.

3. The quantity $\Delta t_{drif}\, v_s/d$ is calculated and the result is added to a variable corresponding to that cluster. If the sum is greater than one, the cluster is moved downwards one lattice spacing and the new value of the variable becomes the remainder of the sum modulus one.

4. If the cluster is moved, we check for overlapping with other clusters, in which case the moving cluster is taken back to its original position and the overlapping clusters are merged. Afterwards, we go back to the starting situation to calculate Δt_{dif} and Δt_{drif}.

5. The cluster now moves one lattice spacing on a random direction, with probability $(\Delta t_{drif}/\Delta t_{dif})(D/D_{max})$.

6. We then check for overlapping with other clusters, etc., as in point 4.

b) If, on the other hand, $\Delta t_{dif} \leq \Delta t_{drif}$, we follow exactly the same procedure as in a) with the exception that everywhere we find Δt_{drif}, this is changed to Δt_{dif}.

The simulations were stopped just before gelation. One volume fraction $\phi = 0.01$ was considered, for which three values of K were chosen: $K = 10^{-8}$ (DLCA simulation), 10^{-3} (the drift is felt predominantly by clusters of size $\sim$ 1000 or larger) and 10^{-1} (the drift is felt by clusters $\sim$ 10 or larger). For each value of K, 50 simulations of 156 250 initial particles were performed in order to have enough statistics for the evaluation of all relevant quantities.

RESULTS

In Fig. 1 (top) we are plotting on a log-log scale the radius of gyration R_g vs. the number of particles N, for all the clusters formed in the 50 simulations with $K = 10^{-8}$. We fitted a straight line to the data points with N higher than 50 (such as to consider the asymtotic behavior for large clusters), shown in the figure, from which the estimate $d_f = 1.888 \pm 0.002$ was made. Although it appears that the straight line is not passing through the middle of the data points, this is due to the fact that the points are denser below the straight line. We note that the d_f estimate is in excelent agreement with a previous DLCA estimate of d_f as a function of ϕ [6], evaluated at $\phi = 0.01$. In Fig. 1 (middle) a similar plot for $K = 10^{-3}$ was made, where we can clearly see a change in behavior. We note two different slopes for the plot, with a breaking point at around $N \approx 1000$. We fitted two straight lines to the data points in this case. In the first one we considered clusters with N ranging from 50 up to 1000, while in the second one we fitted clusters with N greater than 1000. For the first straight line we obtained the estimate $d_f^1 = 1.871 \pm 0.002$, while for the second one we found $d_f^2 = 2.265 \pm 0.013$. The first estimate is again very close to the known DLCA value for $\phi = 0.01$, while the second indicates another regime: that of big drifting clusters sweeping smaller ones on their way to the bottom. That the fractal dimension of these clusters is higher indicates that the small clusters may fit in some of the holes of the big clusters, making them more compact. Finally in Fig. 1 (bottom), for $K = 0.1$, we find a single straight line for all N greater than 50. In this case the d_f estimate was $d_f = 2.216 \pm 0.001$, which is close to the second estimate for $K = 0.001$. It is worth noting that our d_f values for this "sweeping regime" are also close to the value obtained in [3]: 2.32 ± 0.05. Although the authors explain this value by a restructuring and rearrangements within the clusters mechanism, including bond breakage, that is something we didn't have to invoke; just the sweeping of the small clusters, including single particles, by the big ones.

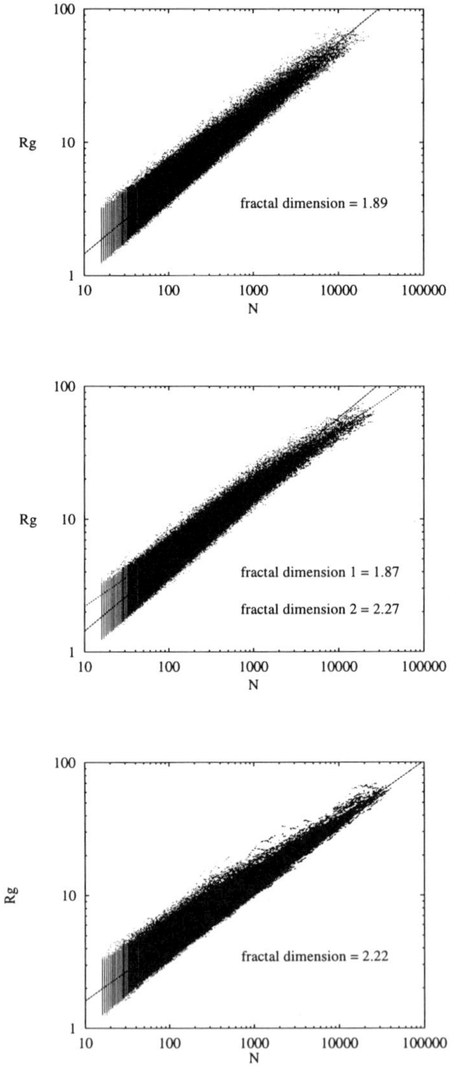

Fig. 1. Top: A log-log plot of R_g vs. N for all the clusters formed with 16 or more particles, in all the 50 simulations performed for $K = 10^{-8}$. Middle: Same as in the figure at the top, but now for $K = 10^{-3}$. Bottom: Same as in the top figure, this time for $K = 10^{-1}$.

In Fig. 2 we plotted on a log-log scale the weight-average cluster size vs. time, averaged over all the 50 simulations, for the three mentioned cases. For $K = 10^{-8}$ we find a straight line (shown between the arrows), defining the exponent z of growth. Note that after the second arrow the graph of S_w curves again due to finite size effects. For the case of $K = 10^{-3}$, the curve splits from the 10^{-8} one in the zone of the straight line, making it much shorter. For the $K = 10^{-1}$ case, the splitting occurs well before the zone of the straight line, indicating a sharp speeding up of the aggregation. The exponent z obtained from this graph for $K = 10^{-8}$ is equal to 1.42 ± 0.04, which is just a little different from the value 1.32 found in [6], applied to the case $\phi = 0.01$.

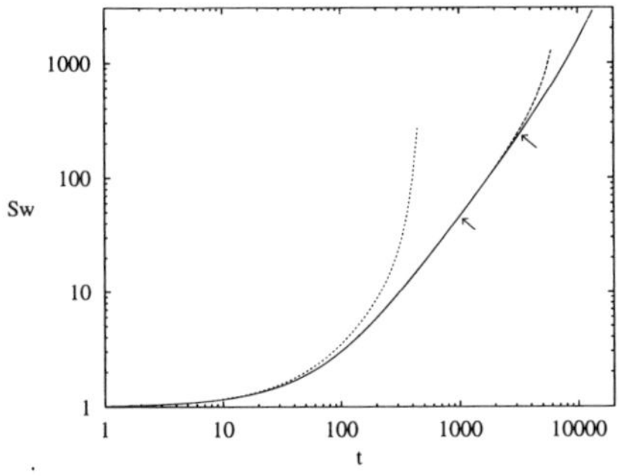

Fig. 2. A log-log plot of the weight-average cluster size S_w vs. time t, for $K = 10^{-8}$ (solid curve), $K = 10^{-3}$ (dashed curve), and for $K = 10^{-1}$ (dotted curve).

In Fig. 3 we show, for the $K = 0.1$ case, the cluster size distribution function $N_s(t)$ vs. s, coming from all 50 simulations. All the times in steps of 0.1 in the $ln(time)$ variable were considered in the plot. That is, for $ln(time) = 0.0, 0.1, 0.2, ..., 6.1$ (= final $ln(time)$). As we can see, for the final times we lose the bell shaped curves characteristic of DLCA and an exponent τ now appears. The 8 straight lines passing close to the 8 last curves all have a slope of 1.16, indicating this value for the exponent τ. The envelope exponent is now 2.15, which is a little different from the standard value of 2 found in DLCA or RLCA simulations. We are currently investigating if there is any relation between this odd value for the envelope exponent and the value found for the exponent τ. The widening of the cluster size distribution indicated by the appearance of an exponent τ has an understandable physical origin: there is an apparent instability in which the big clusters start to grow bigger at a very fast pace, the small clusters being uncapable of disapearing at the same rate. We therefore have the coexistence of very small and very big clusters in our system, widening the size distribution.

In summmary, we have shown that the introduction of a gravitational drift changes the the structural and kinetic quantities of the aggregating system. For high strengths of the drift we crossover to a regime different from DLCA, with a larger fractal dimension, a faster rate of growth, and a wider, algebraically decaying cluster size distribution.

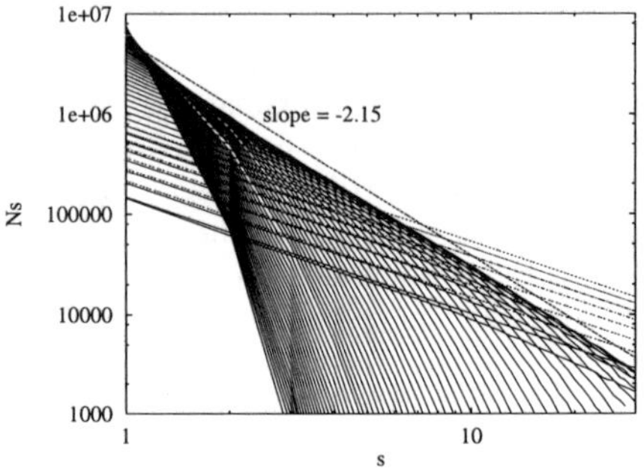

Fig. 3. A log-log plot of the cluster size distribution function N_s vs. s, for the $K = 10^{-1}$ case, for all the times in steps of 0.1 in the $ln(t)$ variable, starting from 0.0 up to 6.1 (final $ln(t)$).

ACKNOWLEDGMENTS

AEG was partially supported by CONACyT grants 4906E and E120.1381, as well as by the Cray Research Grant SC-006096.

References

[1] A. S. Michaels and J. C. Bolger, I&EC fundamentals **1**, p. 24 (1962).

[2] M. Couch, Ph. D. Thesis, Cambridge University, 1993.

[3] C. Allain, M. Cloitre and M. Wafra, Phys. Rev. Lett. **74**, p. 1478 (1995).

[4] R. Jullien and R. Botet, Aggregation and Fractal Aggregates, World Scientific, Singapore, 1987.

[5] T. Vicsek, Fractal Growth Phenomena, World Scientific, Singapore, 1989.

[6] M. Lach-hab, A. E. González and E. Blaisten-Barojas, Phys. Rev. E **54**, p. 5456 (1996).

[7] R. C. Ball, D. A. Weitz, T. A. Witten and F. Leyvraz, Phys. Rev. Lett. **58**, p. 274 (1987).

[8] P. Meakin, Phys. Rev. Lett. **51**, p. 1119 (1983).

[9] M. Kolb, R. Botet and R. Jullien, Phys. Rev. Lett. **51**, p. 1123 (1983).

[10] A. E. González, Phys. Rev. Lett. **71**, p. 2248 (1993).

[11] A. E. González and G. Ramírez-Santiago, Phys. Rev. Lett. **74**, p. 1238 (1995).

[12] A. E. González and G. Ramírez-Santiago, J. Coll. Interf. Sci. **182**, p. 254 (1996).

PATTERN FORMATION AND UNUSUAL $A+B \to 0$ REACTION KINETICS BETWEEN CHARGED REACTANTS IN LOW DIMENSIONS

E.A. KOTOMIN[a,b], V.N. KUZOVKOV[a,c], and W.von NIESSEN[c]
[a]Institute of Solid State Physics, 8 Kengaraga Str., Riga LV-1063, Latvia
[b]Institute of Physics and Astronomy, Aarhus University, Aarhus-C, DK-8000, Denmark
[c]Institut für Physikalische und Theoretische Chemie,
Technische Universität Braunschweig, D-38106 Braunschweig, Germany

Abstract

The effect of non-equilibrium charge screening in the kinetics of the one-dimensional, diffusion-controlled $A+B \to 0$ reaction between charged reactants in solids and liquids is studied. Incorrectness of static, Debye-Hückel theory is shown. Our microscopic formalism is based on the Kirkwood superposition approximation for three-particle densities and the self-consistent treatment of the electrostatic interactions defined by the nonuniform spatial distribution of similar and dissimilar reactants treated in terms of the relevant joint correlation functions. Special attention is paid to the **pattern formation** due to a reaction-induced non-Poissonian fluctuation spectrum of reactant densities. This reflects a formation of loose domains containing similar reactants only. The effect of asymmetry in reactant mobilities ($D_A = 0$, $D_B > 0$) contrasting the traditional symmetric case, i.e. equal diffusion coefficients, ($D_A = D_B$) is studied. In the asymmetric case concentration decay is predicted to be *accelerated*, $n(t) \propto t^{-\alpha}$, $\alpha = 1/3$ as compared to the well-established critical exponent for fluctuation-controlled kinetics in the symmetric case, $\alpha = 1/4$ and/or the prediction of the standard chemical kinetics, $\alpha = 1/2$. Results for the present microscopic theory are compared with the mesoscopic theory.

1 Introduction

Bimolecular $A+B \to 0$ reactions are quite common in condensed matter physics and physical chemistry; e.g. they occur between primary radiation defects of two types, A and B, which recombine when they approach each other during diffusion walks to within some critical distance r_0. These particles (called Frenkel defects in solids and/or electrons and radicals in liquids) could be neutral or charged.

Many-particle effects caused by the spatial fluctuations of the reactant densities have been intensively studied in recent years in the kinetics of bimolecular chemical reactions, including the above-mentioned $A+B \to 0$ reaction. A number of quite different techniques and methods were developed for this purpose, including direct computer simulations, a mesoscopic approach, the scaling, as well as microscopic theory - see review articles [1, 2], a monograph [3]) and proceedings of the conference [4]. These studies clearly demonstrated that the kinetic laws established long ago in standard chemical kinetics [5] could be violated, usually at high particle concentrations/long reaction times. In particular, the asymptotic ($t \to \infty$) concentration decay rate turns out to be $n(t) \propto t^{-d/4}$, where $d \leq 4$ is the spatial dimension, i.e. it is *slower* than the one in standard chemical kinetics, $\alpha = 1/2, 1$ and 1 for $d = 1, 2$ and 3, respectively. This effect, called sometimes 'abnormal kinetics' - abnormal from the standard point of view - is directly related to the reaction-induced non-uniform reactant distribution wich is in contrast to the main prediction of chemical kinetics that all reactants are well stirred and the reaction volume is homogeneous. As a result, modern chemical kinetics uses the language of *critical exponents, correlation lengths, etc* similar to the physics of critical phenomena.

Presently almost all studies of fluctuation-controlled effects deal with neutral, non-interacting particles thus neglecting effects caused by their interaction. In this paper, we study *many-particle effects between charged reactants* focusing on the *dynamical* Coulomb potential screening in the $A+B \to 0$ reaction.

Mat. Res. Soc. Symp. Proc. Vol. 463 © 1997 Materials Research Society

2 Kinetic equations

The basic equations of our microscopic theory for interacting particles have been derived and discussed recently for $d = 3$ for the cases of Coulomb [6] and elastic ($U(r) \propto 1/r^3$) [7] interactions. Now, based on results of our review article [2], we generalize these equations for arbitrary space dimension d. This helps to show peculiarities in the transition to low dimensions.

Use of the Kirkwood superposion approximation [8] for decoupling the infinite hierarchy of equations for the correlation functions leads to a minimum set of variables describing the fluctuation-controlled reaction kinetics. They are: macroscopic concentrations, $n_A = n_B = n(t)$, and three kinds of *joint correlation functions* [1, 2] - two for similar particles, $X_\nu(r,t)$, $\nu = A, B$ and a third one for dissimilar particles, $Y(r,t)$, where r is the relative distance between two particles. These functions describe a spatial distribution of pairs AA, BB and AB, respectively and are analogous to the radial distribution function in statistical physics of dense gases and liquids [8]. The physical meaning of these correlation functions is the following [2, 6]: $C_A^{(a)}(r,t) = n(t)X_A(r,t)$ and $C_B^{(a)}(r,t) = n(t)Y(r,t)$ are mean densities of particles A and B, respectively at the relative distance r provided that a probe particle A is in the coordinate origin. Introducing for simplicity a new function $X(r,t) = (X_A(r,t) + X_B(r,t))/2$ the basic set of kinetic equations read:

$$\frac{dn(t)}{dt} = -K(t)n^2(t) \ , \quad K(t) = \gamma_d \, r_0^{d-1} \, |\mathbf{j}(r_0,t)| \ , \tag{1}$$

$$\partial Y(r,t)/\partial t = \nabla \mathbf{j}(r,t) - 2n(t)K(t)Y(r,t)J_d[X] \ , \tag{2}$$

$$\mathbf{j}(r,t) = (D_A + D_B)\left\{\nabla Y(r,t) + \beta \nabla U'_{AB}(r,t)Y(r,t)\right\} , \tag{3}$$

$$\partial X_\nu(r,t)/\partial t = \nabla \mathbf{j}_\nu(r,t) - 2n(t)K(t)X_\nu(r,t)J_d[Y], \tag{4}$$

$$\mathbf{j}_\nu(r,t) = 2D_\nu\left\{\nabla X_\nu(r,t) + \beta \nabla U'_{\nu\nu}(r,t)X_\nu(r,t)\right\} \ . \tag{5}$$

In Eqs.(1) to (5) the black-sphere recombination model is assumed implicitly: any AB pair recombines instantly when two reactants during their diffusive walks approach each other to within some critical distance r_0 [1, 2]. This fact is incorporated into the (Smoluchowski) boundary condition for the correlation function of dissimilar particles; $Y(r \le r_0, t) = 0$ in Eq.(2). This correlation function defines the quantity of primary importance - the *reaction rate* $K(t)$ which is a flux of particles over the recombination sphere's surface ($\gamma_d = 2, 2\pi, 4\pi$ for $d = 1, 2, 3$, respectively). For a finite r_0 the reaction rate reads

$$K(t) = \gamma_d r_0^{d-1} \partial Y(r,t)/\partial r|_{r=r_0}.$$

The non-linear terms in Eqs. (2), (4) containing the functionals $J_d[Z]$ arise directly from the Kirkwood approximation [3, 8]. Their expressions for $d = 1, 2, 3$ are given in ref.[2]. In particular,

$$J_1[Z] = (Z(r + r_0, t) + Z(|r - r_0|, t))/2 - 1 \ . \tag{6}$$

Expressions for the flux densities $\mathbf{j}$(r,t) and $\mathbf{j}_\nu$(r,t) ($\beta = 1/k_B T$) are also non-linear since the effective potential energies $U'_{\lambda\mu}(r,t)$ have contributions of both direct ($\lambda\mu$ pair) and indirect, lateral particle interactions through surrounding particles. The technique for their calculation in the case of a short-range potential has been discussed in ref.[7], whereas that for the Coulomb potentials in ref.[6].

Low-dimensional ($d = 1, 2$) systems with Coulomb interaction reveal a peculiarity which allows us to reduce the number of independent variables and to simplify the kinetic equations. Namely, we can do the limiting transition $r_0 \to 0$ retaining the finite reaction rate. Physically this means that the reaction rate is governed by the effective radius. This radius is the largest one of the scale lengths in the system. For the Coulomb systems such a scale is called the *Onsager radius*, $R = e^2/\varepsilon k_B T$ [6]. This is a distance at which the thermal energy equals the attraction energy; when approaching to within R two reactants cannot separate and thus inevitably recombine. Usually $R >> r_0$ and thus R determines the reaction rate. Neglecting many-particle effects, the latter has a very simple form: $K = 4\pi D R_{eff}$ [2, 5]. In the limiting case of $r_0 \to 0$ the functional $J_d[Z]$ in eqs.(2),(4) is greatly simplified, $J_d[Z]$=Z(r,t)-1.

The recombination kinetics is defined by the following dimensionless parameters: (i) the initial particle concentration $n(t=0) = n(0)$, (ii) the *partial* diffusion coefficient $\kappa = D_A/D$ (note that dimensionless diffusion coefficients are related by the condition $D_A + D_B = 2$, i.e. $D_A = 2\kappa$, $D_B = 2(1-\kappa)$), (iii) the capillary radius r_c (in the $1d$ case). For more details see [3, 10].

3 Results

3.1 Concentration decay

The kinetics of the concentration decay has been calculated for high initial concentration $n(0) = 1$ and long dimensionless time, $t = 10^8$. At this moment particle concentration drops by three orders of magnitude. (Further concentration decay hardly could be measured experimentally.) To make results more obvious, in Fig.1 we plotted not the very kinetic curves, $n = n(t)$, but their slopes on a logarithmic scale which defines the so-called *current critical exponents*

$$\alpha(t) = -\frac{d \ln n(t)}{d \ln t}. \tag{7}$$

To demonstrate the importance of the effect of *non-equilibrium* charge screening neglected in many previous studies, we present results for three different approximations as follows.

(i) The traditional, Debye-like treatment of the reaction kinetics with unsreened Coulomb interaction [8].

Many-particle effects are neglected, the kinetic equations arise due to linearization of equations for the correlation functions. As a result, the equation for the correlation function of *similar* particles $X_\nu(r,t)$ no longer affects the kinetics. In fact, the latter is defined entirely by the joint correlation function of dissimilar particles obeying the simple kinetic equation

$$\frac{\partial Y(r,t)}{\partial t} = \frac{\partial}{\partial r}\left\{\frac{\partial}{\partial r}Y(r,t) + Y(r,t)\frac{\partial}{\partial r}U(r)\right\}, \tag{8}$$

where $U(r) = -1/r$ is the unscreened Coulomb potential. After linearization of a set of kinetic equations, their solution no longer depends on the partial diffusion coefficient κ (solid and dashed lines in Fig.1). At long times the solution of Eq.(8) is practically defined by the diffusion length $\xi = \sqrt{t}$, i.e. the decay kinetics obeys the 'classical' algebraic law, $n(t) \propto t^{-\alpha}$, $\alpha = 1/2$.

(ii) The complete set of Eqs. (1) to (6) incorporating many-particle effects (via non-linear terms) but with linearized potentials, $U_A(r,t) = U_B(r,t) = -U(r,t) = 1/r$.

In this intermediate approximation the kinetics under study begins to depend on the mobility parameter κ but asymptotically it still follows the kinetics known for neutral, non-interacting particles with $U(r) = 0$.

(iii) The complete set of kinetic equations is combined with non-equilibrium treatment of charge screening making now no linearization.

The dimensionless capillary radius was chosen as $r_c = 0.1$. (Its reduction to the value of 0.01 results in a small, logarithmic correction which does not affect the critical exponent.) Curve 1 in Fig.1 shows that in the time interval considered the critical exponent rather rapidly approaches

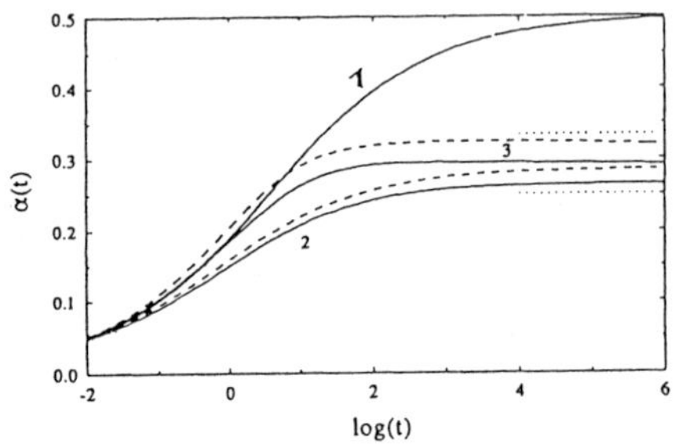

FIG. 1. The critical exponent characterizing the algebraic concentration decay, Eq. (7), as a function of dimensionless reaction time. Solid curves—symmetric reactant mobilities, $D_A = D_B$; dashed curves—asymmetric mobilities, $D_A = 0$. Dotted lines show the two expected asymptotes: $\alpha = \frac{1}{4}$ and $\frac{1}{3}$. Curves 1 correspond to the Debye theory, curves 2 to a solution of the kinetic Eqs. (1)–(6) incorporating spatial reactant correlations but neglecting dynamical charge screening, and curves 3 to the case when all screening effects are incorporated.

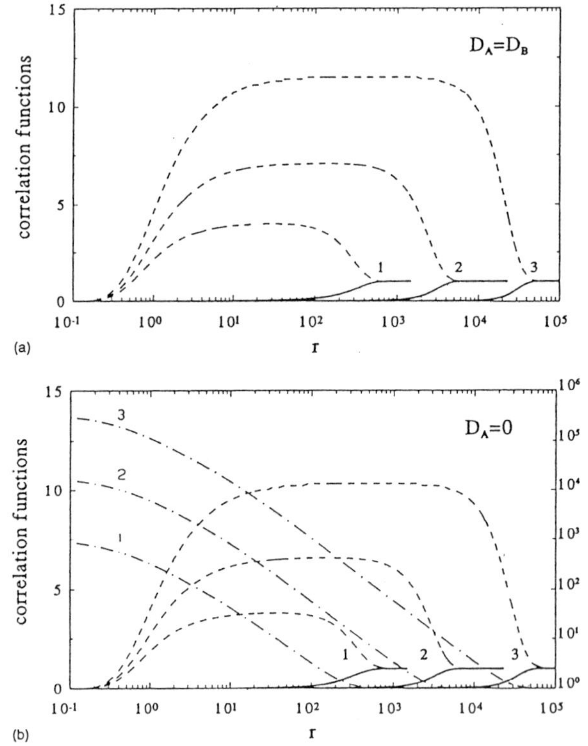

FIG. 2. The joint correlation function of dissimilar particles, $Y(r,t)$ (solid curve), and that of similar particles, $X_A(r,t)$ (dot-dashed curve) and $X_B(r,t)$ (dashed curve). Curves 1–3 correspond to the dimensionless times 10^4, 10^6, and 10^8, respectively. (a) Symmetric mobility case, $D_A = D_B$. (b) Asymmetric case, $D_A = 0$. Note that in case (a) $X_A(r,t) = X_B(r,t) = X(r,t)$; in case (b) $X_A(r,t)$ is plotted on a logarithmic scale.

its limiting 'classical' value of 1/2. Curves 2, incorporating many-particle effects, approach their *quasi*-steady-state after nearly the same time but their further approach from above to another asymptote wich $\alpha = 1/4$ has a logarithmically slow character. For example, for the symmetrical mobilities $\alpha(t = 10^8) = 0.264$. In the asymmetric mobility case the deviation from the asymptote is larger, $\alpha(t = 10^8) = 0.280$. Such a behaviour results from the long-range nature of the Coulomb interaction between particles.

Strictly speaking, the $\alpha = 1/4$ law of the fluctuation-controlled kinetics in $d = 1$ is proved only for non-interacting particles [1, 2, 4]. It was generalized for interacting particles *provided* that their interaction potential is short-range and does not lead to the similar-particle collapse [3]. In fact, this law has been proved for the case when the largest length parameter in the problem is the diffusion length, which is the case if the interaction potential was a Debye- Hückel like. However, the uncreened Coulomb potential, $U(r) = -1/r$, has an *infinite* interaction radius and thus defines the asymptotics of the correlation functions at large distances. The approach to the asymptotic character is very slow, it has diffusion-controlled character. Moreover, in the asymmetric case there is no mechanism of smoothing the fluctuations of the immobile particle distribution at all. This is why the results of our case (ii) are far from trivial.

Lastly, curves 3 in Fig.1 show a considerable difference for the symmetric and asymmetric mobilities which is more pronounced than that in curves 2. However, due to a very slow approach to their limiting values it is not clear whether and by how much the relevant critical exponents differ as $t \to \infty$. Analytical arguments are given in ref. [10] that in the symmetric case $\alpha = 1/4$ (as for non-interacting particles), whereas for asymmetric mobilities the critical exponent is *larger* and the reaction occurs respectively faster, $\alpha = 1/3$. Note that a similar *reaction acceleration* between charged particles with asymmetric mobilities was predicted earlier in the $3d$ case [6]. We found there that $\alpha = 5/4$, to be compared with $\alpha = 1$ known in the standard chemical kinetics, and $\alpha = 3/4$ in the fluctuation-controlled theory. Analogously, in $2d$ [9] we predict $\alpha = 1/2$ and 3/4 for symmetric and asymmetric mobilities, respectively. Consider now briefly the kinetics of the pattern formation in the particle spatial distribution.

3.2 Spatial reactant distributions

Figure 2 shows the time development of the joint correlation functions (note the logarithmic scale on the x axis and the same scale for immobile similar particles X_A). A key role of the diffusion length $\xi(t) = \sqrt{t}$ is evident here: the characteristic relative distance ξ' at which no AB pairs exist ($Y(r < \xi', t) << 1$) increases in time as $\sqrt{t}$: ξ' increases by an order of magnitude as time increased by *two* orders of magnitude, $\xi' = \xi(t)$.

Irrespective of the κ value, the correlation functions of mobile particles, $X_\nu(r, t)$, have a plateau at the same scale $r < \xi'$ and decrease rapidly to zero at $r < 1$. (This comes from the repulsion of similar particles at the relative distances which are short compared to the Onsager radius). In the asymmetric case the correlation functions of similar immobile particles have singularities at short r, where $X_A(r, t)$ drops by several orders of magnitude in a narrow interval $r \in (0, 1)$.

A comparison of these results with earlier findings for non-interacting particles [11] shows their remarkable similarity. The main difference lies in the depletion in the correlation functions of similar mobile particles at short relative distances caused by particle repulsion, whereas for neutral particles the correlation functions are finite as $r \to 0$. For non-interacting particles and symmetric diffusion such a behaviour of the correlation functions led to the conclusion that the *pattern formation* occurs in a form of alternating *domains* of similar particles, A or B, with linear size $\xi(t)$ [1, 2, 3]. This reaction-induced reactant-structure greatly differs from the basic assumption of standard chemical kinetics about well-stirred and homogeneous reactant distribution. In the domain structure reaction occurs only at the boundaries of the domains of particles of different type. In the asymmetric diffusion case for both non-interacting and interacting particles mobile B reactants remain randomly distributed within their domains, whereas immobile A reactants form compact clusters - a kind of "raisins in dough" [6, 11]. The $2d$ case will be discuss in detail in ref. [9].

4 Conclusions

We compare in the conclusion the main results of the mesoscopic [3] and the present, microscopic formalism for the diffusion-controlled $A + B \rightarrow 0$ reaction between charged particles in the $1d$ case. The former theory claims that the critical exponents in the concentration algebraic decay is the same for charged and neutral particles *provided*:

(i) similar particles (AA, BB) repel each other, and (ii) the pair interaction potential is not divergent (e.g. the Debye- Hückel potential).

Our microscopic theory generalizes this result (valid for the case of symmetric particle mobilities) for the *unscreened* (divergent) Coulomb potential. Moreover, we have studied here the case of asymmetric mobilities ($D_A = 0$, $D_B > 0$) and predicted reaction *acceleration*, i.e. an existence of a new critical exponent, $\alpha = 1/3$. We have also demonstrated that this peculiarity is a direct consequence of the *specific spatial distribution of reactants* studied by us in terms of the joint correlation functions for both similar and dissimilar reactants. A large discrepancy between the two approaches takes place for the accumulation kinetics under permanent particle source [9, 10].

This research was partly supported by the Deutsche Forschungsgemeinschaft (via a fellowship for V.N.Kuzovkov). E.K. thanks the Institute of Physics, Aarhus University (Denmark) for financial support and warm hospitality during his visit.

References

[1] V.N.Kuzovkov and E.A.Kotomin, Rep. Prog. Phys., **51**, 1479 (1988).

[2] E.A.Kotomin and V.N.Kuzovkov, Rep. Prog. Phys., **55**, 2079 (1992).

[3] E.A. Kotomin and V.N. Kuzovkov, *Modern Aspects of the Diffusion- Controlled Reactions: Many-Particle Effects* (vol. **34** in Comprehensive Chemical Kinetics; Elsevier: Amsterdam, 1996).

[4] Special issue of J. Stat. Phys., **65** (5/6) (1991).

[5] H.Eyring, S.H.Lin and S.M.Lin, *Basic Chemical Kinetics* (Wiley, New York, 1980).

[6] V.N.Kuzovkov and E.A.Kotomin, J. Stat. Phys., **72** , 145 (1993).

[7] V.N.Kuzovkov and E.A.Kotomin, J. Chem. Phys., **98** , 9107 (1993).

[8] R.Balescu, *Equilibrium and Non-equilibrium Statistical Mechanics* (Wiley, New York, 1975).

[9] V.N.Kuzovkov, E.A.Kotomin and W.von Niessen, J. Chem. Phys., **105** (1996).

[10] V.N.Kuzovkov, E.A.Kotomin and W.von Niessen, Phys. Rev. E, **54** (1996).

[11] E.A.Kotomin, V.N.Kuzovkov, W.Frank and A.Seeger, J. Phys. A: Math. Gen., **27**, 1453 (1994).

KINETICS OF PERIODIC PATTERN FORMATION IN DIFFUSION-REACTION IN A GEL

C-S. KUO *, R. BANSIL ** and E.L. CABARCOS +
*Chemistry Department, Boston University, Boston, MA 02215, cskuo@bu.edu
**Physics Department, Boston University, Boston, MA 02215, rb@buphy.bu.edu
+Departmento de Fisico Quimica, Complutense University, 28040, Madrid, Spain

ABSTRACT

A video photographic study of the kinetics of the precipitation reaction Na_2HPO_4 + $CaCl_2 \rightarrow CaHPO_4$ in agarose gels is presented. Periodic precipitation, known as Liesegang bands, are observed when one of the reactants is incorporated in the gel and the other allowed to diffuse in. The time evolution of the first Liesegang band's profile provides a direct confirmation of the supersaturation model, with a decrease in intensity as the monomer concentration is depleted and the formation of a broad plateau behind the sharp band from large particles left after the incorporation of the smaller mobile particles into the rapidly growing aggregate front. The time dependence of leading edge of the precipitate front agrees with the predictions of the reaction-diffusion-supersaturation model of Liesegang band formation.

INTRODUCTION

The study of non-linear chemical reactions exhibiting features such as temporal oscillations, chaos and pattern formation is currently attracting considerable attention. There has been much recent interest in the study of diffusion controlled reactions in gels or other unstirred media. The simplest case is the reaction $A + B \rightarrow C(solid)$ [1-3] in systems where the initially separated reactants after diffusing are allowed to react irreversibly and the products C are inert particles which may diffuse [4] or not [5]. Such a reaction is usually conducted by allowing one electrolyte front to diffuse and react with another electrolyte which is embedded in a gel forming a slightly soluble product that precipitates discontinuously in bands parallel to the diffusion front's surface. This phenomenon, called Liesegang bands [6], has been widely studied because it offers a model to explain a great variety of formations in nature, ranging from agate rocks and gold veins to gallstones [6-8]. The patterns appear to follow some general laws. First, the position x_n (measured from the gel surface) of the nth band is related to the time t_n of its formation by the so called time law $x_n \sim {t_n}^{1/2}$ which is the hallmark of the diffusion mechanism. Second, the ratio between the positions of adjacent bands , $x_n/x_{n-1} = p$, approaches a constant value $p > 1$ for large n; although the phenomenon of reverse banding, i.e. $p < 1$ has sometimes been observed. Third, the width of the bands ω_n grows with the distance according to the relation $w_n \sim (x_n)^{\alpha}$ with α smaller than one.

Many theoretical and simulation studies have been made to explain the formation of Liesegang patterns [9-10]. No single model can explain all the observed results, such as the presence of precipitate particles between the bands, the reverse spacing ($p < 1$) phenomena or the formation of complex patterns (double banding). However, there is considerable evidence to support the supersaturation theory of Ostwald and Prager [4-5] which considers band formation as a spatially discontinuous nucleation process. The formation of sharp rings is preceded by the onset of a turbidity front, implying that colloidal particles are continuously distributed initially and the sharp band formation is a post nucleation process involving aggregation and growth mechanisms [11].

Recently we have performed precipitation reactions in agarose gels by diffusing two salts from opposite ends of a tube containing the gel and allowing them to react and form an insoluble precipitate in the gel. From such a study we observed that a crossover from a confined precipitate to Liesegang bands occurs when the reactant concentrations change from equal at the two ends to unequal [12].

Mat. Res. Soc. Symp. Proc. Vol. 463 © 1997 Materials Research Society

In an attempt to reproduce the observed behaviour we performed numerical simulations with the diffusion-reaction model [12]. We assumed that in the reaction $A + B \rightarrow C$, the product C is a neutral molecule capable of diffusing and aggregating. The aggregation is modelled by an additional reaction $C \rightarrow D$ where D, although chemically identical to C, represents the solid phase. Following Dee's [4] simplification we considered that the aggregation term consists of two parts: the first one accounts for nucleation and has the form $K_n c$ for $c > c_n$, where c_n is the the supersaturation threshold at which nucleation is no more hindered by kinetic barriers. Nucleation is assumed to be the only mechanism responsible for the triggering of the precipitation in the gel because cluster-cluster aggregation is inhibited by the trapping of mesoscopic particles in the gel structure. The second part of the precipitation term is in general a complicated integro-differential expression [4] of the concentration of C and D. It is based on the fact that in presence of aggregation, precipitation of nuclei is no more hindered by kinetic barriers and occurs for concentration $c > c_s$, where c_s, the saturation threshold is $< c_n$. In order to simplify the equations, we followed the approach of Chopard et al. [9] and retained the non linear term $K_p\, c\, d$ so that the simplified equations become:

$$\delta_t a = D_a \nabla^2 a - Kab \tag{1}$$

$$\delta_t b = D_b \nabla^2 b - Kab \tag{2}$$

$$\delta_t c = D_c \nabla^2 c + Kab - K_n c H(c - c_n) - K_p c d H(c - c_s) \tag{3}$$

$$\delta_t d = K_n c H(c - c_n) + K_p c d H(c - c_s) \tag{4}$$

where $H(x)$ is the Heaviside step function. By changing the initial distributions of the reactants and the boundary conditions, this model can also be used to describe Liesegang band formation when one of the reactants is incorporated at a uniform concentration in the gel and the other allowed to diffuse into the gel from one end.

In this paper we report the early stages of the kinetics leading to the formation of the first Liesegang band for a precipitation reaction carried out by incorporating one of the reactants in the gel and allowing the other to diffuse in. We also examine whether the phenomenon depends on which of two reactant species is incorporated in the gel. Previous studies of the kinetics [7,8] have focussed on the temporal evolution of the successive bands; our focus is on the time evolution of the first band.

EXPERIMENT

The experiments were carried out in capillary tubes of 2mm diameter and 12.5 cm length. Aqueous solutions of Na_2HPO_4 and $CaCl_2$ were used as reactants. Solutions of both electrolytes at several different concentrations were prepared at room temperature. A pre-gel mixture containing 1.5% agarose and one of the two reactants, either Na_2HPO_4 or $CaCl_2$, (i.e. the inner electrolyte) at a concentration of 20 mM was made at room temperature. The agarose was dissolved by by heating at 90°C for about half an hour. The bottom half of the capillary tube was filled with the hot pre-gel solution. The samples were allowed to stand undisturbed at room temperature (22°C) for approximately one hour until a colorless gel was formed. The top half of the capillary tube was then filled, using a syringe, with the other electrolyte solution, leaving the gel at the bottom. The capillary tubes were sealed to avoid evaporation of the solvent during the experiment. All the experiments were performed at room temperature (22°C). Care was taken to keep the tubes in a horizontal position during the experiment (over 30 days).

The time evolution of the precipitate was recorded using a CCD camera and the images were stored on a video tape. The video signal was transformed using a PC-vision image processing card to a digital image. The profiles were grabbed from the original images using a frame grabber and analysed using a digital image analysis program.

RESULTS

Figure 1 shows the periodic precipitate pattern formed in the tube twenty hours after the reaction had been initiated. The inner and outer electrolyte ratio is varied for the different cases shown. Classic Liesegang bands are observed in both cases (a, b) where the inner electrolyte (i.e. inside the gel) is Na_2HPO_4 and the outer electrolyte is $CaCl_2$ at a higher concentration. As seen in the Figure decreasing the relative concentration of the outer/inner electrolyte from 4 to 2 decreases the number of bands from 4 to 3. After waiting for over one month tube (a) shows 6 bands while tube (b) shows 8 bands. Similar bands were observed when the $CaCl_2$ was embedded in the gel and the Na_2HPO_4 was allowed to diffuse into the gel. A comparison of the experiments where the concentrations of the outer and inner electrolyte concentrations are the same but the type of electrolyte is reversed shows that the extent to which the outer electrolyte has diffused and reacted is roughly the same in the two cases ($CaCl_2$ diffuses about 15% faster than Na_2HPO_4), implying that the flux is quite similar for the two electrolytes. The formation of same number of bands irrespective of whether the cation or the anion is the inner electrolyte shows that the number of bands is insensitive to the influence of the surface charge of the growing colloidal droplet and any charges that may be immobilised by the agarose gel which is known to carry some residual sulfate. The slight variation in the spacing of the bands reflects the roughly 15% variation in the diffusion constants of the two species.

Figure 2 shows the time evolution of the initial stages of the precipitation process leading to the formation of the first band in the case of the reaction shown in Figure 1b. At the earliest time only a sharp narrow precipitate peak is observed. The position of this peak shifts into the gel with time while the width of the precipitate band increases. A slight asymmetry is evident even at the earliest time of observation. This asymmetry becomes more pronounced as the peak moves further into the gel leaving behind a broad plateau which widens. The intensity of the sharp peak decreases with time. At late time, around 2.5 hours the peak position gets fixed and is clearly visible at the leading edge of the plateau; a second Liesegang band has also appeared by this time, as shown in the 3-dimensional plot. Figure 3 shows a trace of the density along the flow direction, obtained by averaging the density over the cross-section of the tube, 14 hours after the initiation of the reaction. Four Liesegang bands are clearly visible.

This profile represents the moving reaction front of the precipitated particles. As shown in Fig. 4, at early times the distance travelled by the leading edge of the precipitate front $R(t) \sim A\, t^{1/2} - Bt^{3/2}$ where t is the elapsed time, as predicted by the reaction-diffusion-supersaturation model [13] at short times. The $t^{1/2}$ term dominates at the earliest times, while the $t^{3/2}$ term becomes important later. The systematic deviation at the latest times implies that the full solution in terms of the error function is needed to model the position of the edge as the band asymptotically settles into its fixed position. The time-evolution of R(t) over the entire duration of the experiment is very similar to the numerical results obtained in ref. 13 (c.f. Fig. 3 of [13]).

The decrease in the intensity of the peak at the leading edge (clearly visible in the 3-d plot of Fig. 2) is a direct visualisation of the Ostwald ripening mechanism. The precipitated particles nucleate when the concentration of C particles exceeds c_n; however the rapid growth into a sharp band represents the aggregation process which starts at $c > c_s$ ($c_s < c_n$). The depletion of the small particles by their incorporation into the rapidly growing aggregate particles D leaves behind the broad plateau, while the moving peak is the rapidly growing aggregate which continues to move until the region in front of it is depleted of monomer. This depletion of the monomer is responsible for the decrease in the intensity at the leading edge and the slowing down in the distance travelled by the advancing precipitation front. The broad plateau results from the largest particles which are left behind in the Ostwald ripening process, since their mobility is very small. Similar broad tails can be seen preceding each of the subsequent bands (c.f. Figs. 2 and 3). The width of these broad tails appears to decrease with increasing band number, suggesting that as the reaction proceeds in time, the concentration of the outer electrolyte may also

Figure 1. Photograph showing the precipitates of $CaHPO_4$ produced in a 1.5% agarose gel containing one of the reactants in the gel and allowing the other to diffuse in, after 20 hours of reaction. In both (a) and (b) the inner electrolyte is Na_2HPO_4 at a concentration of 20mM. In (a) the outer electrolyte is 40mM $CaCl_2$, while in (b) the $CaCl_2$ concentration is 80 mM. Similar results were observed by reversing the inner and outer electrolytes.

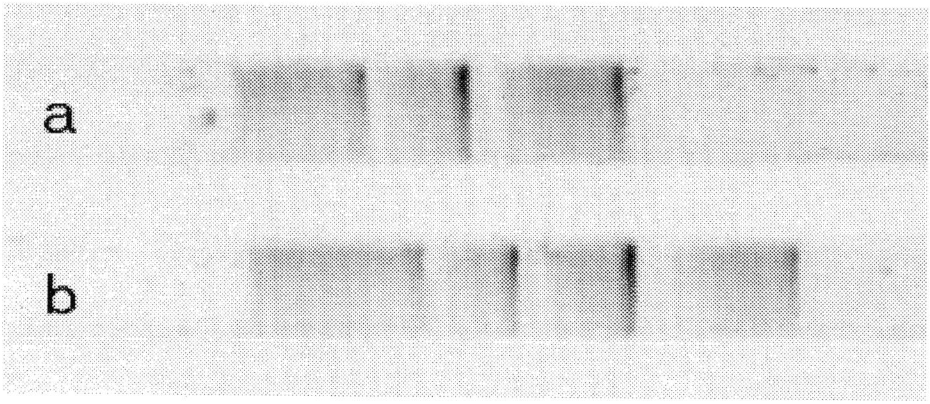

Figure 2. A 3-dimensional plot of the time evolution of the precipitate profile for the reaction shown in Fig. 1b. The first two Liesegang bands are identified as 1 and 2 at the latest times.

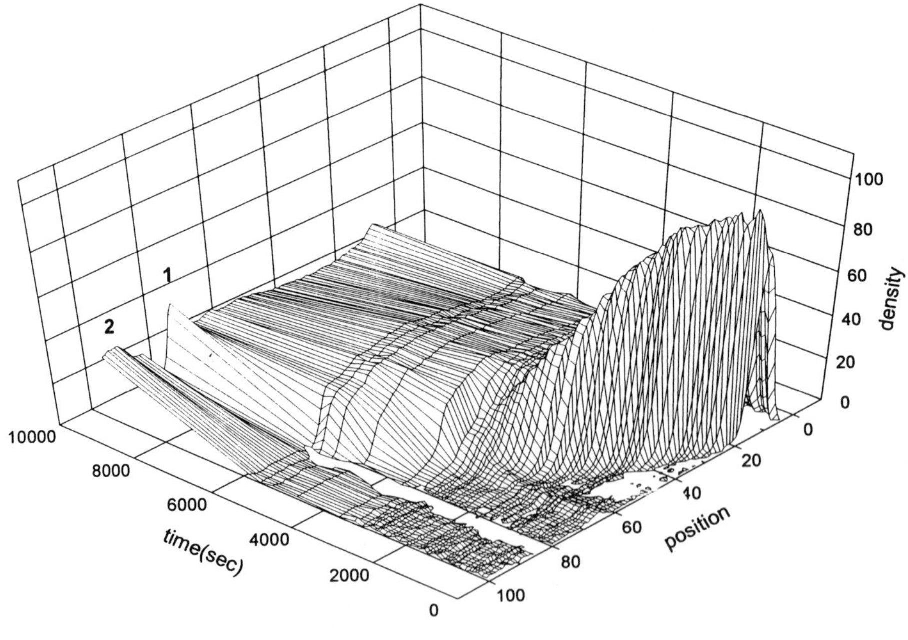

Figure 3. A density profile of the reaction in Fig. 1b after 14 hours of reaction. The graph shows 4 Liesegang bands. Note the broad plateau trailing the sharp frst band and broad tails preceding each of the subsequent Liesegang bands. The position axis is in arbitrary units. The bar in the figure shows the actual length scale of the experiment. The y-axis is proportional to the density of the precipitate.

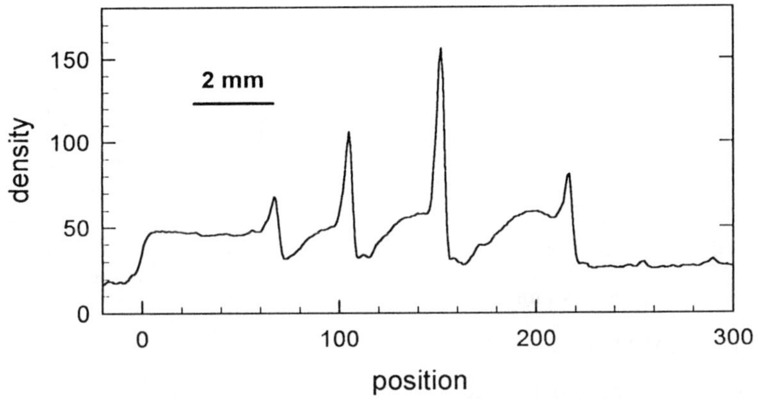

Figure 4. The time dependence of the position of the leading edge of the precipitate front, R(t) for the first Liesegang band of the reaction in Fig. 1(b). The fitted line describes the short time limit of the reaction-diffusion-supersaturation model as described in ref. 13.

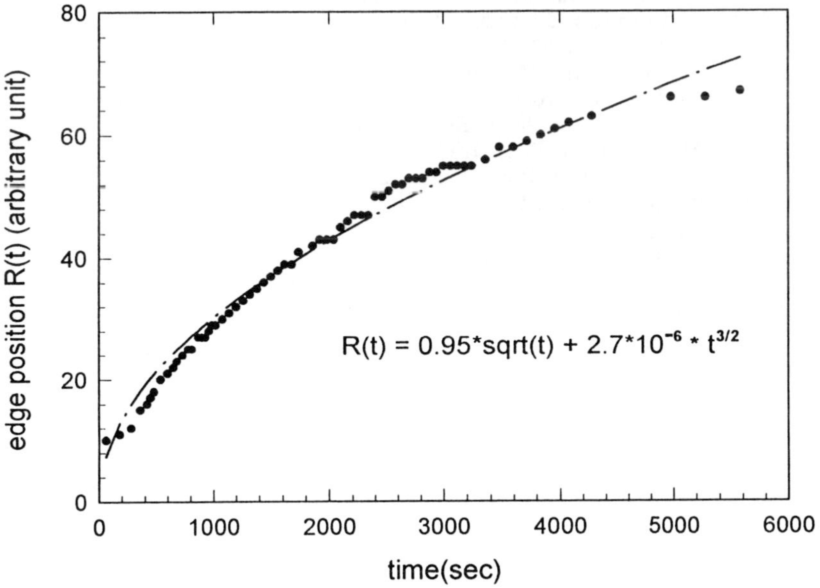

be decreasing. The characteristic profile of a broad plateau or tail with a sharp band at the leading edge is also similar to the profile observed in our experiment on confined precipitate produced by bringing the reactants from opposite ends of the gel [12]. In that experiment the confined precipitate's profile is a broad band in the middle with a sharp edge at either side. The simulation in that case [12] had shown that the sharp band is produced where the concentration of the precipitated particles equals c_s, the saturation threshold; whereas the broad band is centered where the concentration equals the nucleation threshold, c_n.

CONCLUSIONS

We have obtained direct confirmation of the supersaturation mechanism as being responsible for the formation of the discrete bands in a Liesegang band precipitation experiment in a gel. The time evolution of the first Liesegang band is in excellent agreement with the predictions of the reaction-diffusion-supersaturation model. The density of precipitated particles decreases as the supersaturation mechanism comes into play, with the condensation of the smaller particles onto the rapidly growing aggregates. We observed that band formation is independent of which electrolyte is inside the gel and which is outside the gel. This symmetry indicates that the kinetic barriers and the diffusion rates of the growing particles are not strongly dependent on the interaction of the ions, the surface charge of the colloidal particles and the residual charges, if any, of the gel.

ACKNOWLEDGMENTS

This research was supported by an NSF-DMR grant to R.Bansil. E. L. Cabarcos acknowledges DGICYT and the Complutense University for granting a sabbatical year at Boston University.

REFERENCES

1. L. Gálfi and Z. Ráacz, Phys.Rev.A **38**, 3151 (1988).
2. E. Ben-Naim and S. Redner, J.Phys.A **25**, L575 (1992).
3. G. T. Barkema, Phys.Rev.E **53**, 2017 (1996).
4. G. T. Dee, Phys.Rev.Lett. **57**, 275 (1986).
5. D. A. Smith, J.Chem.Phys. **81**, 3102 (1984).
6. R. E. Liesegang, Naturw. Wochenschr. **11**, 353 (1896). For a recent review see R. A. Schibeci and C. Carlsen, J. Chem. Edu. **65**, 365 (1988).
7. S. Kai, S. C. Müller, and J. Ross, J. Chem. Phys. **76**, 1392 (1982).
8. M. E. LeVan and J. Ross, J. Phys. Chem. **91**, 6300 (1987).
9. B. Chopard, P. Luthi and M. Droz, Phys.Rev.Lett. **72**, 1384 (1994).
10. A. Büki, E. Kárpáti-Smidrózki and M. Zŕinyl, J.Chem.Phys. **103**, 10387 (1995).
11. G. Venzi and J. Ross, J.Chem.Phys. **77**, 1308 (1982).
12. E. L. Cabarcos, C-S Kuo, A. Scala and R. Bansil, Phys. Rev. Lett. **77**, 2834 (1996).
13. J. B. Keller and S. I. Rubinow, J. Chem. Phys. **74**, 5000 (1981).

PHASE TRANSITIONS, PATTERNS AND STATISTICAL MECHANICS OF FRONT PROPAGATION IN A DYNAMIC RANDOM IMPURITY MODEL FOR STRIP, UNUSUAL TREES AND OTHER GEOMETRIES

N. VANDEWALLE, M. AUSLOOS,
SUPRAS, Institut de Physique B5, Université de Liège, B-4000 Liège, Belgium

ABSTRACT

A dynamic random impurity model is studied for describing the evolution of an advancing interface through a multiphase random medium containing mobile impurities. A short range repulsion between the front and the impurities leads to an aggregation process along the front, and to the trapping of aggregates behind the front. The patterns of trapped impurities are found to be self-organized. Some theoretical treatment is performed through a transfer matrix technique in various geometries: trees with loops instead of branches, chains of squares and triangles joined by vertices. Arguments are given for applications of such concepts, techniques and models in various cases: chemistry, biology, trafic, sociology, economy,...

INTRODUCTION

Pattern formation is a common phenomenon in nature [1]. Various physical processes leading to spatial and/or temporal self-organization have been investigated: coalescence, aggregation, convection, fragmentation, ... However, less attention has been paid to the trapping of a collection of particles pushed by an advancing interface. This case is e.g. of high interest in crystal growth where the decoration of a crystal by impurities [2] could improve the mechanical and electrical properties of the crystal. The trapping of bubbles during solidification [3] and the behavior of biological cells on a ice/water interface [4] are also addressed with this problem.

While the pushing/trapping of a single particle by a smooth interface has been experimentally and theoretically investigated [5], there is a lack of studies about the more complex phenomenon of collective motion/trapping of many pushed particles near an interface. We recently introduced a model [6] (so-called "dynamic epidemic model") which provides a tool for studying this phenomenon. Unexpected results were obtained and some of them are collected in this paper.

MODEL

The model is constrained to be the most simple as possible. The model was first defined on a two-dimensional half-space of width W. Periodic boundary conditions are used. Each site can receive a liquid unit (phase L), a solid unit (phase B) or a particle (phase A). A single site is assumed to contain at most one phase. Initially, all sites are turned into the liquid phase (L) except for a fraction x of sites which each contain an A-particle. The initial spatial distribution of particles is supposed to be random.

The growth starts from a homogeneous seed line made of the B-phase at the bottom of the lattice. At each simulation step, all liquid sites in contact with the so-called B-cluster are selected. One of them is randomly chosen and is turned into the phase B following the trivial reaction L$\rightarrow$B. Such a growth rule is equivalent to the classical Eden one [7].

The repulsive dynamical mechanism occuring between the front and the particles is then introduced as follows. If a particle is touched by the newly added solid unit, the particle makes a random move towards a nearest neighbour liquid site reducing the contact with the solid front. If the particle cannot reduce its number of nearest neighbouring solid units by such a jump, the position of the particle remains unchanged and later becomes trapped by the front in the cluster. One should note that a particle can be trapped following two different ways: (i) an A-particle can be trapped directly by the front because the former one cannot reduce its number of nearest neighbour B-solid units, or (ii) the displacement of an A-particle can be forbidden by the presence of other particles on neighbouring sites leading further to the trappingsince the pushing of aggregates is not considered herein. The selection, growth and particle motion

Mat. Res. Soc. Symp. Proc. Vol. 463 © 1997 Materials Research Society

process described here above is repeated a large number of times if possible. Indeed, the growth can sometimes stop if there is no liquid site in contact with the cluster. The growth is irreversible (i.e. far from equilibrium) and history dependent (i.e. non-Markovian).

Applications to various types of interfaces (two unmixing fluids, solid/liquid, ...) are obvious. Non-solid particles can be also envisaged like non-coagulating liquid droplets or vesicles. Many other spreading phenomena like epidemia, dielectric breakdown, gelation, ... are also concerned by the present study. Extensions of this model for attractive interactions front/particles but also for interactions between particles themselves are of interest. The dissolution of the pushed particles was also considered in a parallel work for the application to superconducting materials in isothermal undercooling conditions [8].

NUMERICAL INVESTIGATIONS IN TWO DIMENSIONS

Figure 1 presents a typical simulated pattern for x=0.20. One should first note that the distribution of A-particles in the B-matrix is inhomogeneous behind the front. In fact, the A-particles rejected by the interface are aggregating and the aggregates are then trapped behind the front [6,9].

For $x \approx x_c = 0.56 \pm 0.01$, a kinetic phase transition between the ever growing B-cluster case and a blocked B-cluster case takes place. Indeed, for $x < x_c$ the B-cluster are growing for ever; while for $x > x_c$, the growth of B-clusters is stopped due to a large amount of hindrances (A-impurities) along the front. One should remark that the x_c value is much higher than the value $x_c = 0.41$ obtained in the case of static particles, i.e. in the simple percolation problem [10] or usual epidemic model [11]. The high x_c value is explained as resulting from the aggregation phenomenon occuring along the advancing interface [6] and kink front instabilities.

Figure 1 - A typical growth patterns of width W=64. The solid-liquid interface and the trapped particles are denoted by black dots. The fraction of particles is x=0.20.

Figure 2 presents snapshots of the trapped particle distributions behind the growing front for eleven different concentrations. Mobile particles, i.e. particles which can minimize their contact with the B-phase by a move, are not represented. Patterns are illustrated for various particle fractions, i.e. from x=0.05 to x=0.55 by steps of 0.05. The distribution of trapped particles is clearly inhomogeneous in all cases except for very low values of the particle fraction x where the distribution consists in isolated trapped particles (see the x=0.05 case). As the fraction x of particles increases, the patterns tend to be made of trapped aggregates. Filamentary-like structures are well seen for a fraction x close to 0.25. Such filaments are roughly oriented parallel or at 45° to the direction of the front propagation. For higher values of x, larger aggregates are trapped showing branched structures. This is well seen for e.g. the x=0.40 case. For a particle fraction close to x=0.50, some "voids" appear in the distribution of trapped particles. Indeed, liquid regions of finite size cannot be reached by the growth since such regions are blocked by surrounding particles forming static hindrances at that high fraction of particles. One should recall that mobile particles are distributed in these voids and are not drawn in Figure 2. For $x \approx x_c$, it can be shown [6] that the voids in the trapped particle distribution have all sizes. It is not possible to specifically observe the scale invariance in the last snapshot of Figure 2 because of the finite size (40×40 lattice cells) of the picture. However, the scale invariance was numerically demonstrated considering cluster density profiles in ref.[6]. The fractal dimension of the trapped patterns was found to be $D_f \approx 91/48$ at $x \approx x_c$, i.e. the fractal dimension of epidemic clusters with static particles [11].

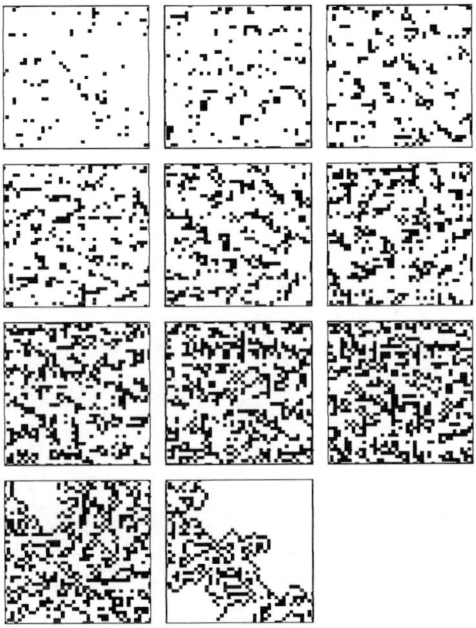

Figure 2 - Snapshots of some typical trapped particle distribution much behind the front. Eleven different particle fractions x are illustrated ranging from 0.05 to 0.55 by steps of 0.05. The size of the snapshots is 40×40 lattice cells. Particles which are still mobile in the "liquid" phase are not drawn.

THEORETICAL INVESTIGATIONS IN ONE AND INFINITE DIMENSIONS

The above model has been analytically studied on simple chains [12] and simple trees [12,13], i.e. on one and infinite dimensional lattices. For solving this problem, a transfer matrix method has been elaborated and its usefulness has been demonstrated for hierarchical lattices [13]. The main results are collected herein. The transfer matrix method provided two main results: (i) the threshold x_c for the transition and (ii) the critical exponents describing this transition. The exact solution of the model has been obtained on various chain and tree geometries as those shown in Figure 3. The chains and trees of Figure 3 are decorated by loops or not. The "loop" nature of a lattice allows the trapping of particles behind the front since the front is able to surround and bypass the small aggregates. This is not the case of loopless lattices (Figures 3a and 3d) where the front is blocked by small aggregates.

On chains (d=1), the critical fraction xc is zero meaning that it is not possible to grow indefinitely the B-phase in presence of A-impurities. On trees ($d=\infty$), it was found non-zero x_c values [?]. We gave [?] the exact solution $x_c=(z^2-1)/z^2$ for mobile particles on the Cayley tree. This is in contrast with the $x_c=(z-1)/z$ result in the case of static particles on the Cayley tree.

Moreover, the values of the critical exponents (ν and γ) describing respectively the divergence of correlation length ξ at x_c and the divergence of the cluster size S at x_c have been also calculated. In infinite dimensions (on trees), the critical exponents $\nu=1$ and $\gamma=1$ are those of percolation (the case of static particles). However, in one dimension (on chains), the values of the critical exponents depend surprisingly on the loop nature of the chain. Indeed, chains with loops have the exponents $\nu=3$ and $\gamma=3$ while simple chains have the exponent values $\nu=1$ and $\gamma=1$, i.e. those of percolation. This was unexpected.

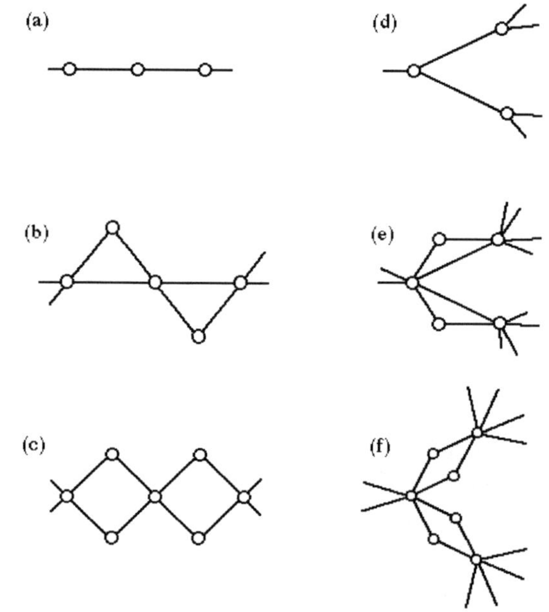

Figure 3 - The various chains (a-c) and trees (d-f) for which the problem of pushed particles was exactly solved [12,13]. In contrast with other lattices (b,c,e,f), the chain (a) and the Cayley trees (d) do not contain loops.

CONCLUSION

The dynamic epidemic model considers the growth of a cluster in a medium containing a fraction x of mobile "particles" which are pushed by a propagation front. It can be the source of many investigations and interesting predictions when being generalized to more realistic constraints as in [8] for superconducting ceramics.

ACKNOWLEDGEMENTS

Part of this work was financially supported through the Impulse Program on High Temperature Superconductors of the Belgium Federal Services for Scientific, Technical and Cultural (SSTC) Affairs under contract SU/02/013. This work was also supported through the ARC (94-99/174) contract of the University of Liège. N.V. thanks the Belgium Research Funds for Industry and Agriculture (FRIA, Brussels).

REFERENCES

1. M.C.Cross and P.C.Hohenberg, Rev. Mod. Phys. **65**, 851 (1993)

2. C.-J.Kim, S.H.Lai, and P.J.McGinn, Materials Lett. **19**, 185 (1994)

3. T.O'D.Hanley and T.Vo Van, J. Cryst. Growth **19**, 147 (1973)

4. C.Körber, Quart. Rev. Biophys. **21**, 229 (1988)

5. D.R.Uhlmann, B.Chalmers and K.A.Jackson, J. Appl. Phys. **35**, 2986 (1964)

6. N.Vandewalle and M.Ausloos, J. Phys. A **29**, 309 (1996)

7. R.Julien and R.Botet, J. Phys. A **18**, 2279 (1985)

8. N.Vandewalle, Ph.D. thesis (University of Liège, Belgium, 1996), unpublished

9. N.Vandewalle and M.Ausloos, Phys. Rev. Lett. **77**, 510 (1996)

10. A.Bunde, H.J.Herrmann, A.Margolina and H.E.Stanley, Phys. Rev. Lett. **55**, 653 (1985)

11. D.Stauffer and A.Aharony, *Introduction to Percolation Theory*, (Taylor & Francis, London, 1994) 2nd printing

12. N.Vandewalle and M.Ausloos, Phys. Rev. E **54**, 3499 (1996)

13. N.Vandewalle and M.Ausloos, Physica A **200**, 1 (1996)

Fluctuations, Lindemann Criterion and Liquid–Solid Transition in Thin Films

Alexei V. Tkachenko and Yitzhak Rabin
Department of Physics, Bar–Ilan University
Ramat–Gan 52900, Israel

Abstract

A simple model of confinement–induced freezing in thin liquid layers is presented. Under boundary conditions of vanishing displacement field at the walls, confinement suppresses the amplitude of fluctuations in a solid film below a critical value determined by the Lindemann criterion of melting. The resulting increase of the melting temperature compared to that of a bulk solid, is estimated and the connection with experiments is discussed.

1 Introduction.

Recent experimental and theoretical studies[1]–[6] show that when the thickness of a thin liquid film confined between two solid plates is reduced to the point where only a few molecular layers can fit between the plates, the film acquires shear rigidity characteristic of a solid. Computer simulations and some simple models suggest that this effect is related to the existence of wall–induced epitaxial ordering in the confined film[4]–[6]. Although this mechanism appears plausible, experiments[2, 3] have shown that the phenomenon occurs even when the diameter of the molecules of the liquid is much larger than the typical size of the constituents of the walls, i.e., under conditions which rule out the epitaxial mechanism. Indirect evidence against the epitaxial mechanism comes from the fact that, in the liquids where confinement–induced freezing is observed, solidification does not take place in films spread on a single solid surface[7] (the other surface is in contact with air). Other possible scenarios such as pressure–induced solidification or density increase due to strong attractive interactions between the walls and the molecules in the thin layer, cannot explain recent experiments which were performed under zero normal stress conditions and which reported an abrupt transition into a solid phase when some critical film thickness (6 to 8 molecular layers, depending on the molecular constituents) was reached[3].

Mat. Res. Soc. Symp. Proc. Vol. 463 © 1997 Materials Research Society

2 Description of the Model

The difficulty in constructing a theory of confinement–induced freezing (CIF) starting from first principles is that there exists no reliable theory of the liquid–solid transition even in the simpler, bulk case. Instead of following this path, we will present a phenomenological approach, based upon the theory of elasticity of solids and the empirical Lindemann criterion of melting[8]. Note that this approach allows us to avoid delicate issues related to the structure (crystalline vs. amorphous) of the solid phase and define the film as solid if it has non–vanishing shear rigidity, at least on experimentally relevant time scales.

What is the minimal model which could describe CIF? It is known that a liquid near a smooth solid surface has a layered structure [9] and one could expect that such layering in the confined geometry may stimulate freezing under conditions when the bulk is liquid. If this was true, it would be sufficient to ensure that the solid walls restrict molecular motion normal to their plane but do not affect motion in the parallel directions. However, even in the extreme limit of a single molecular layer confined between two planes with purely hard–core molecule–plane interactions, we have essentially a two–dimensional liquid whose freezing temperature (under low pressure conditions) is lower than that of the corresponding three–dimensional one.

Since confinement–induced freezing is observed at temperatures for which the bulk phase is still in a liquid state, one concludes that the presence of a hard wall which suppresses only molecular motion normal to its plane is not sufficient to produce CIF, and that the observation of the phenomenon implies that the in–plane degrees of freedom are also affected by the wall. In the following, we will assume that the coupling between the molecules in the thin film and the solid surfaces is sufficiently strong to suppress molecular motion near the walls, in all three directions. Instead of discussing the effect of confinement on the freezing of a liquid, we consider a thin solid film confined between two rigid walls and calculate the effect of boundary conditions on the thermal fluctuations of the molecules about their average positions in the solid. Then, we use the Lindemann criterion of melting (for fixed pressure) to estimate the shift of the melting temperature of the thin film with respect to that of the bulk solid.

Consider a solid film confined between two planes, $z = 0$ and $z = L$. We introduce the displacement field $\mathbf{u}$ which describes arbitrary deviations (e.g., due to thermal fluctuations) of the molecules from their average positions in the film. We demand that the displacement field vanishes on the boundaries of the sample due to strong coupling to the rigid walls:

$$\mathbf{u}\big|_{z=0} = \mathbf{u}\big|_{z=L} = 0 \tag{1}$$

Using elastic free energy of an isotropic solid[10]:

$$H = \frac{\mu}{2} \int d\mathbf{r} \left[\partial^i u^j \partial^i u^j + \frac{1}{1-2\sigma} (\partial \mathbf{u})^2 \right] \tag{2}$$

one can obtain [11] a simple and remarkable result that the maximal fluctuation amplitude in the confined film (which corresponds to the midplane of the film) is smaller than its bulk value and that the correction is inversely proportional to the number N of confined molecular layers:

$$\frac{\delta \langle \mathbf{u}^2 (L/2) \rangle}{\langle \mathbf{u}^2 \rangle_0} \simeq -\frac{1}{N} \tag{3}$$

This result is not sensitive to the non-universal behavior of the elastic theory on molecular length scales. This provides an *a posteriori* justification for our use of the continuum free energy, Eq. (2).

The calculated suppression of the amplitude of fluctuations at the center of the layer is a direct consequence of our assumption that fluctuations in all directions vanish at the wall (it can be shown that if only normal fluctuations were to vanish at the wall, the amplitude at the center would actually increase compared to the bulk).

3 Results and Discussion

We now relate our estimate of the confinement-induced suppression of fluctuations to the shift in the melting temperature, by using the Lindemann criterion. According to the strong form of this criterion which states that the value of the maximal fluctuation amplitude at the melting point is independent of N and thus is the same in the bulk and in the confined film (this statement is supported by studies of finite size effects on thermal flucutations in solids under periodic boundary conditions [13]), the confinement-induced temperature shift is given by:

$$\frac{\delta T^{melt} (N)}{T_0^{melt}} \simeq \frac{1}{N} \tag{4}$$

In deriving the above result, we used Eq. (3) and the fact that $\langle \mathbf{u}^2 \rangle_0$ is proportional to temperature. Strictly speaking, the Lindemann criterion should only be used to estimate the limit of stability (spinodal) of the solid phase. We assume, however, that the scaling and the sign of the relative correction to the critical temperature is the same for all the characteristic points of the transition (i.e. for the spinodals of the liquid and solid phases and for the point of equilibrium transition).

Note that according to Eq. (4), the temperature shift is positive for any finite thickness of the confined film. If this was true, confinement-induced freezing could be observed even in macroscopically thick layers. Clearly, this expression can not be valid when the calculated temperature shift is very small and the precision of the Lindemann criterion is not sufficient to describe it correctly. Moreover, the strong version of the Lindemann criterion can not be applied to most experiments in which CIF is produced by rapid expulsion of liquid layers from the film (at constant temperature) [2, 3]. As is clear from Eq. (4), squeezing out a molecular layer from the space between the walls is equivalent to instantaneous cooling of the film. When the thickness of the film falls below a critical value (for a given temperature), the liquid film becomes rapidly supercooled and freezes. This is likely to result in a solid phase characterized by very irregular structure and large unrelaxed stresses, which is less stable than the equilibrium crystalline phase. A disordered non–equilibrium solid film may also be formed due to the presence of impurities or surface roughness. In all these cases, the critical fluctuation amplitude which enters the Lindemann criterion of melting, is expected to be smaller in the confined solid film than in the bulk.

In the light of the preceding discussion, one can formulate a weaker version of the Lindemann criterion for rapidly confined films: the critical fluctuation amplitude at the melting temperature does not dependent on the thickness of the *non–equilibrium film* but is smaller than the critical amplitude in the *equilibrium bulk solid.* This suggests that the confinement–induced shift of the melting temperature of non–equilibrium films can be described by the following expression:

$$\frac{\delta T^{melt}(N)}{T_0^{melt}} \simeq \frac{1}{N} - \frac{1}{N_{\mathbf{max}}} \tag{5}$$

The parameter $N_{\mathbf{max}}$ is the maximal number of molecular layers, for which the confinement–induced freezing can be detected. This number is likely to be universal for pure substances with the same molecular geometry, but may be different for spherical and linear molecules. For $N \geq N_{\mathbf{max}}$, the suppression of fluctuations by the presence of the boundaries is overcome by the destabilizing effect of internal unrelaxed stresses and CIF can no longer be observed.

Eq. (5) suggests that under fixed temperature conditions, there is a sharp liquid to solid transition when the number of confined layers becomes smaller than some critical value. This general result is in qualitative agreement with experimental observations[3, 12]. The critical film thickness at a given temperature T can be determined from Eq. (5) which yields:

$$N(T) = \left(\frac{T - T_0^{melt}}{T_0^{melt}} + \frac{1}{N_{\mathbf{max}}}\right)^{-1} \tag{6}$$

The predicted temperature dependence of the critical number of molecular layers at which CIF is expected, awaits experimental confirmation. A sharp confinement–induced transition to the solid phase was observed in recent measurements of the shear response of the film[3]. The experimental study was based on the following notion. Solids, unlike liquids, possess finite shear rigidity and, as the result, the solidified confined film exhibits an elastic response up to a certain critical stress. Above this stress, mechanical instability leads to fast stress relaxation. If the film is subjected to shear at a constant shear rate, the alteration of such elastic and inelastic responses leads to the well-known stick-slip phenomenon[14]. Conversely, the observation of a stick-slip response in confined liquids is the signature of confinement-induced freezing.

The available experiments do not allow one to determine which physical processes are responsible for the onset of slip observed when the shear stress reaches a critical value. One possibility is fracture of the entire film, similar to the process which determines the yield stress of bulk solids. Another explanation is that the mechanical instability takes place at the film-wall interface (wall slip) and that the magnitude of the critical shear is determined by the maximal frictional force between the walls and the confined film. The latter scenario appears more plausible in most cases, since the coupling between the wall and the molecular layer of the film adjacent to the surface is probably weaker than that between two internal layers.

4 Conclusion

Our model suggests several possibilities for further experimental studies of CIF. One can perform measurements of the melting temperature of thin solid films obtained by CIF and test our prediction for the shift of the melting temperature, Eq. (5). In principle, one can also start with a confined liquid film and reach its freezing point by slowly changing the temperature while keeping constant the spacing between the walls. In this case the structure of the resulting solid is expected to be close to that of the equilibrium bulk crystal, and the stress-free result, Eq. (4), is likely to apply.

Finally, we would like to comment on the generality of our results. Note that the proposed mechanism is expected to apply to situations in which the difference between the temperature of the experiment and the bulk melting temperature is not too large. CIF was indeed observed in OMCTS and cyclohexane whose melting points (290 and 280 degrees K, respectively) are close to room temperature, but only liquid behavor was reported for ultra-thin layers of toluene [15] whose melting point is 178 degrees K. Note also that although our boundary conditions on the displacement field, Eq. (1), are not equivalent to the non-slip boundary conditions for fluid flow, the two types of boundary conditions have a similar physical origin. Therefore, we expect no CIF in cases where a finite slip at the boundary is observed [15].

ACKNOWLEDGMENT. We thank Jacob Klein for valuable discussions and providing us with experimental data. This research was supported by grants from the Israeli Academy of Science and Humanities and the Research Authority of the Bar–Ilan University.

References

[1] J. N. Israelashvili, P. M. McGuigan and A. M. Homola, *Science* **240**, 189 (1988).

[2] J. Van Alsten and S. Granick, *Phys. Rev. Lett.* **61**, 2570 (1988).

[3] J. Klein and E. Kumacheva, *Science* **269**, 819 (1995).

[4] M. Schoen, C. L. Rhykerd, D. Diestler and J. H. Cushman, *Science* **245**, 1223 (1989).

[5] P.A. Thompson and M.O. Robbins, *Science* **250**, 792 (1990).

[6] P.A. Thompson, M.O. Robbins and G. S. Grest, *Israel Journal of Chemistry* **35**, 93 (1995).

[7] J. Klein, private communication.

[8] D. Tabor, *Gases, Liquids and Solids*, 3rd ed. (Cambridge University Press, Cambridge, 1993).

[9] J. N. Israelashvili, *Intermolecular and Surface Forces*, 2nd ed. (Academic Press, London, 1991).

[10] L.D. Landau and E. M. Lifshits, *Theory of Elastisity*, 3rd ed. (Pergamon Press, Oxford, 1986).

[11] A. V. Tkachenko and Y. Rabin, submitted to *Phys. Rev. Lett.*

[12] J. Klein and E. Kumacheva, to be published.

[13] M. O. Robbins, G. S. Grest, K. Kremer, *Phys. Rev.* **B42**, 5579 (1990).

[14] M. L. Gee, P. M. McGuiggan, J. N. Israelashvili and A. M. Homola, *J. Chem. Phys.* **93**, 1895 (1990).

[15] S.E. Campbell, G. Luengo, V. I. Srdanov, F, Wudl and J. N. Israelashvili, *Nature* **382**, 520 (1996).

ANNULAR RINGLIKE ARRAYS FROM MIXTURES OF METAL NANOPARTICLES

P.C. OHARA, J.R. HEATH, W.M. GELBART
Department of Chemistry & Biochemistry, University of California, Los Angeles, CA 90095-1569, pohara@ucla.edu

ABSTRACT

We describe a theory for hole nucleation in volatile, wetting, thin liquid films. Hole nucleation in this system can occur due to two independent driving forces: evaporation (volatile hole nucleation) and by disjoining pressure effects (non-volatile hole nucleation). Although a concerted combination of these two can be in effect, the cases are treated separately here and will be treated jointly in the context of a simulation elsewhere. Growing holes -- dry areas of the solid surface which are otherwise wet by solvent -- are conjectured to be responsible for the formation of close-packed 2D annular ringlike arrays from dilute solutions of nearly monodisperse, large (3-5nm diameter), organically passivated metal particles [1]. When organic solutions of these nanoparticles are evaporated on a solid carbon substrate (a TEM grid), the resulting sub-monolayer structures are annular ringlike arrays due to the pinning of the rim of an opening hole by a sufficient collection of particles.

INTRODUCTION

Recently, there has been much interest in ordered, non-templated arrays of monodisperse semiconductor [2-5], metal [6-8], and metal oxide [9] nanocrystals. Attention is placed on both the formation of the superlattices, as well as their physical properties [10-11]. Most experimental work has involved extended 2- or 3-dimensional *close-packed* structures. Our work has focused on 2D annular ringlike arrays (0.1-1micron diameter) of organically passivated metal nanoparticles (3-5nm diameter) [1]. These rings are different in origin, and in size -- being *meso*scopic -- compared to other rings [12-13], which are typically *macro*scopic. The macroscopic rings generally originate from single, sessile, non-wetting, drops placed on a substrate. Because fluid flow must compensate for evaporation near the three-phase regions associated with the contact lines, particles are drawn by solvent fluxes to the edges of the pinned drop, producing a thick ring [12]. Our situation is fundamentally different, resulting in the spontaneous formation of *many meso*scopic rings of *meso*(nano)particles arising from a *single* macroscopic film of solution. We believe that these rings form when holes nucleate in the evaporating, wetting thin liquid film (as opposed to the non-wetting sessile drops). The rings result from the pinning of the contact lines of opening *holes* in the thin liquid film, due to the presence of particles.

While a great deal of work has been done on hole nucleation in *de-wetting* films below a critical thickness [14], only Elbaum and Lipson [15] appear to have investigated the corresponding problem for *wetting* layers. Their efforts, however, focus on *volatile* hole nucleation in ≈50-500nm films involving *wet* holes, whereas our problem involves thinner films (a few nanometers thick) and *dry* holes. Our films also contain suspended colloidal particles, and their effect on the inherent instabilities of these thin evaporating films must be explicitly accounted for.

There is also much theoretical work on the non-volatile behavior of liquids on solid surfaces [16]. For a wetting liquid, the liquid film never wants to thin too much because of the disjoining

Mat. Res. Soc. Symp. Proc. Vol. 463 © 1997 Materials Research Society

pressure contribution to the free energy per unit area, which has the form A_H/t^2, where $A_H = (A_{sl} - A_{ll}) > 0$ is the relevant Hamaker constant. At the same time, it wants to avoid forming holes or otherwise de-wetting the substrate, because of the positive spreading coefficient, $S = \gamma_{sv} - (\gamma_{sl} + \gamma)$, where γ is the liquid-vapor interfacial tension and $\gamma_{sv}(\gamma_{sl})$ the corresponding value for the solid-vapor (sold-liquid) interface. The equilibrium thickness $t_e=(3A_H/S)^{1/2}$ [16] reflects a compromise between these competing tendencies of the thinning, wetting film.

THEORY

We formulate here separate theoretical scenarios for volatile and non-volatile hole nucleation in wetting, thin liquid films. For the first case, we assume for simplicity that the rate of evaporation far exceeds the rate of fluid flow, such that hole nucleation is only driven by the volatility of the liquid. Furthermore, we assume that evaporation at the rim of a hole is faster than uniform evaporation -- keeping the film thickness unchanged.

In this case, we start with a drop of nanoparticle solution that spreads across the carbon substrate as a thin film. The liquid film is in a large box (large compared to the volume of liquid), and evaporates because the equilibrium vapor pressure of solvent, P^o, greatly exceeds the actual pressure above the film, P. Because of this condition ($P \ll P^o$), the film would need to thin, via evaporation, to unphysically small dimensions in order for the liquid to be in equilibrium with its vapor - equilibrium is reached when all of the liquid has evaporated. Before reaching this point, the liquid evaporates steadily as a thin film until it becomes unstable with respect to the nucleation and growth of holes.

The formation of holes in an evaporating thin film is analogous to ordinary bubble nucleation [17] in a bulk superheated liquid, with the difference that holes in thin films are "open" to the vapor *above* the film. For a system with total pressure $P_{atm} \approx 1$atm, a given liquid thickness t, a fixed area of the substrate A_s and number of liquid molecules N, the Helmholtz free energy can be written as: $A_{initial} = N\mu_l(P_{atm}) - P_{atm}A_s t + A_s(\gamma_{sl} + \gamma + A_H/t^2)$, where μ_l is the chemical potential of the liquid. A (assumed) cylindrical hole of radius R then nucleates, such that N_g liquid molecules have evaporated. We can write the free energy of this system as $A_{hole,volatile} = (N-N_g)\mu_l(P_{atm}) + N_g\mu_g(P_{atm}) - P_{atm}(A_s t-\pi R^2 t) - \rho_l \pi R^2 t k_B T + (A_s - \pi R^2)(\gamma_{sl} + \gamma + A_H/t'^2) + \pi R^2 \gamma_s + 2\pi R t\gamma$, where R can be obtained by using the liquid number density ρ_l, where $N_g = \pi R^2 t\rho_l$. The change in free energy ΔA for the nucleation of this hole is then given by $\Delta A_{volatile} = A_{hole,volatile} - A_{initial}$:

$$\Delta A_{volatile} \approx \pi R^2\{\rho_l t k_B T[\log(P/P^o) - 1] + (S - A_H/t^2)\} + 2\pi R t\gamma \qquad (1)$$

If this free energy difference is monitored as the hole grows, the barrier associated with hole nucleation can be obtained. When this barrier $\Delta A_{max,vol} = \pi\gamma t R_{max}$ is sufficiently small, on the order of $k_B T$, holes of critical radius $R_{max} = \gamma t/[t\rho_l k_B T\{\log(P^o/P_l + 1\} - (S - A_H/t^2)]$, will nucleate and grow. Equation (1) presented above is similar to that presented by Elbaum and Lipson [15], with the primary difference that our holes are dry, due to the lack of an equilibrium film thickness for our system. For our system from which annular rings are observed, typically $P \ll P^o$, such that holes can nucleate at thicknesses greater than t_e. This is where the importance of time scales (which can be accomplished via controlled evaporation experiments) becomes evident -- if the fluid flow is faster than evaporation, these holes will close. Holes will only nucleate and open when the solvent thickness is less than the equilibrium thickness, which will be discussed next. This is also the case when the vapor pressure P is close to the equilibrium vapor pressure, which may result if the system size is significantly reduced.

In addition to the above case, hole nucleation can also occur in *non-volatile* wetting fluids, where the film becomes unstable at thicknesses below a critical value ($<t_e$) [16]; holes open in order to restore the film to its equilibrium thickness $t_e=(3A_H/S)^{1/2}$. Elsewhere [1], we chose to develop a mechanism for ring formation in a slowly disappearing (effectively *non-volatile)* fluid, based on the assumption that evaporation is much slower than fluid movement -- which gives rise to a "local equilibrium" -- and has allowed us to relate equilibrium ideas to our system. Given the same initial condition as above, allow a hole of radius R open in this non-volatile system. Let $\Delta A_{hole,nv}$ denote the free energy associated with the formation of this hole. It is then straightforward to show that this quantity can be written approximately as

$$\Delta A_{hole,nv} = [\gamma t]\, 2\pi R + \{S - A_H/t^2 - A_H/tt' - A_H/t'^2\}\pi R^2 \qquad (2)$$

where t' is the final thickness, determined by conservation of liquid volume. Fluctuations corresponding to small holes will grow when barrier $\Delta A_{max,nv} = \pi\gamma t R_{max}$ is brought down to $\approx k_B T$, where $R_{max} = \gamma t/[A_H(1/t^2 + 1/t'^2 + 1/tt')-S]$. This happens when the film thickness t is sufficiently small (a few per cent) compared to the "pancake" thickness t_e. Below this thickness, the thin film becomes unstable against the nucleation and growth of holes, in its drive to achieve the optimum thickness t_e.

As holes open up and grow, solvent and particles are "pushed out" toward the bulk film of solution (see Fig. 1). The rim continues to grow unless the outward thickening force is balanced by frictonal effects due to particle-substrate interactions [19]. The mechanism for ring formation due to hole nucleation and growth, and the effects of particle size and concentration, are discussed in detail elsewhere [1]. The observed inverse dependence of ring size with concentration has provided the crucial argument that holes are important in our system. In the absence of a sufficient number of large particles collected along the hole rims, these holes will percolate to form drops -- resulting in compact structures [6].

In determining the effect of the particle on the film, we noted that capillary forces [20] between particles -- arising from the deformation of the liquid surface due to the presence of particles -- are weak because the film thickness is small enough for disjoining pressure to dominate over hydrostatic pressure (gravity). For nanometer film thicknesses, the effective capillary length λ, or decay length of the capillary forces, is short-ranged -- on the order of angstroms.

CONCLUSIONS

We have presented here two mechanisms for hole nucleation in volatile, wetting thin liquid films: volatile hole nucleation driven by evaporation; and non-volatile hole nucleation due to disjoining pressure effects. The importance of each mechanism can be varied in an experimental setup, via controlled evaporation experiments, the use of different solvents, or by varying the vapor pressure. Hole formation and the presence of particles in solutions have been related to mesoscopic annular ring formation. Although both forms of hole nucleation may be operative in producing rings, we have presented the two cases separately. The importance of the time scales of fluid flow and evaporation on the resulting annular ringlike configurations are currently being addressed, and these results will presented elsewhere.

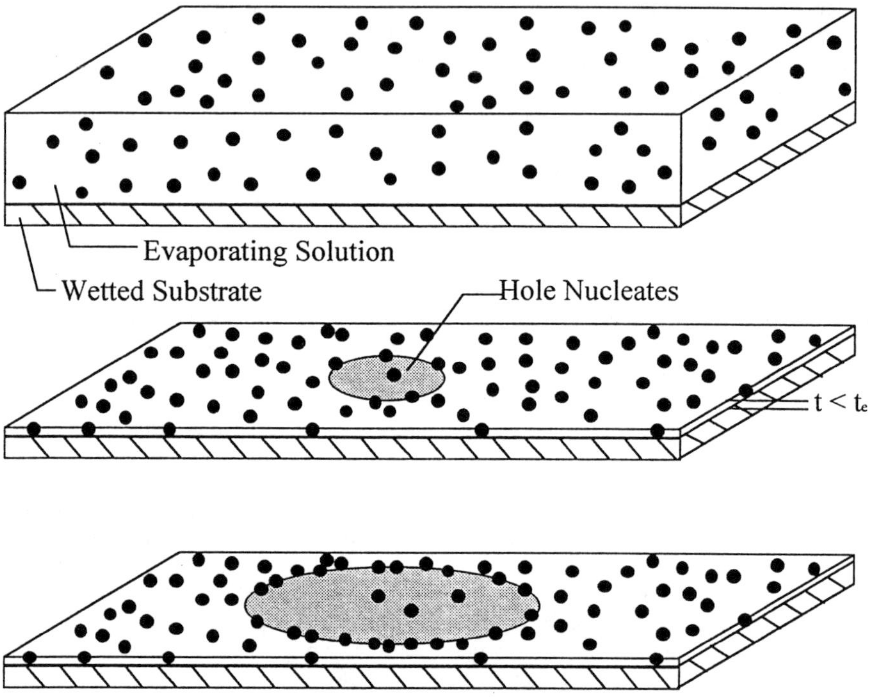

FIGURE 1. This figure shows a pictorial scenario of the evaporating solution wetting the solid substrate (top), and thinning to a thickness $t_{hole}<t_e$, at which a hole nucleates (middle). The hole opens, pushing most particles out along its advancing rim (bottom), which becomes an annular ring.

ACKNOWLEDGEMENTS

W.M.G. acknowledges current funding from the National Science Foundation, J.R.H. from the NSF-NYI program and Packard Foundation, and P.C.O. from the University of California, Los Angeles in the form of a Dissertation Year Fellowship. The authors would also like to thank Prof. Howard Reiss, Prof. Robijn Bruinsma, and Prof. Andrea Liu for many helpful discussions.

REFERENCES

1. P.C. Ohara, J.R. Heath, W.M. Gelbart (submitted).

2. N. Herron, J.C. Calabrese, W.E. Farneth, Y. Wang, Science **259**, 1426 (1993).

3. T. Vossmeyer, G. Reck, L. Katsikas, E.T.K. Haupt, B. Schulz, H. Weller, Science, **267**, 1476 (1995).

4. A. Eychmuller, A. Mews, H. Weller, Chem. Phys. Lett. **208**, 59 (1993).

5. C.B. Murray, M.G. Bawendi, Science **270**, 1335 (1995).

6. P.C. Ohara, D.V. Leff, W.M. Gelbart, J.R. Heath, Phys. Rev. Lett. **75**, 3466 (1995).

7. L. Motte, F. Billoudet, M.P. Pileni, J. Phys. Chem. **99**, 16425 (1995).

8. W.D. Luedtke, U. Landman, J. Phys. Chem. **100**, 13323 (1996).

9. M. Giersig, P. Mulvaney, Langmuir **9**, 3408 (1993).

10. T. Takagahara, Surf. Sci. **267**, 310 (1992).

11. C.R. Kagan, C.B. Murray, M. Nirmal, M.G. Bawendi, Phys. Rev. Lett. **76**, 1517 (1996).

12. R. Deegan, O. Bakajin, T.F. Dupont, G. Huber, S.R. Nagel, T.A. Witten (submitted).

13. E. Adachi, A.S. Dimitrov, K. Nagayama, Langmuir **11**, 1057 (1995).

14. See J.A. Moriarty, L.W. Schwartz, J. Colloid Int. Sci. **161**, 335 (1993) and references contained therein.

15. M. Elbaum, S.G. Lipson, Phys. Rev. Lett. **72**, 3562 (1994).

16. See, for example F. Brochard-Wyart, J.M. di Meglio, D. Quere, P.G. de Gennes, Langmuir **7**, 335 (1991) and P.G. de Gennes, Rev. Mod. Phys. **57**, 827 (1985).

17. P. DiBenedetti, Metastable Liquids, (Princeton University Press, 1996).

18. F. Brochard-Wyart, J. Daillant, Can. J. Phys. **68**, 1084 (1990).

19. See, for example, S.A. Safran, Statistical Thermodynamics of Surfaces, Interfaces, and Membranes, (Addison-Wesley, 1994).

20. See P.A. Kralchevsky, I.B. Ivanov, K. Nagayama, J. Colloid Int. Sci. **151**, 79 (1992) and related work.

Part VIII

Granular Media

Foam Drainage: Extended Large-Q Potts Model Simulation

Yi Jiang*, James A. Glazier
Department of Physics, University of Notre Dame, Notre Dame, IN 46556 USA

ABSTRACT

We study foam drainage using the large-Q Potts model extended to include gravity on a three dimensional lattice. Without adding liquid, homogeneously distributed liquid drains to the bottom of the foam until equilibrium between capillary effects and gravity is reached, while in an ordered dry foam, if a fixed amount of liquid is added from the top, a sharp flat interface between the wet and dry foam develops. The wetting front profile forms a downward moving pulse, with a constant velocity. The pulse decays over time while its leading edge for a brief time behaves like a solitary wave. With continuous liquid addition from the top, the pulse does not decay and we observe a soliton front moving with a constant velocity. Continuously adding liquid to an initially wet foam keeps the liquid profile constant. Our simulations agree with both experimental data and simplified mean field analytical results for ordered foams but predict an unstable interface for disordered foams.

INTRODUCTION

Foams are macroscopically structured cellular materials, of practical importance in applications from brewing to lubrication. A liquid foam, with clearly separated liquid and gas phases, has a characteristic time scale of evolution and distinct rheological properties [1]. Foams are also a model for other nonequilibrium, disordered materials, such as polycrystalline metals, in which the dynamics is driven by the surface energy. Characterizing the structure and evolution of foams challenges current theory and experiments. In three dimensional liquid foams, gravitational drainage of the liquid is an additional complication. Often drainage, rather than diffusion, determines foam stability and properties. Mysels *et al.* were the first to investigate the different types of thin film drainage [2], concentrating on vertical films formed by withdrawing glass frames from pools of surfactant solution. Early experiments, measuring the amount of drained liquid as a function of time, did not offer much immediate insight [3]. Princen [4] has discussed the asymptotic vertical equilibrium profile of drainage by considering the osmotic pressure of foams. Mean field calculations on the microflow in Plateau channels in relatively dry foams suggested a solitary wave solution for liquid flow in foam [5] [6]. Recently Weaire *et al.* measured the velocity of the interface between the dry and wet monodisperse foams in terms of the rate of continuous addition of liquid from the top. AC capacitance and conductance measurements of the vertical liquid profiles for different types of drainage could not distinguish horizontal wetness changes [7] [8] [9]. Because the foam is inhomogeneous and resembles a porous medium, we might expect that the penetration of liquid into the Plateau channels would result in an unstable interface or in viscous fingering. Instead, the mean field theory predicts a flat interface.

Magnetic resonance imaging (MRI) allows a non-invasive visualization of the interior structure of foams. German, McCarthy *et al.* measured one dimensional vertical wetness profiles of draining beer foams, whipped cream, and egg white *etc.* [10]. Prause *et al.* have taken images of three dimensional foams with high resolution [12] and 1-d drainage profiles,

Mat. Res. Soc. Symp. Proc. Vol. 463 © 1997 Materials Research Society

but the full three dimensional drainage experiments remain to be done.

MEAN FIELD CALCULATIONS

Theoretical calculations of the liquid flow during foam drainage assume that the liquid flows through a system of interconnected, randomly oriented Plateau channels, with negligible flow through the much thinner soap films [2]. A monomolecular layer of surfactant on the film surfaces reduces the surface tension and retards liquid flow, resulting in non-slip conditions on the boundaries of channels. Neglecting bubble size distribution, bubble density and liquid viscosity *etc.* produces a mean field theory.

The capillary pressure, i.e. the pressure across the liquid surface, is: $P_{capillary} = P_l - P_g = \frac{2\gamma\cos\theta}{r}$, where γ is the surface tension, θ is the contact angle between liquid and gas, and P_l and P_g are the pressure of the liquid and gas, respectively.

For an incompressible liquid, the derivative of equation (1) yields the between the change of gas pressure δP and the change in the channel area δs:

$$\delta P = \gamma \alpha s^{-3/2} \delta s, \tag{1}$$

where α is a constant determined by the curvature of the lateral surface of the channel. For ideal Plateau channels, the radius of curvature of the lateral surface r can be related to the area of the Plateau channel s by $s = (\sqrt{3} - \pi/2)r^2$, yielding $\alpha = (\sqrt{3} - \pi/2)^{\frac{1}{2}}$.

Assuming the motion of the liquid in Plateau channels to be microscale flow, we apply the continuity equation to the channel flow:

$$\frac{\partial s}{\partial t} + u\frac{\partial s}{\partial x} + x\frac{\partial u}{\partial x} = 0, \tag{2}$$

where u is the velocity of the flow.

Following the derivation in [5], we consider the solution of the Navier-Stokes equation for massive fluid in a tube of arbitrary cross section:

$$u = -\frac{s}{\beta\eta}\left[\frac{1}{3}\frac{\partial P}{\partial x} - \rho g\right], \tag{3}$$

where η is the viscosity of the fluid, the factor 1/3 comes from averaging over the directions of motion of the liquid, and β is a numerical coefficient depending only on the shape of the cross section of the channel: e.g. $\beta = 8\pi$ for a tube of circular section, $\beta = 49.1$ for a cross section in the form of a Plateau triangle [5]. This equation can also be viewed as the force balance between the capillary force, the pressure gradient and gravity [6].

If we eliminate p and u from the above equations and introduce $x_0 = \sqrt{\gamma/\rho g}$ and $t_0 = \beta\eta/\sqrt{\gamma\rho g}$, the equations are reduced to a nonlinear PDE:

$$\frac{\partial S}{\partial \tau} - \frac{\partial}{\partial X}\left[\alpha S^{1/2}\frac{\partial S}{\partial X} - S^2\right] = 0, \tag{4}$$

with dimensionless variables $X = x/x_0$, $\tau = t/t_0$ and $S = s/{x_0}^2$.

Solving equation(4) in a moving frame, we get the following three solutions:

$$S = \begin{cases} v \\ v\tanh^2\left[\frac{\sqrt{v}}{\beta}(X - v\tau)\right] \\ v\coth^2\left[\frac{\sqrt{v}}{\beta}(X - v\tau)\right] \end{cases} \tag{5}$$

The first solution to the channel area describes a steady state, in which the drainage reaches an equilibrium with the influx of liquid. The second solution is not physical as it diverges at $X - v\tau = 0$. The last one, a solitary wave traveling at a constant velocity, was observed in experiments and was proposed in [5] and [6].

SIMULATIONS

The large-Q Potts model partitions the foam on a three dimensional square lattice. A bubble, σ, is the collection of lattice sites $\vec{i} = (x, y, z)$ with spin $\sigma(\vec{i}) = \sigma$, σ ranging from 1 to Q ($Q > 5000$ in this study). While it was originally used to simulate foam coarsening [13] in static one-phase systems, it can be easily extended to include drainage. We treat the two phases in the system, liquid and gas, as two bubble types, with the liquid represented as a single bubble subject to a gravitational field. The extended Potts Hamiltonian of the system is then:

$$\mathcal{H}_{Potts} = \sum_{\vec{i},\vec{i}' neighbors} J_{\tau(\sigma),\tau(\sigma')}[1 - \delta_{\sigma(\vec{i}),\sigma(\vec{i}')}] + \sum_{liquid} g\ x_{liquid}. \tag{6}$$

where $J_{\tau(\sigma),\tau(\sigma')}$ is the coupling constant between bubbles, where $J_{gas,gas} > J_{gas,liquid} > J_{liquid,liquid}$; g is the force of gravity per unit density and x_{liquid} is the average height of the liquid component of the foam. The second term in $\mathcal{H}_{Potts}$ applies to liquid only.

At each time step, a site is selected at random, and a spin change to a neighboring spin proposed. Only the changes that lower the total energy of the system are accepted, corresponding to zero temperature dynamics.

Boundary conditions are periodic in the horizontal direction but rigid in the vertical direction. We use a very ordered initial foam which affects the shape of the wetting interface. We monitor the mean liquid fraction as a function of vertical position. Free drainage has no input of liquid and the liquid begins with a homogeneous vertical distribution. In pulsed drainage, the foam is initially dry and a fixed amount of liquid placed above the dry foam drains under gravity. In forced drainage, liquid is added from the top of the foam at a constant rate, different initial conditions (wet or dry foams) resulting in different profiles.

DISCUSSION

Figure 1a shows the time evolution of the liquid profiles in free drainage. With no liquid input, the initially homogeneously distributed liquid drains downwards until gravity is balanced by capillary effects, to produce an equilibrium profile. Drainage profiles from MRI experiments by McCarthy [10] and our group(figure 1b) are very similar, though it is difficult to compare them quantitatively. Efforts to quantify the slope of the linear part of the profile have not had much success.

In figure 2, we plot the liquid profiles for pulsed drainage. The pulse is monitored as it descends. The leading edge of the pulse, over short intervals such that decays is negligible, is quantitatively identical to the soliton in [9] [8]. In figure 2, we fit the leading edges of profiles to the same soliton solution. A longer time simulation (long distance for the pulse) allows us to see liquid transfer from the peak of the soliton to the tail, resulting in pulse spreading. The wave front of the soliton moves at a constant velocity, in good agreement with [9]. The high frequency structure of the solitary wave is due to the ordered structure of the simulated foams, shown in the vertical cross sections in figure 3.

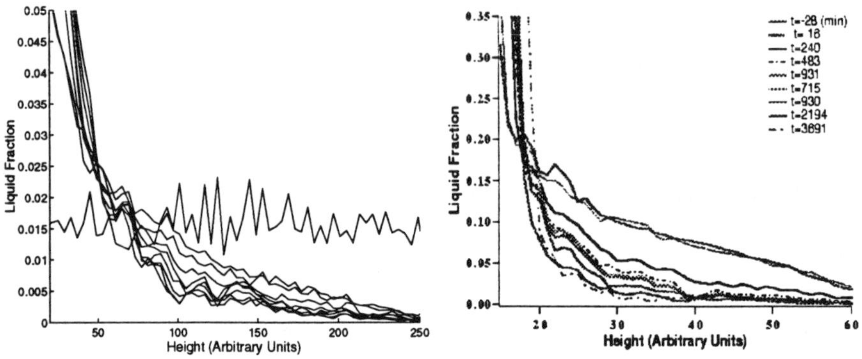

(a) Potts Model Simulation (b) MRI experimental data

Figure 1. Liquid profiles for gravitational drainage as a function of time

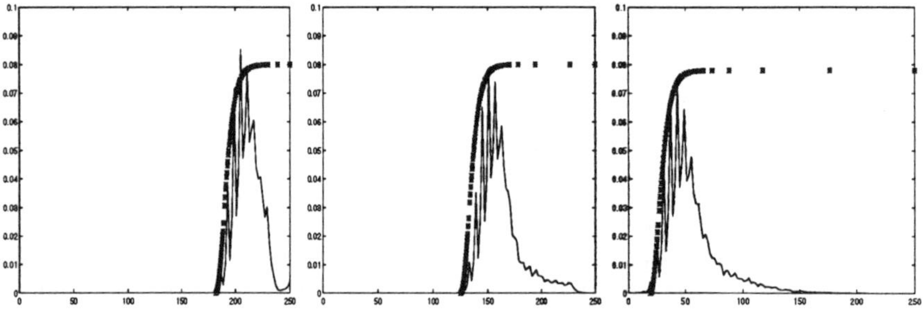

(a) 3000 (b) 5000 (c) 9000

Figure 2. Liquid profiles for pulsed drainage in Monte Carlo steps (MCS)

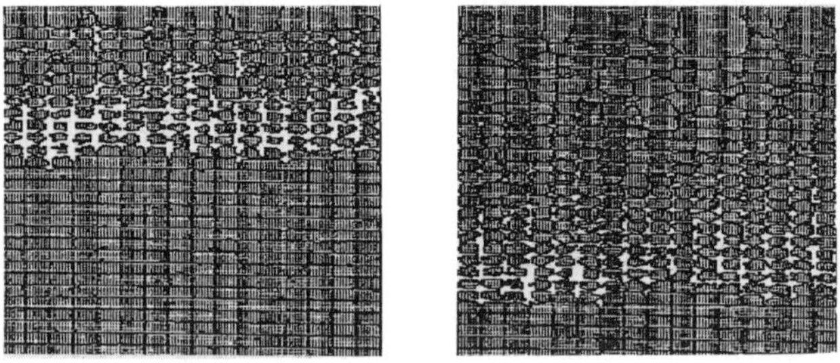

(a) 3000 MCS (b) 8000MCS

Figure 3. Snapshots of vertical cross sections of pulsed drainage

Naturally, for a continuous liquid supply, the soliton has a non-decaying fixed profile (Figure 4a), matching both the analytic solution and the experiment results. In addition, the width of the wave front spreads over time, as predicted in [7].

On the other hand, if the foam is initally wet, continuously adding liquid does not change the shape of the profiles except to increase the foam wetness (Figure 4b), indicating equilibrium between the liquid input, gravity and capillary forces, corresponding to the steady state solution $S = v = constant$.

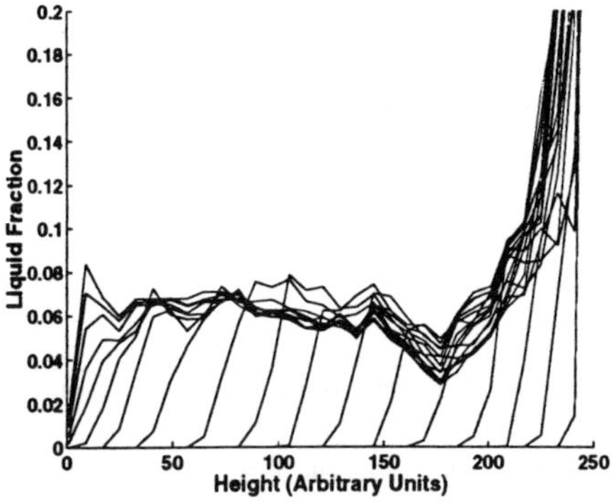

(a) Initially dry foam

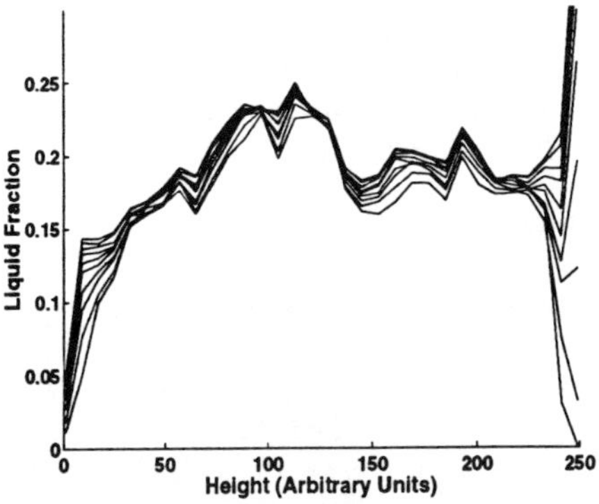

(b) Initially wet foam

Figure 4. Liquid profiles for forced drainage

When the rate of coarsening is comparable to that of drainage, the interface between the wet and dry foam is diffuse due to the disordered structure. The liquid forms clusters in the middle of the foam (Figure 5). The penetrating of liquid into relatively dry foam through the Plateau channels resembles the front instability of viscous fingering. Thus mean field theory fails for disordered foams. When drainage is much faster than coarsening, the interface is flat, within a band thinner than the dimension of a bubble. Thus drainage of a disordered foam is actually a porous medium problem with evolving pore sizes and shapes.

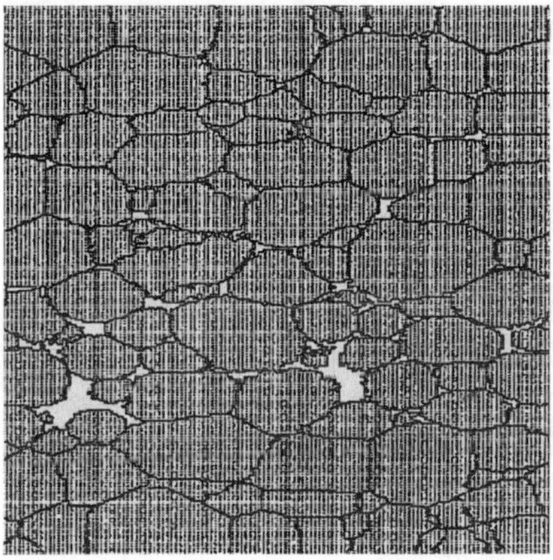

Figure 5. Snap shot of drainage in a disordered foam.

CONCLUSIONS

The extended large-Q Potts model simulations produce results comparable to experiments and also validate the mean field drainage theory for ordered foams. The model allows us to continuously vary the liquid content, as well as the strength and range of interactions between like and unlike phases. With experimental data for foams of varying compositions and growth characteristics, we will develop our simulation to predict the behavior of foams of arbitrary viscosity and wetness, leading to quantitative simulations of the properties of industrial foams.

ACKNOWLEDGMENTS

The authors acknowledge support from the Exxon Educational Foundation, ACS/PRF & NSF/NYI DMR-9257011.

REFERENCES

* Electronic address: yjiang@charpentier.phys.nd.edu

[1] A. Kraynik, Ann. Rev. Fluid Mech., **20**, 325 (1988).
[2] K. J. Mysels, K. Shinoda, S. Frankel, *Soap Films: Studies of Their Thinning and a Bibliography*, Pergamon Press, Oxford (1959).
[3] J. J. Bikerman, *Foams: Theory and Industrial Applications*, Reinhold, New York, (1953).
[4] H. M. Princen, Langmuir **2**, 518 (1986).
[5] I. I. Gol'dfarb, K. B. Kann, and I. R . Sreiber, Fluid Dyn. **23**, 244 (1988) and references therein.
[6] G. Verbist and D. Weaire, Euro. Phys. Lett. **26**,631 (1994).
[7] D. Weaire, N. Pittet, S. Hutzler and D. Pardal, Phys. Rev. Lett. **71**,2670 (1993).
[8] D. Weaire, S. Findlay and G. Verbist, J. Phys. Condens. Matter **7**, L217 (1995).
[9] S. Hutzler, G. Verbist, D. Weaire and K. A. van der Steen, Europhys. Lett. **31**, 497 (1995).
[10] J. B. German and M. J. McCarthy, J. Agric. Food Chem. **37**, 1321 (1989); M. J. McCarthy and E. Perez, Biotechnol. Prog. **7**, 540 (1991).
[11] C. P. Gonatas, J. S. Leigh, A. G. Yodh, J. A. Glazier and B. Prause, Phys. Rev. Lett. **75**, 573 (1995).
[12] B. A. Prause, J. A. Glazier, S. J. Gravina and C. D. Montemagno, J. Phys. Condens. Matter **7**, L511 (1995).
[13] M. P. Anderson, D. J. Srolovitz, G. S. Grest and P. S. Sahni, Acta metal. **32**, 783 (1984) J. A. Glazier, M. P. Anderson, and G. S. Grest, Philos. Mag. B **62**, 615 (1990).
[14] J. A. Glazier, Ph.D. thesis, Univ. of Chicago (1989) unpublished.
[15] J. Stavans and J. A. Glazier, Phys. Rev. Lett.**62**, 1318 (1989); J. A. Glazier, S. P. Gross, and J. Stavans, Phys. Rev. A **36**, 306 (1987).

DYNAMICS OF GRAINS IN DRIVEN GRANULAR MEDIA

NARAYANAN MENON, DOUGLAS J. DURIAN
Department of Physics and Astronomy, University of California, Los Angeles, CA 90095-1547

ABSTRACT

We use diffusing-wave spectroscopy (DWS) to study microscopic dynamics in the interior of 3-dimensional granular systems. We study two granular systems where particle motions are excited by different driving mechanisms -- gravity-driven channel flow and a gas-fluidized bed. In both instances we obtain detailed information about short-time collisional dynamics such as rms velocity fluctuations, mean free paths, and collision frequencies. We also observe a slow crossover from short-time ballistic motions to long-time, grain-scale diffusive motions.

INTRODUCTION

In the absence of external driving forces, a granular system comes to rest because of dissipative interactions such as friction and inelastic collisions between individual grains. However, when subject to geophysical processes, human handling or manipulation in industrial situations, granular systems are driven into motion [1]. We do not, however, have an established 'fluid mechanics' for sand to describe macroscopic response to external sources of energy. The major effort toward such a continuum description are in the form of kinetic theories [2] based on the intuition that sand may be described as a gas of inelastic particles. In contrast to molecular fluids, the scale of velocity fluctuations is non-thermal and in these models is introduced as a temperature (referred to as the 'granular temperature') which is determined by the state of flow. While this qualitative picture appears reasonable (at least in the limit of a dilute, highly-driven system), it has never been established by experiment or been quantitatively probed in real 3-dimensional flows.

In this article we present a direct probe of local, short-time dynamics of grains in dense, fully three-dimensional systems with Diffusing Wave Spectroscopy (DWS) [3] a multiple light scattering technique which measures mean-square particle displacements as a function of time down to 10^{-8} secs with a resolution of 1Å. For two different driving mechanisms -- gravity-driven flow and gas-fluidization -- we deduce important microscopic inputs to a description of transport : the rms velocity fluctuations, the mean free path and mean collision frequencies. Since granular systems are opaque due to scattering at grain surfaces, most previous experimental work has been confined to quasi-two-dimensional flows. MRI techniques [4] and (less directly) video imaging of tracer beads [5] can yield information about 3-dimensional flows. However, as we shall show here, the short-time collisional dynamics are at too short length (.01 - 1 μm) and time (10^{-6}-10^{-4} s) scales to be directly accessible to these techniques. Furthermore, we shall also show that the relationship between the short-time collisional motions and long-time diffusive and convective processes is more complex than previously believed, rendering it impossible to infer the collisional parameters from measurements at scales greater than the particle size.

EXPERIMENT

Materials

We use cohesionless, smooth, spherical glass beads with diameters, d = 49±5, 95±15, 97±11, and 194±17 μm. The beads are prepared by baking them and cooling in a dry, inert atmosphere before use. In channel flow, we monitor the effect of humidity on the beads and

Mat. Res. Soc. Symp. Proc. Vol. 463 © 1997 Materials Research Society

repeat this procedure as necessary. In the fluidized bed, a continuous flow of N_2 gas prevents such effects.

Gravity-driven channel flow:

Sand flows from a large reservoir into a vertical, rectangular channel (30 cm high, 10 cm wide and 0.3 - 1 cm thick), and then flows out through a grid of finely-spaced holes at the bottom. The flow velocity in the channel is varied between .03 and 3 cm/s by controlling the size and spacing of the grid. The flow is stationary, and if a uniform grid is used, may be characterized by a single average density and flow velocity, V_f, everywhere in the channel. Spatial gradients are found to be small in all three directions by long-range video microscopy, light transmission, and DWS. We have also studied non-uniform flows, where we create macroscopic velocity gradients using a non-uniform grid at the outlet. The arrangement of beads in flow shows no evidence of density inhomogeneities or crystalline clusters. The effect of the ambient air is small: the Bagnold number, given by the ratio of grain inertia to the Stokes drag, is greater than 10^4.

Fluidized bed

The bed is contained in a glass cell of square cross-section (5.9 cm x 5.9 cm) and supported on a frit which also functions as a gas distributor. The fluidizing gas is N_2, metered into a chamber below the distributor at a regulated pressure through a needle valve. A flowmeter measures the volume flow rate of gas. The pressure drop across the frit is measured by a differential pressure transducer as a function of gas velocity and is subtracted from the total pressure drop to obtain the drop across the bed.

Light scattering

We illuminate the sample using an Ar^+ laser with wavelength, λ=488 or 514 nm and 3mm beam waist. Incident photons perform random walks through the sample due to multiple scattering by the beads, and interfere producing a speckled interference pattern. A description of diffusive photon transport [3] through the medium involves two length scales: a transport mean free path, l^*, and an absorption length l_A. These lengths are determined by measuring the fraction of light transmitted through the sample, T, as a function of its thickness and fitting to the solution of the diffusion equation treating the sample as a slab of infinite extent and thickness, L:

$$T = \frac{\sinh(\sqrt{y}) + \sqrt{y}\cosh(\sqrt{y})}{(1+y)\sinh\left(\frac{L}{l^*}\sqrt{y}\right) + 2\sqrt{y}\cosh\left(\frac{L}{l^*}\sqrt{y}\right)} \quad (1)$$

where $y = 3l^*/l_A$ (we assume here and in equation 2 that the extrapolation length ratio [3], z_e=1). In Fig. 1 we show the results of such a fit applied to transmission through our samples.

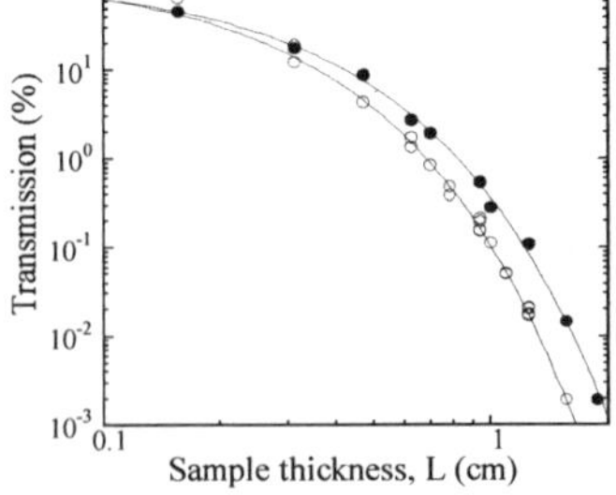

Fig. 1 Transmission vs. thickness, L, in cm for 194μm beads (open circles) and 95μm beads (solid circles). The lines are fits to eqn. 1

DWS involves measuring intensity fluctuations of speckles due to grain motions. From the measured intensity autocorrelation function, we compute the normalized electric-field autocorrelation function $g_1(\tau)$ using the Siegert relation [3]: $g_1(\tau) = [\langle I(\tau)I(0)\rangle / \langle I(\tau)\rangle^2 - 1]/\beta$, where β is a constant related to the collection optics. Using diffusion theory with the

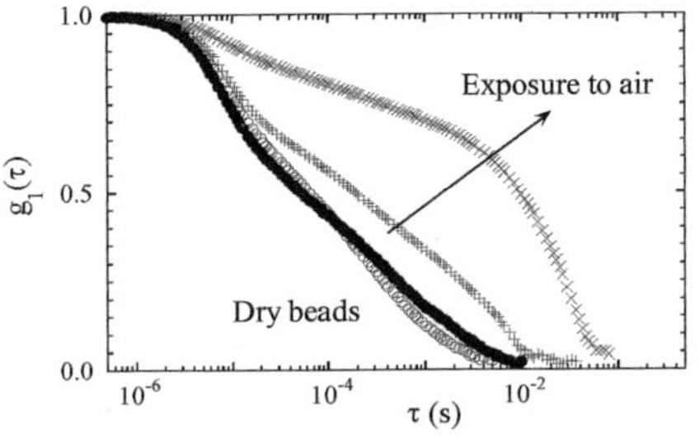

Fig.2 Electric field autocorrelation function, $g_1(\tau)$ vs. τ in secs. Circles (open and solid) represent 95μm beads, prepared dry, the plus (+) and cross (x) symbols represent increasing exposure to atmosphere.

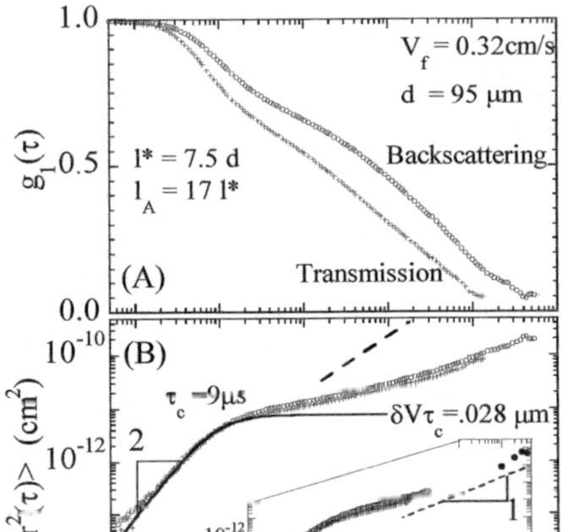

Fig. 3 a) Electric field autocorrelation function, $g_1(\tau)$, vs τ, in transmission (+) and backscattering (o). (b) $\langle\Delta r^2(\tau)\rangle$ vs τ. The solid line is a fit to $(\delta V\tau)^2/(1+(\tau/\tau_c)^2)$, showing ballistic motion with a random velocity, δV, in a cage of size $\delta V\tau_c$. The inset shows $\langle\Delta r^2(\tau)\rangle$ vs τ for 194 μm sand ($l^*/d=5.3$, $l_A/l^*=5$). The open symbols are from DWS and the solid circles are video measurements in 1-d, perpendicular to the flow. In both the inset and main figure, there is sub-diffusive motion (diffusion is indicated by dashed lines) over 3-4 decades in time for $\tau > \tau_c$.

appropriate boundary conditions for the specific sample and scattering geometry in use and including absorption effects [3], $g_1(\tau)$ may then be inverted to obtain the mean-squared displacement $<\Delta r^2(\tau)>$ of the scattering sites. The relevant expressions are: $g_1(\tau) = f(x+y)/f(y)$ where in transmission,

$$f(x) = \frac{\sinh(\sqrt{x}) + \sqrt{x}\cosh(\sqrt{x})}{(1+x)\sinh\left(\frac{L\sqrt{x}}{l^*}\right) + 2\cosh\left(\frac{L\sqrt{x}}{l^*}\right)} \quad (2)$$

and in back-scattering, $f(x)=\exp(-2\sqrt{x})$
where $x=x(\tau)\equiv k^2<\Delta r^2(\tau)>$, $k=2\pi/\lambda$, and $y=3l^*/l_A$.

RESULTS

Gravity-driven channel flow:

Typical data for normalized electric-field autocorrelation functions, $g_1(\tau)$, are shown in Fig. 2 for light transmitted through the sample. The agreement of the solid and open circles, which show different experimental runs on identically prepared beads, demonstrates reproducibility. The shape of $g_1(\tau)$ is very different for sand that is aged by exposure to air -- presumably by adsorbing moisture -- in spite of the fact that the flow velocity, V_f, is not appreciably affected by aging. Aside from presenting a useful diagnostic to test the preparation of beads, this suggests that the features in $g_1(\tau)$ are determined by relative motions of beads rather than their average flow (which decorrelates $g_1(\tau)$ at times of 0.1-10s)

In Fig. 3a we show two more examples of the electric field autocorrelation function $g_1(\tau)$ measured in transmission and backscattering on 95 μm sand with a flow velocity V_f = 0.32 cm/s along with the grain displacement, $\langle\Delta r^2(\tau)\rangle$, obtained by the prescription of eqns 2 a and b. The dynamics inferred from very different $g_1(\tau)$ in Fig. 3a coincide, showing the robustness of the analysis. This also implies that the dynamics of grains are uniform across the thickness of the channel, since photons transmitted through the sample are scattered by sand grains through the bulk of the cell whereas photons backscattered from the sample are scattered mainly by sand grains within a few l* of the walls. The short-time motion of sand grains is ballistic, i.e. $\langle\Delta r^2(\tau)\rangle=(\delta V)^2\tau^2$, where δV is a randomly directed velocity. The mean free time τ_c represents the duration of this ballistic flight, which is terminated by a collision with a neighbouring grain. The mean free path, $s=\delta V\tau_c$, corresponds to the mean distance between surfaces of neighbouring beads. Further changes in the relative positions of grains of grains requires many collisions. The slow increase in $\langle\Delta r^2(\tau)\rangle$ at times longer than τ_c is interpreted as a gradual distortion of the cage of neighbours. The solid line in Fig. 3b shows ballistic motion of grains with a randomly directed velocity δV within a fixed cage of size s.

In Fig. 4 we display the dependencies of the mean velocity fluctuation (δV), the mean free path (s) and mean free time (τ_c) on the flow velocity, V_f. Transmission measurements taken on cells of different thicknesses as well as in back-scattering have the same dependence on V_f showing that velocity gradients perpendicular to the walls of the channel are negligibly small. (Gradients in the other two directions are directly confirmed to be small by scanning the beam over the sample). Fig. 4a shows again that δV is comparable in magnitude to V_f and that $\delta V \propto V_f^{2/3}$ over this range. The mean free time τ_c, (Fig. 4b) has only a weak dependence on V_f. The collision frequency ranges from about 500 kHz (for d=95 μm) to 10 kHz (for d=194 μm). The mean free path, s, is small compared to the particle diameter, d. Thus, the dilation is tiny (~0.1 μm) ranging from 0.01 to 0.1% of d. An optical or MRI image therefore cannot even identify the flow as being in a collisional regime (as opposed to particles being in constant contact). The physical picture of collisional dynamics represented by the data in Fig. 4 are consistent with energy and momentum balance at the level of orders of magnitude as is illustrated by the lines in Fig. 4c which show the dilation required to produce a free fall velocity of δV, scaled to the particle diameter. Naive energy and momentum budgets do not, however, recover the dependence of the measured parameters on the driving velocity V_f. The rate of inelastic losses, $1/2m(\delta V)^2(1-e^2)/\tau_c$, is comparable to the rate of gravitational work, mgV_f, but balancing these two terms with the assumption that τ_c is independent of V_f (as suggested by Fig. 4b) gives an incorrect scaling of $\delta V \propto V_f^{1/2}$. Likewise, assuming that velocities of neighbouring grains are uncorrelated and that a collision completely decorrelates the velocity autocorrelations yields an estimate $me\delta V \sim mg\tau_c$.

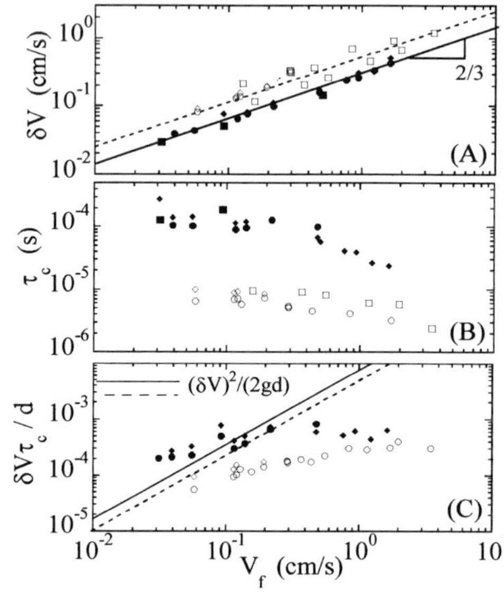

Fig.4 Microscopic scales of motion vs. macroscopic flow velocity V_f. The open symbols are for 95 μm beads (squares and circles are transmission in 0.32 and 0.625 cm thick cells; diamonds are back-scattering). The solid symbols are for 194 μm beads (circles and squares are transmission in 0.625 and 0.92 cm thick cells; diamonds are back-scattering). (a) δV vs V_f, both in cm/s. The lines are power-law fits with exponent 2/3 to the 95 μm (dashed) and 194 μm (solid) data. (b) Mean collision time, τ_c (sec) vs. V_f. (c) Mean free path, s ($=\delta V\tau_c$), scaled to the particle diameter, d. The lines show $(\delta V)^2/2gd$, the fraction of its own diameter a particle must fall to attain a speed δV

This also leads to an incorrect scaling $\delta V \propto V_f$. A treatment of the particle motions as uncorrelated thus seems insufficient, suggesting that effects such as velocity correlations between particles or slowly decaying velocity autocorrelations, may be important. Long time motions, which we discuss next, give further indication of such complex dynamics.

At times longer than τ_c, the relative displacement of particles is characterized by sub-diffusive motion over several decades in time, as displayed in Fig. 3b and its inset. Indeed, we do not obtain the diffusive limit, $\langle \Delta r^2(\tau) \rangle \propto \tau$, even at the end of the range available to DWS. The four data points (solid circles) at the extreme long time end of the inset in Fig. 3b show data from long-range video microscopy of self-diffusion of sand grains in the direction transverse to V_f and in the plane of the channel. Our data indicate that the parameters of collisional dynamics such as δV and s are not simply related to long-time diffusion and therefore cannot be deduced from measurements at long time- and length-scales.

The data of Fig. 4a also demonstrate that velocity fluctuations can be large even in the absence of a macroscopic velocity gradient, contrary to the simplest models of granular hydrodynamics. In Fig. 5 we plot $\langle \Delta r^2(\tau) \rangle$ taken at various points in a cell with spatially varying velocity gradients (dashed lines). For comparison, we also plot (solid lines) data taken at two different values of V_f in a uniform flow. It is quite clear that the imposition of a macroscopic velocity gradient alters the transition from collisional dynamics discussed above, but has little influence on the ballistic regime itself. In particular, this shows that δV is quite insensitive to gradients, a fact that is not captured in kinetic theories of flow.

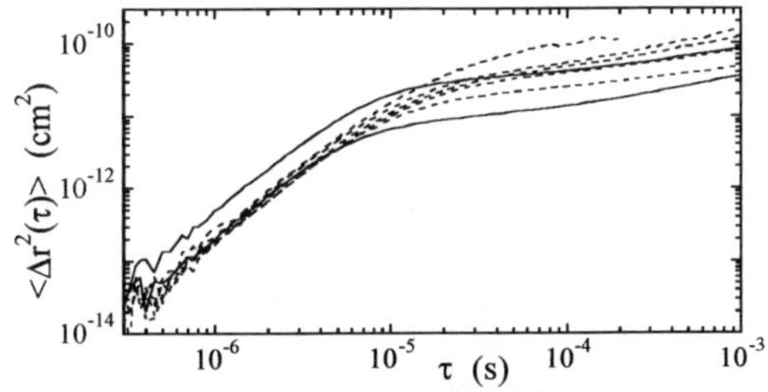

Fig. 5. Effect of velocity gradients. The solid lines are in a uniform flow with V_f=0.5 and 1.1 cm/s, the dashed lines are in velocity gradients, the local flow velocity is 0.4 - 0.5 cm/s.

Fluidized beds

We have taken DWS data for particle motions in beds composed of 49 and 97 μm glass beads. Beads of this size and density fall in the Geldart A category, in which increasing gas velocity causes the static bed first to fluidize uniformly and then admit macroscopic bubbles at a higher flow rates. The hysteretic nature of these transitions has led to the belief that particles in the bed are able to attain less compact packings in the uniformly fluidized state and do not actually lose contact with neighbouring particles. For a narrow regime of gas flow rates above the minimum velocity for fluidization, the DWS signal indicates that the particles are static. This, for the first time, directly confirms the hypothesis that interparticle contacts may be sustained above minimum fluidization. As bubbling commences, we visually observe large-scale convective motions in the bead pack. Microscopic dynamics induced by this overall convection are obtained in the DWS signal as shown in Fig 6 where we plot $\langle \Delta r^2(\tau) \rangle$ for a range of flow rates from just above the uniform

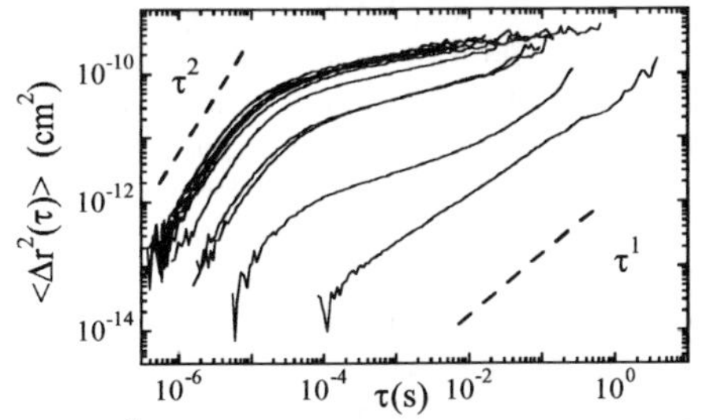

Fig. 6. $<\Delta r^2(\tau)>$ vs. τ at several gas flow rates (increasing from right to left) for 49μm beads fluidized by N_2 gas. The dashed lines represent diffusion (τ^1) and ballistic motion (τ^2)

fluidization threshold to flow rates well in excess of the bubbling threshold (flow rates increase from right to left). Just above the bubbling threshold we observe diffusive motions (as indicated by the dashed line with a slope τ^1). As the gas flow is further increased, the collisional regime is revealed (ballistic motion is indicated by the dashed line with slope τ^2.) and we can obtain the 'granular temperature'[6], $(\delta V)^2$, the collision frequency, and the mean free path just as in the case of channel flow. δV grows rapidly with increasing gas flow just above the bubbling threshold but appears to turn over to a slower dependence on gas velocity at higher rates of gas flow as may be seen in the crowding of the curves at the extreme left of Fig. 6

CONCLUSIONS

We have created a simple granular flow and described a technique which allows a full characterization of the macroscopic and microscopic behaviour of the system. This should serve as an experimental bench-mark for any theory of flow, which we feel is vital in a field where theory has been advanced well beyond experiment in the absence of such constraints. What have we directly learnt that is new? We believe that for the first time, we have been able to study a real, three dimensional flow and present both a qualitative and quantitative picture of the collisional regime. While hitherto unproven, the fact that collisional dynamics do occur will not surprise many. What is surprising and new are the scales of motion, the fact that a 'temperature' occurs in the absence of any shear gradient (and indeed, seems to be only weakly affected by a gradient), and a well-defined puzzle regarding the dependence of this 'temperature' and the other scales of the collisional dynamics on the average driving velocity. Yet another new and unexpected fact regards the relationship between the short-time collisional dynamics and the long-time behaviour. The large separation in time scale is reminiscent of dynamics in viscous liquids and dense colloids and suggest that there may be analogous cooperative dynamics underlying the problem of flow. The interplay between macroscopic velocity gradients and the nature of the cross-over from ballistic to diffusive regimes is also not captured in existing models of grain flow.

We have also demonstrated the utility of DWS techniques in characterizing particle motions in gas-fluidized beds. We are able to show directly that particle contacts persist in the uniformly fluidized regime. At higher flow rates we are able to track both diffusional and collisional dynamics. DWS thus is a versatile, non-intrusive and sensitive probe of particle motions in granular systems at length scales much smaller than the grain size and is potentially a diagnostic of wide utility even in real industrial settings.

REFERENCES

1. H. M. Jaeger, S.R. Nagel, R.P. Behringer, Phys. Today, **49**, 32 (1996).
2. R.A. Bagnold, Proc. R. Soc. Lond. A **225**, 49 (1954);; J.T. Jenkins, S.B. Savage, J. Fluid Mech, **130**, 187 (1983); P.K. Haff, *ibid.,* **134**, 401 (1983)
3. D.J. Pine, D.A. Weitz, *Dynamic Light Scattering: The method and some applications*, ed. Wyn Brown, pg. 652 (Clarendon, Oxford, 1993)
4. M. Nakagawa *et al.*, Exp. Fluids **16**, 54 (1993); E.E. Ehrichs *et al.*, Science, **267**, 1632, (1995)
5. V.V.R. Natarajan, M.L. Hunt, E.D. Taylor, J. Fluid Mech., **304**, 1 (1995); O. Zik, J. Stavans, Europhys. Lett., **16**, 255 (1991).
6. G.D. Cody et al. Powd. Tech. **87**, 211 (1996) infer δV from the acoustic response of the container to discrete collisions of particles.

AVALANCHE SEGREGATION OF BINARY MIXTURES OF GRANULAR MEDIA

J. KOEPPE, M. ENZ and J. KAKALIOS
The University of Minnesota, School of Physics and Astronomy, Minneapolis, MN 55455

ABSTRACT

When an initially homogeneous binary mixture of granular media such as fine and coarse sand is poured near the closed edge of a "quasi-two-dimensional" Hele-Shaw cell consisting of two vertical transparent plates held a narrow distance apart, the mixture spontaneously forms alternating segregated layers. Using digital images taken with a CCD camera and Fourier transformed to obtain the structure function of the banding pattern, the wavelength selection mechanism and degree of segregation are studied as the plate separation of the Hele-Shaw cell is systematically increased. For a given flow rate, the degree of segregation, as reflected in the amplitude of the peak in the structure function, decreases as the plate separation increases. For wider plate spacings the peak in the structure function shifts to lower wavelengths. These results are compared to numerical simulations of the avalanche segregation using a stochastic sandpile model.

INTRODUCTION

A striking property of granular media is segregation when a homogeneous mixture of granular materials of different sizes or masses is shaken or rotated.[1-11] The former case is known as vertical segregation,[4,5] wherein convective rolls due to interactions between the smaller particles and the side walls of the container cause the larger particles to be entrained and brought to the top surface,[5] while the second case is termed axial segregation.[6-11] If a cylinder partially filled with a binary mixture of granular media is tipped on its side so that its axis of symmetry is horizontal and rotated like a drum about that axis, the individual species will segregate into alternating bands of relatively pure single concentrations along the axis of rotation. This phenomenon has been discussed in terms of a model whereby the dynamic angle of repose of the mixed phase is greater than that of one of the homogeneous phases during rotation.[9-11] Recently a third type of segregation effect[12, 13] has been described by Makse and co-workers,[12] where an initially homogeneous mixture of fine and coarse sand will form alternating segregated layers when poured near the closed edge of a "quasi-two-dimensional" Hele-Shaw cell. This new form of de-mixing, which we term "avalanche segregation" may be the physical mechanism responsible for striation patterns in sandstone.[14-19] In this paper we confirm this avalanche segregation effect and report on the influence of systematically varying the plate separation of the Hele-Shaw cell on the wavelength selection and segregation process.

EXPERIMENTAL TECHNIQUES AND RESULTS

Our experimental setup is similar to that employed by Makse and co-workers.[12] Two vertical Plexiglas sheets 6 mm thick of area 8 inches by 10.5 inches are mounted parallel to each other onto a plexiglas base plate. One sheet is bolted to the base plate, while the second vertical plate is attached at one edge to the fixed plate. By employing spacers which extend the full height of the vertical sheets (so that one edge remains closed), the separation between the plates can be varied from 3 mm to 24 mm. A 50/50 mixture by volume of sand and sugar is poured against the closed edge of this Hele-Shaw cell[20] using a titrating bulb with a rotating stopcock. The granular material is pre-mixed by shaking and stirring prior to and once in the flow bulb. The sand is approximatetly spherical with an average diameter of 0.2 mm while the sugar crystals are approximatetly cylindrical with a length of ~ 0.8 mm. The sand has an average density of 1.5 mg/mm^3, while the sugar's density is only ~ 0.5 mg/mm^3. The sand was dyed blue/black for

Mat. Res. Soc. Symp. Proc. Vol. 463 © 1997 Materials Research Society

contrast in the photos, this dying does not influence the segregation process. Digital photos of the segregated pattern are taken with a monochrome CCD camera (Cohu 4910) in conjunction with a Scion LG-3 frame grabber and a Power Macintosh 7100/80. Data image analysis is performed using the public domain program Image from the NIH.

Figure 1 shows a CCD image of the Hele-Shaw cell after avalanche segregation of the sand/sugar mixture has occured. The spacing between the vertical plates is 4 mm and the flow rate was 0.78 gm/sec. The mixture was poured against the closed edge of the cell, on the far right of fig. 1a. There are several features of note in fig. 1a. There is a "dead-zone" of mixed sand and sugar in the bottom corner which extends vertically up ~ 10 cm before the segregated bands appear. The volume of this dead-zone depends on the plate separation. A second feature of note is that the segregation is weaker toward the bottom and top of the pile (toward the left in fig. 1). Both of these features were reported by Makse and co-workers.[12] As shown in fig. 1b, we have also observed identical segregation patterns when the sand is poured in the center of the cell, the "downhill segregation", with the denser material (sugar) congregating near the bottom and the sand collecting near the top of the pile is not an edgewall effect. Fig. 1c shows the results of numerical simulations of a stochastic sandpile model, as described later.

The influence of varying the plate separation d of the Hele-Shaw cell on the avalanche segregation effect has been investigated. Two trends are observed as the plate separation is increased: the degree of segregation decreases and the spacing between segregated bands decreases. Both trends can be quantified through calculation of the structure function. Figure 2 shows a digital image of the banding pattern near the top of the sandpile, as indicated by the white box in fig. 1a. The flow rates, plate separation and sand/sugar mixture were the same as in fig. 1. The program Image is used to convert the digital image in fig. 2 into a a plot of pixel intensity against position. Since the sand is dark and the sugar is white, the intensity on the image in fig. 2 corresponds roughly to the concentration of sand near the transparent vertical plate in the Hele-Shaw cell. A high pixel value corresponds to a high concentration of sand and a low pixel intensity corresponds to a high concentration of sugar, and an intermediate pixel value reflects a mixture of the two. One then records a series of peaks and valleys of the digital pixel values against position, which are then Fast Fourier transformed to yield the structure function for the banding pattern. The location of the peak in the plot of the FFT amplitude against wavevector in fig. 2 gives the average wavelength of the segregated bands, while the amplitude of the peak provides a quantitative measure of the degree of segregation. Figure 3 shows a log-log plot of this FFT amplitude and average wavelength against plate separation. The solid line in figs. 3a and 3b are fits to a power law expression, with power law exponents of ~ 1.6 and 0.8, respectively. These exponents are fairly reproducible, however, equally good fits to the data in figs. 3a and b can be obtained to an exponential dependence on plate separation. Consequently, without further theoretical insight as a guide, the results in fig. 3 should only be interpreted as demonstrating a clear decrease in segregation amplitude and banding wavelength with plate separation.

We next address the physical mechanism for the avalanche segregation effect. As the binary granular mixture is poured against the closed edge of the Hele-Shaw cell, a heap or pile is formed which will eventually exceed the maximum angle of repose (which determines the stability condition) and the introduction of additional granular material will cause an avalanche event.[12,13,16] As the granular material flows down the top surface of the pile, only the particles within a few grain diameters of the free surface can move. There is consequently a considerable shear near the top surface, due to this discontinuity in velocities.[16,17] It is well known that granular materials dilate when sheared, so that this flowing top surface will be expanded relative to the stationary bulk of the pile.[16] Within this dilated region, the smaller or denser granular material will sieve or percolate to the lower region of the flowing layer.[8,17,21] The segregation process is stopped once the avalanche is complete, until the addition of further material at the top

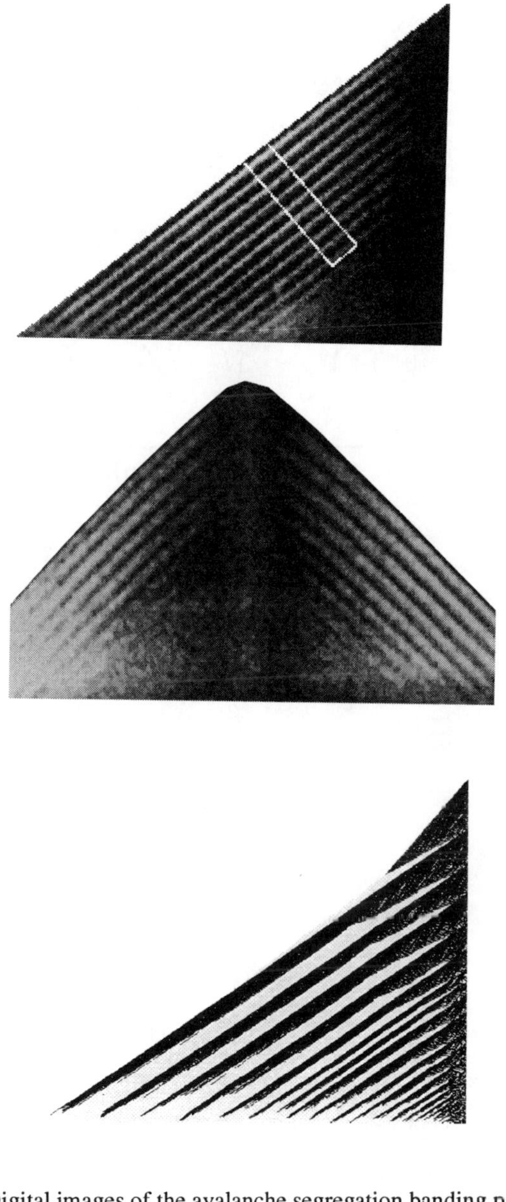

(a)

(b)

(c)

Fig. 1: Digital images of the avalanche segregation banding pattern for a 50/50 mixture of sand and sugar poured along the closed edge of a Hele-Shaw cell (fig. 1a) and in the center of the cell (fig. 1b). Fig. 1c is the resulting segregation pattern obtained from a numerical simulation of a stochastic sandpile model as described in the text.

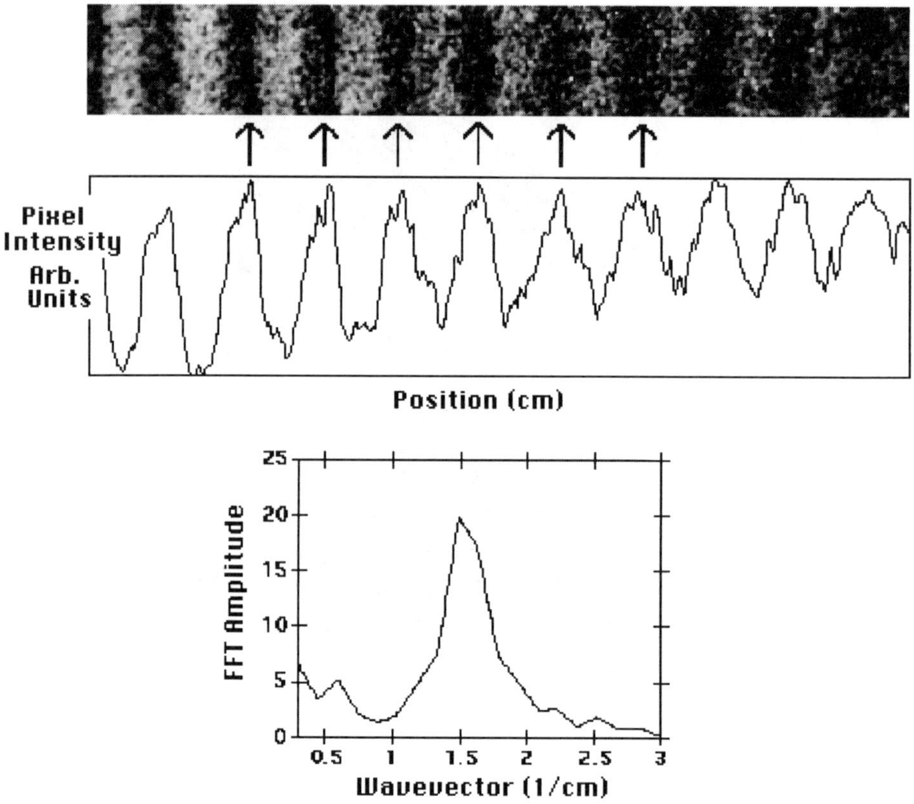

Fig. 2: Digital image of the segregated banding pattern near the top of the sandpile (indicated by the white box in fig. 1a). Also shown is the average pixel intensity against position, obtained using the graphics program Image. The high pixel values correspond to the dark regions of the image (high sand concentration) and the low pixel values indicate regions of high sugar (white) concentration. This data is then Fourier transformed to yield the structure function of the banding pattern.

of the pile builds up to once again exceed the maximum angle of stability. Close examination of the mixture poured into the Hele-Shaw cell confirms that segregation occurs only during the avalanching event. The wavelength or width of the segregated bands reflects the balance between the flow rate at which granular material is poured into the cell at the top of the pile, and the average time between avalanches, which depends on the details of the granular material (the exact value of the maximum angle of repose) as well as the plate separation. A key aspect of this model, that the segregation process is triggered by avalanches of the granular material determined by the local maximum angle of repose, can be tested via numerical simulations using the stochastic sandpile model of Makse and co-workers,[12] as described below.

NUMERICAL SIMULATIONS

The simulation, following Makse et al,[12] represents the granular material as two different rectangles having different heights and a constant width of unity. The granular particles arrive in colums from the edge of the two dimensional lattice, the probability of a granular particle falling onto a column decreases with a Gaussian distribution multiplied by an arbitrary thickness dx, so that the total sum of the rectangles under the Gaussian curve is properly normalized. The local slope between two adjacent columns determines the motion of the granular material. If the height difference between two columns is greater than some predetermined arbitrary maximum value, representing the maximum angle of repose, then the granular particle is moved to the next, lower colum. If a particle reaches the horizontal base of the lattice, then the entire pile is at the maximum angle of repose, which causes an avalanche event wherein particles move as described above based upon the smaller local height difference. The simulation was written in C++ and implemented with the GNU C compliler on a Pentium PC running the Linux operating system. The simulation results in fig. 1c are for a run consisting of 100,000 granular particles with equal

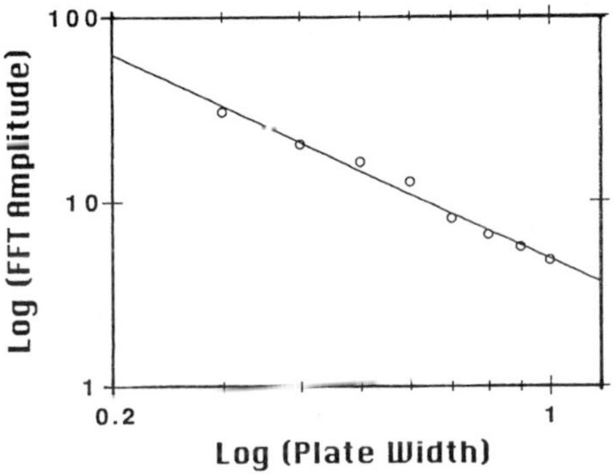

Fig. 3a: Log-log plot of the amplitude of the peak in the structure function for the avalanche segregation banding pattern against plate separation of the Hele-Shaw cell for a mixture of sand and sugar. The solid line is a fit to the data with a power law dependence on plate separation, where the power law exponent is ~ 1.6.

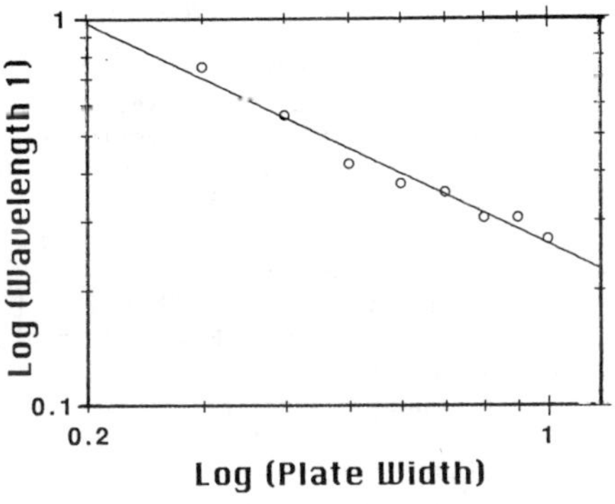

Fig. 3b: Log-log plot the wavelength of the the avalanche segregation banding pattern determined by the location of the peak in the structure function against plate separation of the Hele-Shaw cell for a mixture of sand and sugar. The solid line is a fit to the data with a power law dependence on plate separation, where the power law exponent is ~ 0.8.

volume ratios of the two different sizes. The heights of the grey and black particles were 3 and 2, and the local slopes which would trigger movement were chosen to be 3.1 and 5.1, respectively. The segregation of the different sizes of granular particles into alternating layers in fig. 1c is very similar to the experimentally observed avalanche segregation as in fig. 1a. We find that as the relative height difference between the granular particles is increased, while maintaining a constant arbitrary slope difference to trigger an avalanche event, then the spacingbetween the segregated layers decreases. However the segregated band pattern is very sensitive to the local slopes chosen to trigger an avalanche event. For fixed heights of granular particles, the segregated band widths increase when the larger of the avalanche triggering slopes increases, so that the difference between slopes increases, while regardless of particle size, as the ratio of smaller slope to larger slope decreases, the segregated band width increases. These numerical simulation results will be discussed in greater detail, and directly compared to experimental segregation processes, in a later publication.

We gratefully acknowledge experimental assistance and helpful comments from Kimberly Hill. This work was supported by NSF CTS-9501437 and the University of Minnesota through the Undergraduate Research Opportunity Program.

REFERENCES

1. *Physics of Granular Media,* ed. by D. Bideau and A. Hansen (North-Holland, Amsterdam, 1993).
2. *Disorder and Granular Media,* ed. by D. Bideau and A. Hansen (North-Holland, Amsterdam, 1993).
3. *Granular Matter, An Interdisciplinary Approach,* ed. by Anita Mehta (Springer-Verlag, New York, 1994).
4. James L. Olsen and Edward G. Rippie, J. Pharmaceutical Sci. **53,** 147 (1964).
5. H. M. Jaeger and Sidney R. Nagel, Science 255, 1523 (1992); James B. Knight, H. M. Jaeger, and Sidney R. Nagel, Phys. Rev. Lett. **70**, 3728 (1993).
6. Y. Oyama, Bull. Inst. Phys. Chem. Res. (Tokyo), Rep. **5,** 600 (1939).
7. M. B. Donald and B. Roseman, British Chem. Eng. **7,** 749 (1962); B. Roseman and M. B. Donald, British Chem. Eng. **7,** 823 (1962).
8. J. Bridgwater, Powder Tech. **15,** 215 (1976); J. Bridgwater, N. W. Sharpe and D. C. Stocker, Trans. Instn. Chem. Engrs. **47,** T114 (1969).
9. S. Das Gupta, D. V. Khakhar and S. K. Bhatia, Powder Tech. **67,** 145 (1991); Chemical Eng. Sci. **46,** 1513 (1991).
10. Stuart B. Savage, in *Disorder and Granular Media*, edited by D. Bideau and A. Hansen (North-Holland, Amsterdam, 1993), p.255.
11. K. M. Hill and J. Kakalios, Phys. Rev. E **49**, R3610 (1994); *ibid* **52**, 4393 (1995).
12. H. A. Makse, S. Havlin, P. R. King and H. E. Stanley, Bull. A.P.S. **41**, 384 (1996) and this volume.
13. J. C. Williams, Powder Tech. **2**, 13 (1968).
14. J. R. L. Allen, *Sedimentary Structures: Their Character and Physical Basis* (Elsevier, Amsterdam) (1982).
15. G. V. Middleton, *Mechanics of Sediment Movement* (S.E.P.M. Providence,RI) (1984).
16. R. A. Bagnold, *The Physics of Blown Sand and Desert Dunes* (Chapman and Hall, London) (1941).
17. A. V. Jopling, J. Geophysical Res. **69**, 3403 (1964).
18. S. G. Fryberger and C. Schenk, Sedimentology **28**, 805 (1981).
19. G. V. Middleton, Geo. Assoc. of Canada, special paper no. **7**, 253 (1970).
20. H. S. Hele-Shaw, Nature **58**, 34 (1898).
21. K. Ridgway and R. Rupp, Powder Tech. **4**, 195 (1970).

DRIVEN GRANULAR MEDIA AND DISSIPATIVE GASES: CORRELATIONS AND LIQUID-GAS PHASE TRANSITIONS

D.R.M. WILLIAMS
Institute of Advanced Studies, Research School of Physical Sciences and Engineering
The Australian National University, Canberra

ABSTRACT

We study a simple model of a granular material or powder where the particles are excited by an external noise source and dissipate energy by inelastic collisions. Due to the inelastic collisions between particles there is an effective interaction between them. In one dimension this leads to long-range correlations between the particles in a gas phase despite the absence of long-range forces between the particles. In two dimensions the dissipative effects cause a very sharp liquid-gas phase transition at which the susceptibility has a pronounced peak.

INTRODUCTION

Granular materials or powders may be defined as any material which consists of "grains" or small particles. Examples include many products of importance to the agricultural, mining and food industries, such as rice, wheat, coal, most ores and most breakfast cereals. Although granular materials have long been of interest to engineers they have only attracted significant attention from the physics community over the past decade [1]. To physicists these materials are interesting because they exhibit a wide range of interesting phenomena, many of which remain either unexplained or not fully understood. These include, heaping, clustering, size-segregation and the breakdown of macroscopic hydrodynamics [1–16]. These effects, and others have been much studied experimentally and by computer simulation. However, despite extensive studies a reasonable understanding of granular materials is still lacking. This is presumably in part because the individual grains undergo complex interactions, where short range elastic forces and nonlinearities become important. However, there is clearly also a more fundamental reason. The vast majority of the experiments and simulations on granular materials involve several separate and identifiable influences. These are:
(i) The interactions between the particles are inelastic. Thus when two particles collide, energy is dissipated.
(ii) The systems studied usually include a boundary such as the walls of the container.
(iii) Energy normally flows from the boundaries into the granular media.
(iv) The system is placed in a gravitational field.
Each of these influences obviously complicates the system and it is thus not surprising that many of the experiments and simulations show novel and rich behaviour. There are two obvious analogs here. The first is with the dynamics of simple fluids. In general the flows of such fluids away from fields and boundaries are relatively easy to understand. However, in the presence of boundaries and fields simple fluids exhibit convection and turbulence and are still not perfectly understood. The second analogy is with the theory of magnetism. If one wants to understand ferromagnetism it is simplest to study the Ising model on an infinite lattice in the absence of any external fields, boundaries or imposed gradients. The situation with granular materials is even more complicated. In general influences (ii) to (iv) artificially break the symmetry and induce spatial gradients in the system which make a simple understanding somewhat difficult. In this paper we present a system which has some or all of these undesirable influences removed. Indeed a step in that direction has

Mat. Res. Soc. Symp. Proc. Vol. 463 © 1997 Materials Research Society

already been made [2–4]. The system is a dissipative gas, which is simply a gas in which the particles undergo inelastic collisions. Thus far the studies have concerned "cooling" gases which are not heated in any way. These gases can show interesting dynamics, but they have no non-trivial steady-state behaviour: i.e. all the particles just cool down and stop moving. The simplest system for which there is any non-trivial steady-state is the one we shall study here : a driven dissipative gas. This system was introduced in reference [15] and was studied further in [14] and [16].

In this system the particles undergo inelastic collisions but each particle is heated individually and continuously. It thus includes the important difference between granular and other fluids (inelasticity) whilst ignoring the complicating boundary and field effects. This system is intrinsically much simpler than traditional granular materials. However, as we shall see here, an examination of driven dissipative fluids also reveals non-trivial behaviour and can lead to a better understanding material properties.

CLUSTERING IN ONE-DIMENSIONAL GASES

Clustering is observed when granular media are sheared [9–11]. In general clustering is driven by inelastic collisions. When two particles collide inelastically they dissipate energy, slow down and hence remain close to one another . Here we investigate numerically the effect of such inelastic collisions in a one-dimensional system of point-like particles that are excited by a thermal reservoir. We show that clustering, as described by the two-point correlation function, occurs even in the absence of any other forces between the particles. There have been several studies of systems which are started in a "hot" state and the slowly cool [2–4,6,7], as there is no energy input. In one and two dimensions these can show "inelastic collapse". The novel feature of the collapse is that for coefficients of restitution η below a critical value η_c the kinetic energy is dissipated in a finite time. For $\eta > \eta_c$ the kinetic energy dissipates gradually.

In the model considered here [14,15], we consider a continuous input of energy locally to each particle, as well as dissipative collisions. This might model for instance a system of particles confined to a line on a vibrating plate. For a coefficient of restitution $\eta < 1$, the system eventually settles down to a "steady-state". For $\eta \cong 1$, this steady-state looks, at least superficially, like an ideal gas, in which there are no significant spatial correlations between the particles. However, we show that as η is reduced, even in the absence of any long-range interactions, the system develops a structure factor that is characteristic of an equilibrium system with long-range interactions. In particular, the dissipative interactions lead to a correlation function $g(x)$ that is no longer a constant as it would be for an ideal gas, but shows a peak near the origin. Furthermore, this enhancement of $g(x)$ near $x = 0$ follows a power-law. Thus, the system behaves as if it there were long-range attractive interactions between the particles. Only in the limit $\eta \to 1$ does the correlation function become uniform. As η decreases $g(x)$ becomes more and more sharply peaked about $x = 0$.

We consider N point particles of unit mass, $m = 1$, confined to a line of length $L = 1$. We use periodic boundary conditions, so that the particles lie on a circle of unit circumference. When two particles i and j collide in this one-dimensional system, the final (primed) velocities are given in terms of the initial (unprimed) velocities by

$$v_i' = \frac{1}{2}(1-\eta)v_i + \frac{1}{2}(1+\eta)v_j \qquad v_j' = \frac{1}{2}(1-\eta)v_j + \frac{1}{2}(1+\eta)v_i \tag{1}$$

The main difference between this and previous studies of dissipative gases is that each individual particle is "heated" at a constant rate. This is done by adding a random amount to the velocity of each particle during a time-step Δt via the Langevin equation

$$v_i(t+\Delta t) = v_i(t) + \sqrt{r}\sqrt{\Delta t} f(t), \tag{2}$$

where $f(t)$ is a random number chosen uniformly between $-1/2$ and $1/2$ and r is a number proportional to the heating rate. After the velocities are adjusted the system is transferred to the centre of mass frame, so that $v_i \to v_i - \bar{v}$, where $\bar{v}$ is the average velocity of all the particles in the system. The algorithm (2) ensures that the velocities undergo a random walk, whilst the transfer to the centre of mass frame ensures that the particle speeds do not increase indefinitely. This transfer step is for convenience only, since the properties of collisions do not depend on the absolute speeds, only on the relative speeds. Before the heating step the kinetic energy of the system is $K = \frac{1}{2}\sum_{i=1}^{N} v_i^2$ and after heating it is

$$K' = \frac{1}{2}\sum_{i=1}^{N}(v_i + \delta v_i)^2 = K + \sum_{i=1}^{N} v_i \delta v_i + \frac{1}{2}\sum_{i=1}^{N}(\delta v_i)^2 \tag{3}$$

where $\delta v_i = \sqrt{r}\sqrt{\Delta t} f(t)$. On average the term linear in v_i vanishes and the average of the remaining term, $\frac{1}{2}\sum_{i=1}^{N}(\delta v_i)^2$, is $\frac{1}{2} rtN\langle f^2\rangle = \frac{1}{24} rtN$. Hence the heating rate is $\Omega \equiv \frac{1}{24} r$, which is the energy input per unit time per particle. Provided $\eta < 1$, this system, started with some initial random speeds, rapidly reaches a steady-state configuration.

Qualitatively the system appears to form liquid-like clusters of high density surrounded by a gas-like "phase" of lower density. In order to study this effect quantitatively we use the two-particle correlation function $g(x)$. We select a particle and ask what is the density of particles at distance x from it. This, suitably normalized, is $g(x)$. For an ideal gas of point particles the answer is $g(x)$ = a constant. For a gas which has some attractive potential $U(x)$ acting between the particles $g(x)$ will be peaked about $x = 0$ and will decay to a constant as $x \to \infty$. For the dissipative gas discussed here we find $g(x)$ shows a peak at the origin, even though we have no potential acting between the particles. This peak is caused by the dissipation effect discussed above, and can be thought of a steady-state version of the collapse and clustering seen in cooling gases [2,7,3,4,6]. As $\eta \to 1$ the structure becomes less pronounced and $g(x)$ approaches a constant. However, as η becomes small, $g(x)$ becomes very sharply peaked. Some characteristic results are shown in figure 1. We find that the correlation function depends only on the density and on η and is independent of the heating rate. At least for small x, where the finite size of the system has little effect, $g(x)$ can be approximated by a power law $g(x) \sim x^{-\alpha(\eta)}$. Here $\alpha(\eta)$ is a monotonically increasing function of η. In the limit of a perfectly elastic system $\eta \to 1$ and $\alpha \to 0$. However, for perfectly inelastic systems where $\eta \to 0$ we find $\alpha \to 1/2$

The long-range correlations can be thought of as being induced by an effective potential of mean force, $U(x)$ where $g(x) \propto \exp(-U(x)/kT)$. Here kT can be taken as the kinetic energy of the particles. This potential is approximately logarithmic. Note however that since the only energy occurring in the problem is the kinetic energy of the particles the potential is just proportional to the temperature.

TWO-DIMENSIONAL GASES: LIQUID-GAS PHASE TRANSITIONS

In this section we study a two-dimensional excited dissipative gas. It is well known from the theory of phase transitions that the one-dimensional case is special. It is thus important to study the above effects in two dimensions. A logical question to ask is how do the long-range correlations affect the phases of this system in two or three dimensions. Here we show

Figure 1. A log-log plot of the two-point correlation function versus distance for three values of the coefficient of restitution for $N = 10$ particles. For $\eta = 0.99$ (lowest curve) the correlation function is almost a constant. For $\eta = 0.5$ (middle curve) strong correlations have developed, and for $\eta = 0.01$ (upper curve) the function is clearly a power-law with exponent $-1/2$. The correlations arise because of the effect of inelasticity, i.e. when two particles collide they move more slowly and hence stay near each other, thus causing a correlation. The correlation function is independent of the heating rate Ω. Here we have superimposed data from two heating rates which differ by a factor of 100, $\Omega = 0.0017$ and $\Omega = 0.17$, and each point represents an average over 2×10^5 collisions.

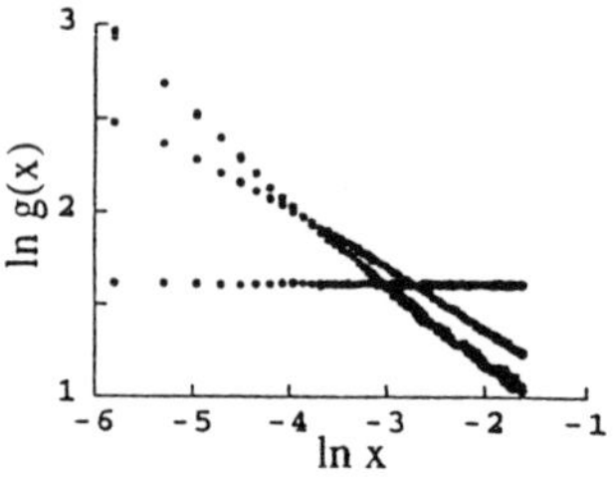

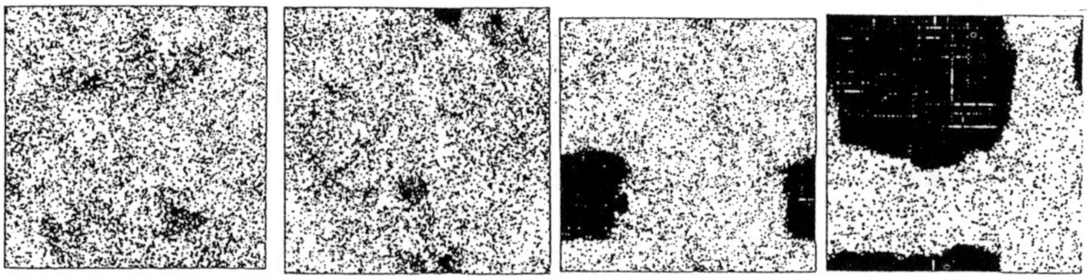

Figure 2. Snapshots of the 2D lattice model for a 200×200 lattice with $\eta = 0$ and $r = 5$. The area fractions covered by the particles are (a) $\phi = 0.25$ (b) $\phi = 0.259125$ (c) $\phi = 0.25925$ and (d) $\phi = 0.5$. At low densities (a) the system is gas-like. Between (b) and (c) the system undergoes a sharp transition from gas to liquid-gas coexistence. In (c) and (d) the existence of a single droplet can clearly be seen. Note that periodic boundary conditions are used so there is only one droplet. The different shades of grey in the pictures correspond to different temperatures. Note that in going from (b) to (c) only five new particles are added, i.e. a change of area fraction of $\Delta\phi = 1.25 \times 10^{-4} = 0.04\%\phi$.

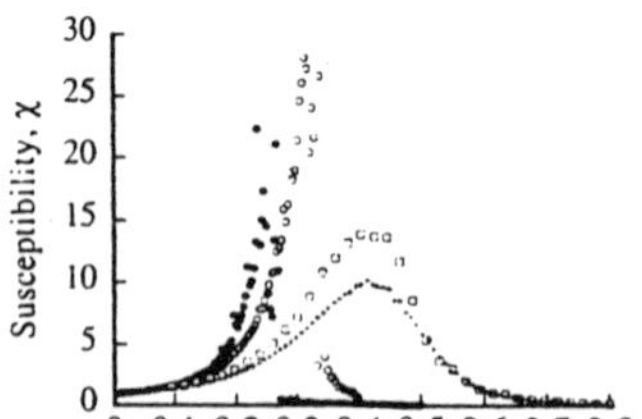

Figure 3. The susceptibility χ versus the area fraction of particles, ϕ, for different $L \times L$ lattices. Here $\eta = 0$ and $r = 5$. The different points are as follows: $L = 30 \rightarrow +$, $L = 40 \rightarrow \square$ $L = 100 \rightarrow \circ$, $L = 200 \rightarrow \bullet$. The position of the beak gives the transition density at which the system changes from a gas to gas-liquid coexistence. Note that as the size of the lattice increases the sharpness of the peak increases. The fact that the peak is higher for $L = 100$ compared to $L = 200$ is probably a result of finite sample times. Near the transition point there are large fluctuations in the susceptibility and an accurate measurement requires many samples.

that they can induce a liquid phase in coexistence with a gas phase. The system we have in mind is a two-dimensional excited powder, i.e. a group of grains laid upon a surface, and gently randomly vibrated. The particles reside on a square $L \times L$ lattice with lattice constant unity, and move in the x and y directions, with velocity components v_x and v_y. The particles undergo inelastic collisions with each other with coefficient of restitution η. For simplicity we assume that the collision occurs only in the x direction or the y direction at any one time. Thus if the x direction is chosen the final (primed) and initial (unprimed) velocities of the two colliding particles, i and j are given by

$$v'_{ix} = \frac{1}{2}(1-\eta)v_{ix} + \frac{1}{2}(1+\eta)v_{jx} \qquad v'_{jx} = \frac{1}{2}(1-\eta)v_{jx} + \frac{1}{2}(1+\eta)v_{ix} \tag{4}$$

$$v'_{iy} = v_{iy} \qquad v'_{jy} = v_{jy} \tag{5}$$

The particles a continuously heated by changing either their x or y velocities as in the one dimensional case. If the x direction is chosen then

$$v_x(t+\Delta t) = v_x(t) + \sqrt{r\Delta t} f(t) \qquad v_y(t+\Delta t) = v_y(t) \tag{6}$$

where r and f have the same meanings as in the one-dimensional model.

We use periodic boundary conditions. The algorithm for particle motion is as follows. (1) A particle is chosen at random and either the x or y direction is then chosen, also at random. The time step is fixed originally at $\Delta t = 1$ and is decremented as the particle moves. (2) If say the x direction is chosen and $|v_x t| > 1$ the particle attempts to move to the next lattice site. If this is unoccupied the time is decremented and step (1) is repeated. (3) If the chosen lattice site is occupied a collision takes place and the x velocities of the two particles are adjusted according to (4), (5) and the direction is again chosen randomly. (4) Once the particle has finished it motion, i.e $|v_x t| < 1$ another particle and direction is chosen at random and heated according to (6). The process then repeats.

The advantage of using a lattice is clear in the collision step. Searching for collisions is trivial on a lattice, but would be very time-consuming in an off-lattice simulation. For off-lattice simulations with no heating an "event-driven" algorithm can be used, but for a heated system the random heating invalidates this algorithm. The particles undergo continual collisions and continuos heating. Eventually these two effects balance out and the system reaches a steady state. At low densities the system behaves like a gas. However, the gas is not ideal. This is because the inelastic collisions between particles cause correlations between them. The simplest way of seeing this is to consider two particles colliding in the centre of mass frame. Their initial speeds are v and their final speeds are $\eta v < v$. Thus the particles recede from each other more slowly than they approached and indeed more slowly than if they had not collided. They thus spend more time together and a correlation is induced. As shown in section (2) , in one-dimension this correlation takes a power-law form.

Some characteristic results are shown in figure (2). At low densities the inelastic correlation causes one to have a relatively uninteresting correlated gas. To the naked eye this looks very similar to an ideal gas, and only by looking in detail at the correlation function can one distinguish the two. However, at higher densities the effective "potential" induced by the inelastic collisions becomes very important. In particular, it leads to a first-order phase transition from a gas phase to a dense liquid phase which coexists with a vapour phase. This is clear in figure (2) where the behaviour of the system suddenly changes when the density is changed by of order 0.04%.

Although the transition is clear to the naked eye in large lattices it is not so clear on smaller lattices. For this reason and to describe the transition quantitatively, it is useful to introduce the susceptibility [17] χ, defined by $\chi = \phi^{-1}\sum_s' s^2 n_s$. where n_s is the number of clusters of s particles and the prime means that the largest cluster is omitted. Here ϕ is the fraction of sites occupied by the particles, which we also call the density.

For small lattices where $L = 30, 40$ the transition is not clear and manifests itself as a pronounced peak in the susceptibility (figure 3). However, for larger lattices the peak is very pronounced, and there is a clear transition density ϕ_t beyond which the susceptibility is zero. This, together with the visual evidence shows that the system undergoes a liquid-gas phase transition at this point.

CONCLUSION

In this paper we have a simple model for a granular material, a driven dissipative "gas". This model avoids many of the complications inherent in previous granular studies, but still shows non-trivial behaviour. This behaviour is entirely due to the inelastic collisions between the particles. In one dimension these lead to long-range correlations which have approximately a power-law form. state. In two dimensions the inelastic collisions lead to a transition from a gas phase to a liquid-gas coexistence.

ACKNOWLEDGEMENTS

The author thanks Fred Mackintosh for useful discussions. The work was funded by a QEII fellowship.

REFRENCES

[1] H.M. Jaeger and S.R. Nagel, Science, 255, (1992) 1523 .
[2] I. Goldhirsch and G. Zanetti, Phys. Rev. Lett., 70, (1993) 1619 . N.Sela and I. Goldhirsch, Phys. Fluids A, 7, (1995) 507.
[3] M.A. Hopkins and M.Y. Louge, Phys. Fluids A, 3, (1991) 47.
[4] S. McNamara and W.R. Young, Phys. Fluids A, 4, (1992) 496.
[5] S. McNamara and W.R. Young, Phys. Rev. E, 50 (1994) R28.
[6] S. Luding, E. Clément, A. Blumen, J, Rajchenbach and J. Duran, Phys. Rev. E., 49, (1994) 1634 .
[7] B. Bernu and R. Mazighi, J. Phys. A, 23, (1990) 5745.
[8] Y.S. Du, H. li and L.P. Kadanoff, Phys. Rev. Lett., 74,(1995) 1268 .
[9] M.A. Hopkins and M.Y. Louge, Phys. Fluids A, 3, (1991) 47.
[10] C.S. Campbell and C.E. Brennen, J. Fluid Mech., 151, (1985) 167 .
[11] M. Babić, J. Fluid Mech., 254, (1993) 127.
[12] J.T. Jenkins and S.B. Savage, J. Fluid Mech., 130, (1983) 187.
[13] P.K. Haff, J. Fluid Mech., 134, (1983) 401 .
[14] D.R.M. Williams and F.C. MacKintosh , Physical Review E (in press)
[15] D.R.M. Williams and F.C. MacKintosh in "Fractal Aspects of Materials", edited by F. Family *et al.* MRS: (1995).
[16] D.R.M. Williams, Physica A (in press).
[17] K.Binder and D.W. Heermann, Monte Carlo Simulation in Statistical Physics, Springer: Berlin (1988).

AUTHOR INDEX

SUBJECT INDEX